Power Integrity

Measuring, Optimizing and Troubleshooting Power-Related Parameters in Electronics Systems

The Faraday Press Edition

Steven M. Sandler

Power Integrity: Measuring, Optimizing and Troubleshooting Power-Related Parameters in Electronics Systems

Other books by Steven M. Sandler

Switched-Mode Power Supply Design with SPICE
Power Integrity Using ADS (with Anto K. Davis)
SPICE Circuit Handbook
Measuring Power: Application Notebook

Cover Design by Guy D. Corp, www.GrafixCorp.com

www.FaradayPress.com
1000 West Apache Trail—Suite 126
Apache Junction, AZ 85120 USA

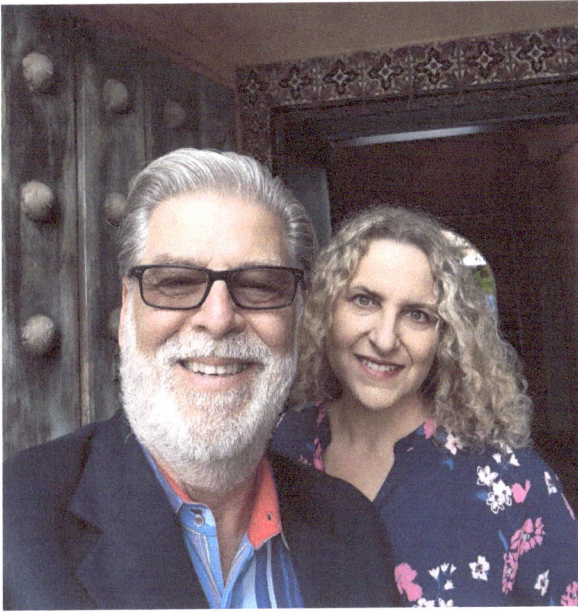

Dedication

THIS BOOK IS dedicated to my wife, Susan. She had the very difficult job of keeping my stress level down and my outlook up. She sacrificed the many hours I spent in my "man cave" making measurements rather than spending time with her. When I (frequently) whined about how much work writing this book was, she never said "I told you so", but gave me the pep talk I needed to keep me motivated. Sweetheart, I love you.

Foreword

WHEN I HEAR the name of Steve Sandler, the first thing that comes to my mind is his long list of excellent publications and books about power converter circuits that go back decades. Second, in most recent years, the impressive range of power integrity test accessories envisioned and made reality for all of us in the industry. Not only do I value and trust Steve as an excellent engineer, I also count him among my friends. For both reasons I am truly honored to write this foreword for his new edition of his book *Power Integrity*.

This book is a very valuable summary of power integrity measurement and test principles and can serve as a useful reference book for all practicing technicians and engineers, who need to decide what should and can be measured—with what—in today's electronic circuits.

What significantly increases the value of the book is the collaborative effort of major players in the test and measurement industry so different test and measurement approaches can be compared as implemented by different companies using different instruments. This fact is also underlined by the long list of people in the Acknowledgement section.

When people want to do careful and thorough designs, the

result has to be validated. Before the circuit or system is built, it can be simulated, but the real test is when we have the hardware built and measure it.

Today we have a large choice of software for power integrity purposes, ranging from the simple and free to the complicated and expensive. Also, software changes fast; bug fixes, feature updates and full revisions come to market regularly, sometime two or three times a year. Unfortunately, often they become obsolete or incompatible equally quickly. As opposed to software tools, hardware equipment, especially professional test equipments, tend to have longer design cycles, but they also stay relevant and useful much longer. This is true also for instruments used for power-integrity measurements.

What usually lags is the development of smaller accessories that may serve specialized purposes and make general-purpose test instruments suitable for the various power integrity tests. As the power integrity discipline gradually became relevant for more and more users in the 1990s, existing instruments, like some low-frequency network analyzers, quickly became the preferred choice to measure the frequency domain behavior of the power delivery network, but for years I had to create my own home-made probes, common-mode transformers, injectors, just to name a few, to do the necessary measurements.

It was a very welcome change when Picotest entered the market and started to make such accessories in high quality and professional form. I have always been happy to take a look at the new accessories before they got finalized, try them out and share my comments, suggestions and feedback. These accessories are essential in power integrity labs, regardless of the manufacturer of the instrument we use them with.

There are many things I like about this book.

For one, the many illustrations. I am a visual type of person, so, for me, seeing what we are talking about is important. As the saying goes "a picture is worth of thousand words." Good pictures like the setup photos in this book, are worth millions of words. They show a lot of additional details that the figure captions and the text going with the figures can not capture.

The illustration photos and resultant screen captures are masterfully arranged to summarize each particular message in a clear way. What makes these illustrations even more valuable is the fact that Steve is a practicing engineer himself, so he understands what is important for fellow engineers to observe.

The Tricks and Tips at the end of each chapter are based on decades of Steve's experience and should serve as important guidance for anyone entering the field of power integrity. So many times we have seen results presented based on measurement-only or simulation-only, lacking any checks and tests of the process used to generate the data. Though we cannot (and should not) always aim for perfection, many times we consciously settle for a 'good enough' result, but the conclusions still have to be based on solid and trusted measurements or simulation data, or, even better, correlated data, data sufficiently above the noise floor that does not include masking noise picked up from the surroundings or are based on invalid calibration or wrong simulation settings, just to name a few of the common pitfalls that are hard to recognize without careful cross checking—even for a trained set of eyes.

And last, but not least, I like the fact that the book brings together the different domains (time, frequency, impedance) and the three major disciplines (power integrity, signal integrity and electromagnetic compatibility) that are all important for us practicing engineers to create successful designs and perform proper measurements and simulations.

When I completed my first book, *Frequency-Domain Characterization of Power Distribution Networks*, none of the great accessories described in this book were available. As a welcome change, now, many years later, there is an ever-growing list of probes, active and passive devices, software, device models and tutorial videos are readily available to support our power integrity work.

Finally, I want to thank Steve for putting in the time and effort to create this book; I am sure it will serve as a valuable reference for many of us for years to come.

—Istvan Novak

Istvan can be found on LinkedIn
https://www.linkedin.com/in/istvan-novak-865792/

Acknowledgments

WRITING THIS BOOK was a major undertaking. Of course writing any book is a lot of work, but the vast range of instruments necessary to perform the measurements and finding ideal examples to measure stepped it up a bit. Many of the concepts in this book may be new and so the information had to be presented very clearly and concisely. This effort required a lot of support from many individuals and companies. Without this support, I would never have been able to complete this book. I also had a record number of peer reviewers for this book. I am forever grateful to the following individuals and companies and apologize profusely in the event I left anyone out.

I want to thank my editor, Michael McCabe, Kritika Kaushik and all of the folks at McGraw-Hill for their work in creating the first edition of this book.

Thanks to my long term friend and business partner, Charles Hymowitz, Vice President of Sales and Marketing for Picotest, and CEO of AEi Systems. He read every page, edited, commented, and offered many helpful suggestions to make this a better book. Thank you does not seem to cover it, but thanks.

Bernhard Baumgartner, Florian Hämmerle, and Wolfgang Schenk of OMICRON Lab for their constant support, for including the noninvasive measurement in their instrument and for being great friends in addition to sales partners. Thanks for the support and the many helpful comments and suggestions.

Mark Roberts, Stacy Hoffacker, Mike Mende, Amy Higgins,

and Tom Lenihan from Tektronix were always ready and willing to help, whether it was to discuss equipment, answer questions, or arrange the shipment of loaner instruments to and from my lab. They also offered some comments and suggestions.

David Tanaka, Yasuhiro Mori, Eileen Meenan, and Hiroshi Kanda from Agilent Technologies for the tremendous knowledge they possess regarding their instruments and their willingness to share some of it with me. I also thank them for arranging the shipment of loaner instruments in and out of my lab.

Dan Burtraw, David Rishavy, and Mike Schnecker from Rohde-Schwarz for arranging the loan of their RTO1044 oscilloscope, for answering the many questions I asked, and for providing helpful comments and suggestions helping to make this a better book.

Bob Hahnke, Steve Murphy, Stephen Mueller, and Kathleen Woods from Teledyne Lecroy for their support of demo equipment and for the helpful comments they provided.

Hawk Shang of PICOTEST Corp for his generous support of our projects, for making excellent general purpose test equipment, and for manufacturing the Picotest signal injectors. Hawk, thank you for all that you do.

Chris Hewson of Power Electronic Measurement for providing the CWT015 probe used in this book as well as answering my questions about Rogowski current probes in general.

Paul Ho, Nazila Arefazar, Cesar Redon, Gordon Leverich, Michael Lui, Shivam Patel, Sahar Sadeghi, Josh Behdad, John Aschennbrenner and Tom Boehler of AEi Systems, and Dave Morrison of How2Power.com for all of your comments and suggestions to make this a better book. Thanks to Tim Guzman, also of AEi Systems, for his photography and image contributions.

Shawn Winchester and Artescapes for enhancing many of the oscilloscope and spectrum analyzer images to make them clearer and easier to see. Shawn also managed daily operations at Picotest, allowing me to focus on writing this book.

To all of you, "thanks" doesn't say enough, but I could not have finished this book without you.

The index was created and revised by Robert Swanson and Zurain Shahzad. Zurain can be found on LinkedIn at: https://www.linkedin.com/in/zurain-shahzad-3a367a204/

DURING A MAY 2013 visit to Austria, while having dinner in a medieval castle with Bernhard Baumgartner, my daughter Rachel Sandler and a few others, we discussed what makes books popular in this generation. The final conclusion was that to be successful, the book must include dragons.

Not just any dragons, but medieval dragons.

I am not superstitious, but just to be safe I have included the following image and hope it will help make this book a best seller.

Image credit: Elena Schweitzer / 123RF.com

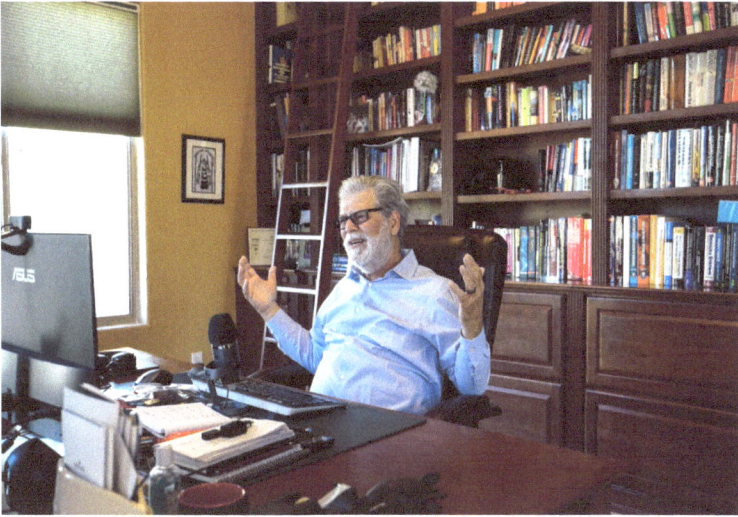

What has changed since this Book was originally Published?

Part 1

Everything!

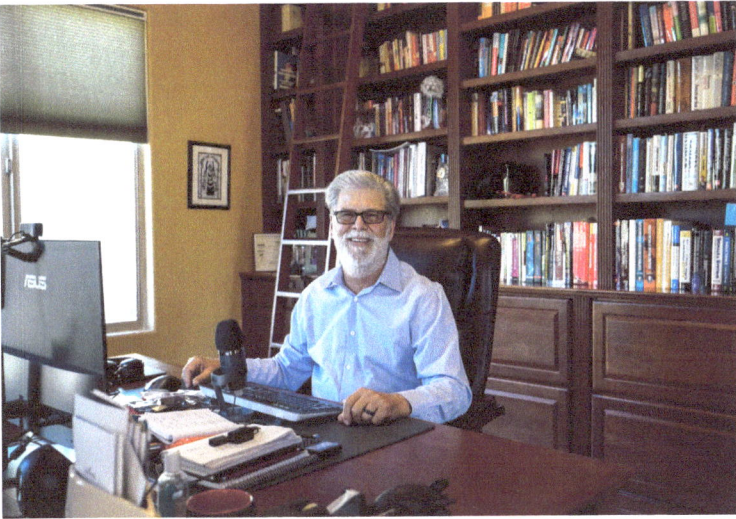

What has changed since this Book was originally Published?

Part 2

WOULD YOU LIKE a less frivolous answer? Among the many things which have changed:

- Impedances got lower. What was once measured in milliohms is now measured in micro-ohms.
- Accurately measuring lower impedances requires better common-mode isolators.
- Probing became much more difficult and a focus of interest.
- The Non-Invasive Stability Margin (NISM) calculation is becoming more mainstream and is now supported by modern software tools and equipment.
- Most oscilloscopes can now directly do impedance and PSRR measurements…and very soon will also directly measure NISM.

About the Author

STEVEN M. SANDLER has been in the power conversion business since 1976. He has worked for companies such as Acme Electric, Lambda Electronics, Venus Scientific, Transistor Devices, Keltec-Florida and Signal Technology. The majority of his work has been involved with the development of power conversion products for commercial, military and space platforms. He was the founder of AEi, a company specializing in the computer simulation and worst case analysis of power electronics. AEi performed a great deal of the computer simulation of the power electronics used on the International Space Station as well as many other platforms.

In 2010, Steve founded Picotest, a company dedicated to providing instruments, accessories and training to support electronic test and measurements, with an emphasis on power integrity solutions. Picotest is headquartered in Phoenix, AZ.

For more information on Picotest, please contact the company at 1-877-914-PICO or visit www.Picotest.com.

Steve can be found on LinkedIn
https://www.linkedin.com/in/steven-sandler-022a7210/

Publisher's Note

THIS BOOK IS an updated reprint of Steve's historic *Power Integrity*. One of the benefits of working with power supplies is that the underlying physics are timeless and eternal. Magnetics, capacitance, Ohm's Law (this should really be called Ohm's relation…a useful relation between voltage and current we call ohms), frequency effects and control loops are not subjective.

The ways applied physics works today is the way it will work forever. What a luxury compared to other technical fields where information and strategies have a half-life measured in years—often just a few years. By half-life, I mean that period when half what the engineer knows becomes obsolete.

Thus, a book like this will always contain useful information. Forever. Maybe this is as close to immortality as we can achieve.

All that said, many things change and will continue to change. In fact, the evolution of our trade is accelerating at a breathtaking pace.

I'm halfway kidding, but it seems like modern loads evolve to maximize the difficulty of creating a robust design. Core voltages are headed toward zero. It's conceivable a future power supply design will require tight tolerance around 0.3V. ±3% would require a total tolerance window of 18mV and that includes initial setpoint, ripple and transient response.

Load currents are increasing. Perhaps this imaginary power supply will have to deliver 2,000W. That's 6667A @ 0.3V.

The frequency content of load transients is increasing. Perhaps this imaginary power supply will need a flat impedance at very high frequency, like 30Ghz.

Power densities are increasing. Remember the good old days when 100W per cubic inch was state-of-the art?

All this reminds me of a question I recently asked Steve…

Ken: Let's say present trends continue and in 10 years we're asked to deliver 0.1 V at 10 kW to serve some monster ASIC. How in the world are we going to do it?

Steve: That's the fun part—we have no idea, yet. When there is a void, rest assured that engineers will find a way to fill it! There are sub-threshold devices now operating at several hundred millivolts. Devices are slow there, and so the current isn't high.

But who knows what the future will bring?

Today there is an explosion of immersive devices running in liquid to keep them cool. That's inefficient, so we push for smaller geometries to reduce the current. But then the devices are faster, so we run them at higher rates and the current climbs back up.

It's a never ending cycle and I love it!

Bring it on.

Steve condensed a lifetime of knowledge and experience into this book. It will, as much as anything can, prepare you for the future.

What's left? To get out there and get it done.

Keep your impedances flat and no higher (or lower) than necessary, my friends.

—Ken Coffman

Ken can be found on LinkedIn
https://www.linkedin.com/in/kencoffman/

Chapter One

Introduction

I CHOSE TO write this book because it has become increasingly clear that much of the data that we need to do our jobs as electronics engineers is lacking. Either the data we need is missing entirely or when we do have data—that we created or received from others—it frequently lacks completeness, fidelity and/or accuracy.

There are a variety of reasons for these shortcomings, and it is my hope that this book will provide useful information and direction to several different audiences.

One goal for this book is to show component and device manufacturers the breadth and fidelity of the data end-users really do need to do their jobs, as well as to help them improve their datasheets accordingly.

Another goal is to provide design and test engineers with methods that enable them to generate higher fidelity measurements with less effort by using appropriate techniques and equipment. It is also my hope that test instrument manufacturers will gain insight into the issues engineers face, as well as how they can improve their equipment capabilities, operating systems, software, and documentation.

Lastly, and maybe most importantly, this book also illustrates the impact power supply performance has on the systems they serve.

What You Will Learn

This book provides useful insights into all aspects of making high fidelity measurements, including power, high-speed, and low-power analog and instrumentation circuits.

Technology continuously, consistently, and rapidly advances. A few new technologies, such as eGaN, GaN, SiC and GaAs, will present new measurement challenges due to the combination of high voltage and ultra-high speed switching. Switching frequencies and edge speeds are increasing, while devices are becoming more highly integrated.

For example, many Point of Load (POLs) switching regulators now include the switching MOSFETs internally. Some devices include the output inductor internally as well. These advances in technology make measurement more difficult and demand a better understanding of measurement fundamentals in order to obtain accurate results. The evolution of high speed FPGAs and CPUs have propelled the measurement of Power Distribution Networks (PDNs) to magnitudes below 1 milliohm and to frequencies above 10 GHz.

Once the need for a measurement is established, there are several important decisions that need to be made. These decisions relate to the measurement domain, the selection of appropriate test equipment, the impact of the connection of the equipment to the device being tested and the interpretation of the acquired data. This book provides the necessary information to evaluate the needs of the measurement and make the best decisions to achieve high fidelity results.

I highly recommend reading the first section of the book

(Chapters One through Six) prior to making any of the measurements discussed. This material provides all of the background information related to equipment selection and fidelity, as well as how test equipment should be interfaced to the device being tested. The relative significance of each type of test at the system level is also discussed.

Once you have reviewed the introductory material, the remainder (Chapters 7 through 15) can be used as reference allowing you the freedom to refer to the material on each type of measurement as needed.

Who Will Benefit from This Book?

This book is written for engineers and technicians of all levels of experience, including those working in field support, design and test engineering disciplines. It is also appropriate for engineering managers, as well as those that are responsible for the leasing or purchasing of test equipment.

In most instances, engineers are underequipped for the measurements they need to perform. More often than not this is due to a lack of understanding of what minimum set of capabilities are actually required to make measurements needed for the particular application. For engineers and management alike it should be noted that there are substantial costs for bad or misleading data.

This book addresses these issues for both the test engineer and the purchaser trying to secure the least expensive solution.

The General Format of This Book

As noted above, this book is written in two sections. The first section is dedicated to the available types of test equipment, measurement fundamentals and interfacing or connecting the test equipment to the Device-Under-Test (DUT).

The second section of the book addresses the specifics of

making particular measurements. Each chapter discusses one or more specific measurement methods. Each measurement method includes a brief discussion of the measurement including why it's important.

Additionally, each measurement method includes setup pictures, pros and cons of the measurement method, tips and/or hints, and example measurements.

Why Measure?

It seems appropriate to start a book about measurement fundamentals by considering the goals we hope to achieve. To that end, this book takes the viewpoint that the end goal of the learning process is to enable the reader to acquire better (as in more-precise, higher-fidelity) data. There are four major reasons for testing:

1. To obtain data that is not available or published or validate data that is.
2. To compare possible devices or circuit topologies for use in a design.
3. To troubleshoot.
4. To validate or verify design performance.

Obtain or Validate Data

In many cases, the manufacturer provides very limited data or data that is not at the operating points that we are interested in. As is often the case, data sheets are more of a marketing tool than technical documents.

Some of the more common examples of this are evident in operational amplifier (op-amps), voltage reference and regulator data sheets. In the case of op-amps, the open-loop gain and phase response curves may not be at the operational voltage we plan to use. The open loop gain and phase curve in Figure 1-1 shows the performance a single 5V power supply and also for a ±15V supply. The phase is shifted significantly between these two voltages. How would the performance differ if our circuit operates with ±5V or a single +12V supply?

Likewise, this figure shows a load capacitance of 100pF. If our circuit does not include a 100pF load capacitance, how do we determine the impact of the load capacitance?

Gain, Phase vs Frequency

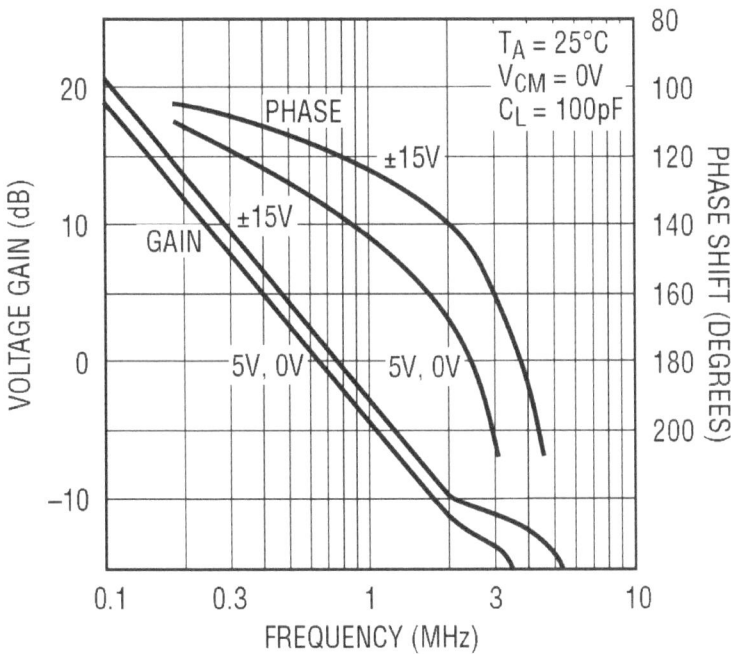

Op-amp Gain and Phase versus Frequency for the LT1014 Op-amp

Figure 1-1

In the cases of voltage references and linear regulators, stability information is generally not provided at all. The load transient of

15

a voltage reference is shown in Figure 1-2. The datasheet does not include a stability plot, but does include the statement that the device is stable with a load capacitance in the range of 0.1-10µF. The step load response shows three rings, indicating less than 'ideal' stability. We'll talk more about this in Chapter 8.

While this reference may not break into a full-blown oscillation under these conditions, it will not perform optimally in terms of regulation, PSRR, or noise. This is a good example of the need to interpret the manufacturer's data. As designers, we have a different idea than the component manufacturers about what constitutes "stable" performance.

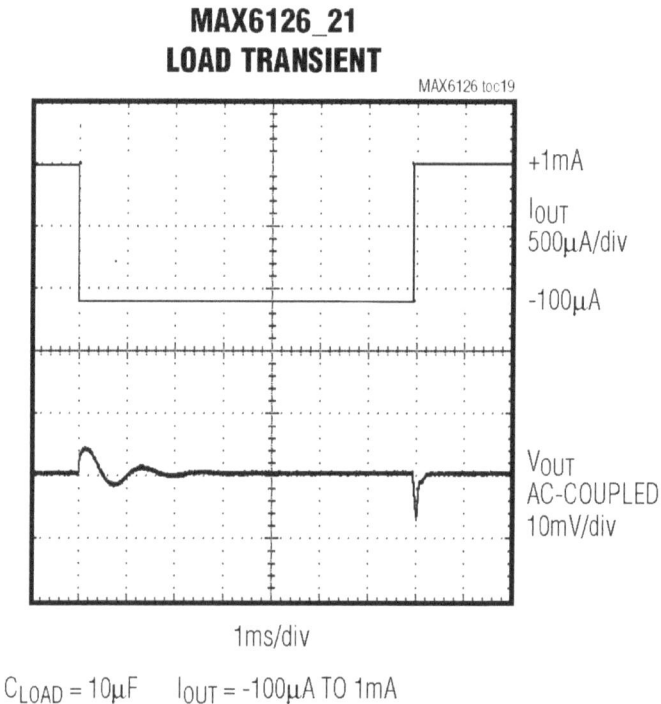

MAX6126_21
LOAD TRANSIENT

Step Load Response of a Voltage Reference

Figure 1-2

Design, Selection and Optimization

We may need to test in order to obtain missing information, as in the case of the op-amp above—or possibly to compare devices from different manufacturers to see which one performs better in the circuit.

An example of this is seen in Figure 1-3 which compares the PSRR of two linear regulators under the same conditions.

If PSRR is the only concern, it is clear that one of these regulators outperforms the other by a large margin. In fact, the better of the two reveals the noise floor of the measurement to be 100dB.

Finally, we might use measurements to optimize component values for a new design.

In some cases, this optimization is performed in production, where adjustments are made either by discrete component value selection or by the adjustment of trimmers in order to more precisely set a particular parameter.

In the semiconductor industry, devices such as voltage references and voltage regulators are routinely laser trimmed during manufacturing to improve their performance.

While this optimization process may or may not be automated, the selection is performed in conjunction with a measurement.

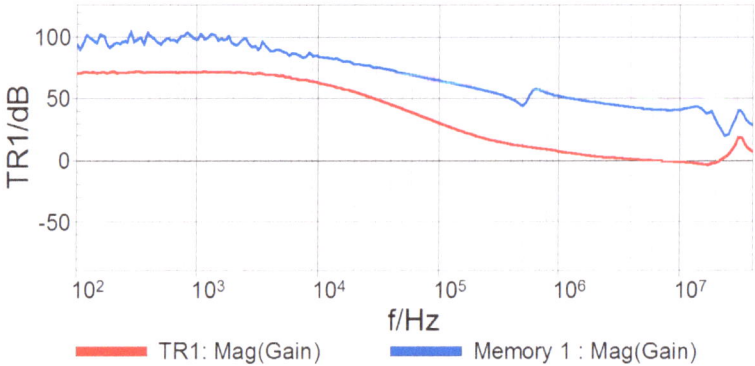

**PSRR of Two Linear Regulators under the Same
Operating Conditions**

Figure 1-3

Troubleshooting

As much as we would like to see all of our designs work perfectly the first time we power them up, it rarely works out that way.

In some cases, there may be an interaction between different devices or subsystems, while in other cases there may be a bad component. Yet in other cases, the design may not perform properly because of printed circuit board influences.

The process of troubleshooting is generally dependent on a series of measurements that finally lead us to the culprit. This is one area where an understanding of high fidelity measurement is crucial as the sources of the problems are generally very good at hiding and the quality of the data can be critical to the search. Generally, it falls to the engineers to find these root causes.

Usually the engineer is expected to do so very quickly and under great pressure. In some cases, the underlying problem is simply that the manufacturer provided incorrect or misleading data.

Output Impedance vs Frequency

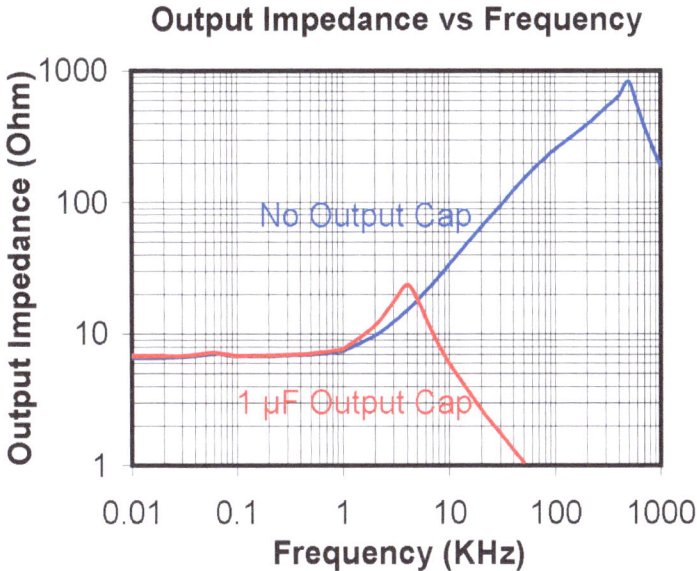

Manufacturer's Datasheet for a Voltage Reference with and without an Output Load Capacitor

Figure 1-4

In the case of Figure 1-4, the manufacturer shows the output impedance of their voltage reference with and without the addition of a 1µF capacitor.

From the 1Ω impedance measurement at 50kHz we can easily calculate the capacitance as:

$$C = \frac{1}{2\pi \cdot (50kHz \cdot 1\Omega)} = 3.2\mu F \qquad 1.1$$

The output capacitor is actually a 3.3µF capacitor and not a 1µF capacitor as stated. These incorrect results are not intentional, nor are they uncommon, offering further evidence for the need to make your own measurements.

In the case of Figure 1-5, we can see that the reference is specified to have an output noise of 9.6PPM or 24µVrms in the bandwidth from 10Hz to 1kHz. Looking at the output impedance plot the device presents a resonant peak at 2kHz to 3kHz, depending on the output capacitor, which is outside of the specified 1kHz bandwidth.

At the peak of the resonance, the output impedance is approximately 600Ω. Simple math tells us that if a noise current of 40nArms is presented to the voltage reference at the resonant frequency, the induced noise equals the specified device noise (i.e. 24µVrms).

This is a simple example of how the interaction between a device and the system in which it resides can present unexpected behavior.

Output Noise	Voltage	$0.1Hz \leq f \leq$ 10Hz	8	PPM_{p-p}
		10Hz	9.6	PPM_{RMS}
		$10Hz \leq f \leq$ 1kHz		

Voltage Reference Output Impedance and Noise Specification

Figure 1-5

Validation or Verification

In many instances, engineers verify the performance compliance of their design by comparing measurements of the circuit with a specification or other requirements document.

Once the design is approved, and in production, the performance is generally assured by testing every product manufactured for various performance characteristics. Often a separate, and more comprehensive test set is also performed on a representative sample from a given production lot. Each of these test processes has a different specific goal, though they are all necessary to verify the quality of the product being manufactured. We might also be interested in validating the computer simulation models we are (hopefully) using to analyze performance. It is essential that we validate the accuracy and fidelity of these models before using them.

This validation process is a check and balance that helps to verify that both the design is built as intended and that the model is correlated. In the case where the measurements and simulations do not agree it can be either because the design is not what we modeled, that the model is not correct, or both.

The process of correlating your models with the test measurements helps to resolve such errors.

Terminology

Before taking the discussion further, let's define some basic terms that are used throughout the book.

Measurement

The result of a test. This can be in the form of a number, a curve, or a dataset. The dataset can be amplitude and time, amplitude and phase, amplitude and frequency or a set of singular numbers, such as 5V, 100kHz, i.e. any data that accurately quantifies a performance characteristic.

High Fidelity

Dictionaries generally relate this term to the reproduction of an audio signal along the lines of a reproduction of an electronic signal that is faithful to the original without any added distortion.

For our purposes, we use it to describe a test or simulation result that presents a faithful reproduction of the signal being measured while presenting sufficient detail and clarity to allow precise values to be determined and precise conclusions related to the outcome to be made.

Precise

A measurement that provides a result without uncertainty or ambiguity. The measurement would generally not be subject to different interpretation by different observers.

For example, if we measure the rise time of a MOSFET, the measurement should be *precise* enough that a selection of engineers would all obtain the same value from the measurement.

Non-Invasive

One of our fundamental goals in making high-fidelity, precise measurements is to observe the measurement without impacting or influencing its result.

We take that to mean two different things. First, the connection of the equipment should not in any way impact the measurement. The impact of the equipment connection is one of the more common sources of measurement error.

A simple example is the measurement of a Pulse Width Modulator (PWM) switching frequency.

Many engineers use a passive scope probe to see the timing ramp on an oscilloscope. The oscilloscope probe capacitance is often large enough to change the frequency. This is an *invasive* measurement. Often the connection of test equipment to the device being tested is a limitation of the fidelity and precision of the measurement.

The second interpretation of non-invasive measurement is that the measurement does not require traces to be cut, lifted, unsoldered or otherwise manipulated.

This is a common issue for high-reliability systems, such as aerospace programs, where such invasive measures are not allowed due to risk, cost and manufacturing constraints.

Indirect Measurement

In some cases, it is desirable to measure performance indirectly, meaning that our measurement is not made on the circuit providing the signal, but on the circuit receiving the signal.

For instance, in high-performance clock oscillators, power supply noise is a major contributor to jitter. Power supply noise can degrade the performance of the clock and the circuit using the clock, such as an ADC.

In such cases, we can often measure the power supply noise more accurately by measuring its effect on the clock than we could by measuring the power supply directly.

In-Situ or In-Circuit

This refers to circumstances surrounding the device under test (DUT) and its interconnections to the circuit driving and loading it.

For instance, the power supply source and load connections influence the performance. On the input side, the input filter interacts with the switching power supply. This is one particular circuit example where the interaction between the wiring to the power supply and the power supply itself interact with one another.

There are many other such examples.

As we saw in Figure 1-4, the addition of the external load capacitor *increased* the output impedance of the voltage reference at 4kHz. This is an example of an interaction on the load side of the voltage reference. Such degradation would also be noted in other aspects of performance, such as PSRR.

For these reasons it is often best to make measurements or with the power supply integrated into the system or in-situ.

Chapter Two

Measurement Philosophy

MOST MEASUREMENT ERRORS can be eliminated by adopting a set of basic guidelines.

Each guideline in this chapter addresses a particular issue that is seen repeatedly. In most cases, the remedial action is minimal in terms of time and cost. These rules might seem obvious; nonetheless, they are frequently overlooked.

Cause No Damage

The first rule of measurement is so simple it appears obvious.

Do not damage the test equipment or the circuitry being tested.

Test equipment is expensive, and in many cases, the circuitry being tested is even more expensive. In a development environment, the circuitry could well be one-of-a-kind. That means that you want to be certain to protect both the test equipment and the devices being tested because mistakes happen.

Consider the voltage and current limits of the test equipment, especially the sensitive high frequency input circuits. Also consider whether the test equipment can generate voltages or currents that can damage the device being tested, especially sensitive devices, such as wideband op-amps, low power BJTs and JFETS.

Many circuits provide test points with protection resistors connected in series to assure no damage, though these can greatly impact the measurement fidelity—as will be discussed in later chapters.

Measure without Influencing the Measurement

Measurement errors are the most common source of error and cannot be completely eliminated. Any connection to a circuit has some influence on performance. You just want to minimize the interaction as much as possible.

There are many ways we can influence the behavior of the DUT. As a simple example, the use of a typical 10X oscilloscope probe introduces 10-15pF of capacitance at the point of measurement, while using a unity gain probe (1X) can easily add 75pF.

A typical high-speed active probe contributes roughly 1pF. Attempting to measure a signal such as the oscillator ramp of a pulse width modulator controller can alter the switching frequency whether you use a 10X or a 1X oscilloscope probe.

The measurement in Figure 2-1 demonstrates this well.

The frequency measured with the 1X probe is 363.6kHz, with the 10X probe is 452.3kHz and with the active probe, 452.9kHz. There is only a slight difference between the readings obtained with the 10X probe and the active probe, while the 1X probe reduces the actual switching frequency by 20%. The active probe has the lowest loading and is, therefore, the most correct measurement.

The 1X probe not only indicates an incorrect result, but if left connected will also influence many other measurements, such as ripple and efficiency, since they are both related to the switching frequency.

In Figure 2-1, the significant loading of the 1X probe produces excessive measurement error.

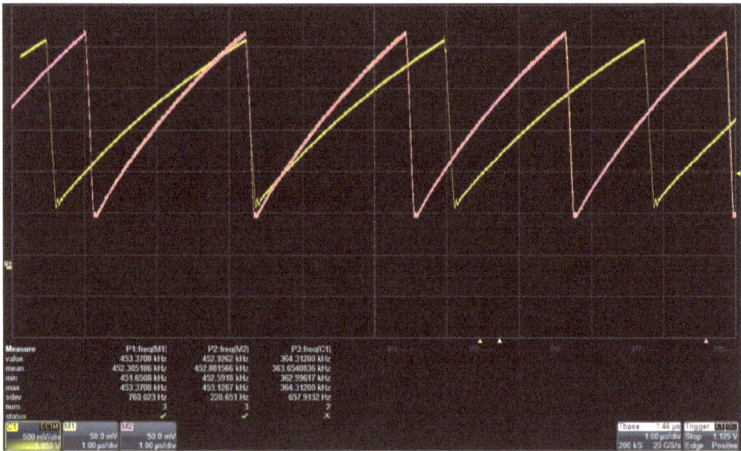

PWM Ramp Frequency Measured with the Standard 1X Probe (C1 in yellow), a Passive 10X Probe (M1 in pale yellow) and a 2.5 GHz Active Probe (M2 in pink)

Figure 2-1

It is important to be aware of how measurement probes impact the DUT performance and not just whether the probe has the fidelity for the measurement.

We'll address this in further detail in Chapter 5 *Interface Cables and Probes.*

Validate the Test Setup and Measurement Limits

Before using a measurement setup to record data, the setup should be used to measure a known quantity to verify that the measurement has the dynamic range, precision and noise floor margin to produce the expected result.

One example is a measurement of the power supply rejection ratio (PSRR) for a linear regulator. The measurement in Figure 2-2 shows the noise floor of the testing environment and the test measurement. The measurement is a minimum of 15dB below the noise floor (magnitude of 5.62).

The impact on the result is a maximum of 15%, calculated as shown in Equation 2.1.

$$Max_{error} = \frac{1}{1 + 5.62} = 15.1\% \qquad 2.1$$

PSRR Measurement of a Linear Regulator along with the Measurement Noise Floor

Figure 2-2

A second example is shown for a noise level measurement, as seen in Figure 2-3.

A 100mV signal source is connected to an Agilent N9020A Spectrum Analyzer through two Picotest J2140A precision cascadable attenuators inserting a total of 100dB attenuation. The correct result should be 1µV.

The measurement is within approximately 5%, validating the measurement.

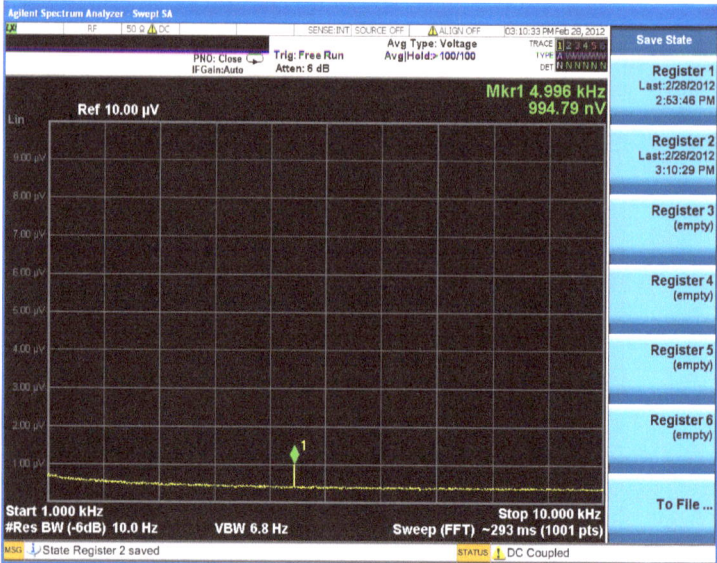

Validating a 1 µV Noise Measurement by Inserting 100dB of Attenuation between a 100mV signal and the Spectrum Analyzer

Figure 2-3

Measure in the Most Efficient and Direct Way

Non-Invasive vs. Invasive Measurement

Ideally, making measurements should not require any lifting or unsoldering of components or wires or other mechanical or electromechanical disruptions to the circuit.

For the purposes of this book, non-invasive measurements are those that do not require any such disruptions. Invasive measurements are those that do require such disruptions.

In-Situ Measurement

In an ideal world, it is preferable to perform all measurements on a circuit while it is in its final configuration, in the production version of the system working under normal operating conditions.

For example, if we want to make an in-situ measurement of a car battery, we would measure the battery while it is in the car operating in typical fashion. This method assures that all connection impedances, operating load conditions and other related characteristics are included in the measurement.

Measurements that are not made in-situ must be carefully evaluated to assure that the measurement is not impacted by the external circuitry or signals, such as input power, loading and interface cables.

In some cases, there may be significant differences between in-situ measurements and on-the-bench type measurements. For example, if we measure the power supply ripple voltage in a computer system you will see not only the ripple created by the power supply, but also the ripple created by the interaction between the dynamic load current and power supply output impedance.

The question then becomes: what measurement result is desired? Is the purpose of the measurement to determine the ripple caused by the power supply or the ripple that exists in the system? These results will often be different.

Indirect vs. Direct Measurement

A direct measurement is a measurement that allows us to measure the characteristic of interest.

For example, if we wish to measure the voltage of our car battery, we can simply use a voltmeter to directly measure the battery voltage. In some cases, we cannot make a direct measurement or it may be simpler to make an indirect measurement. As an example of an indirect measurement, we can often assess power supply noise by measuring the noise floor of an ADC that is connected to the power supply or the jitter of an oscillator connected to the power supply.

The indirect measurement is, therefore, not directly measuring the characteristic of interest, but the end result or impact of the characteristic of interest on the system. The benefits of indirect measurement are that the measurement may be more accessible or may be more sensitive than a direct measurement. The indirect may also be a better assessment or simplify troubleshooting.

For example, while troubleshooting an ADC, do we really want to know the ripple on the power supply or do we want to know why the ADC is noisy?

Document Measurements Thoroughly

No matter how good the quality of the data is, the value of the data is diminished if the measurement is not well documented.

In terms of documenting the measurement consider the possibility that questions will be asked about the measurement and or measurement setup long after the data has been delivered, published, or otherwise distributed. Having data without the detailed annotations and descriptions can be frustrating at best.

The following information should be recorded along with the data.

1. The name of the test engineer and this person's contact information
2. The purpose of the test
3. Simulated or expected results if available
4. The date and physical location of the testing
5. Operational test environment and conditions
6. The model number of each piece of test equipment (including probes) and verification that it is within the

allowable calibration period
7. Setup diagram and/or picture
8. Measurement annotations and comments
9. Any observed anomalies
10. A summary of the results of the test and any follow-on work that may be needed

The Test Engineer and Contact Information

The reason for recording this information should be obvious, but is often overlooked. In the event there are questions, it is always helpful to know who to ask questions of and how to contact the test engineer in order to ask them.

A secondary reason is that each test engineer has his or her own characteristics, including strengths and weaknesses.

Some test engineers are better at some tests than others, and so this can help to identify trends or patterns in results, as well as being useful in determining the possible need for additional training.

The Purpose of the Test

The reason for the test is generally not documented and yet can provide great insight into the focus of the measurement.

For example, if the purpose of a measurement is to observe anomalous behavior at cold temperature, it would be apparent that this data was being gathered to resolve an issue. We would not expect much attention to be paid to other performance attributes that could have been noted or observed nor should we consider this to be a comprehensive test.

The oscilloscope image in Figure 2-4 is provided as a verification of stability in response to a transient load current step.

While this response is quite well behaved and indicative of a stable control loop, the image in Figure 2-5 shows a drastically different and quite unacceptable level of performance. The voltage and time scales in Figure 2-4 are 4mV/div and 2ms/div while the scales in Figure 2-5 are 100mV/div and 200ms/div.

This set of oscilloscope measurements shows why it is so important to understand the purpose of the test.

The image in Figure 2-4 shows that the small signal stability is good, though this image offers no hint that there are voltage excursions in 100's of mV that can be seen in Figure 2-5 and could be a significant issue to the circuit being powered.

In Figure 2-4, The voltage excursions (yellow trace) are less than 10mV in Response to a 25mA load step (blue trace).

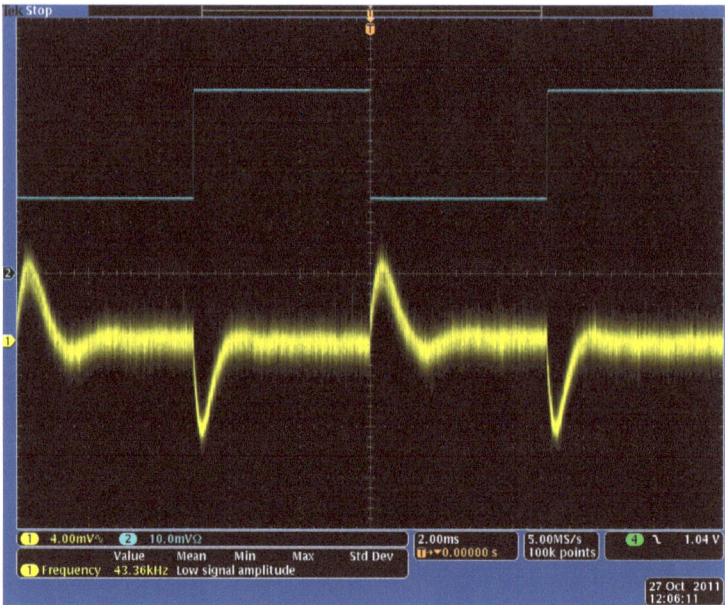

The Small Signal Step Load Response is Well- Behaved

Figure 2-4

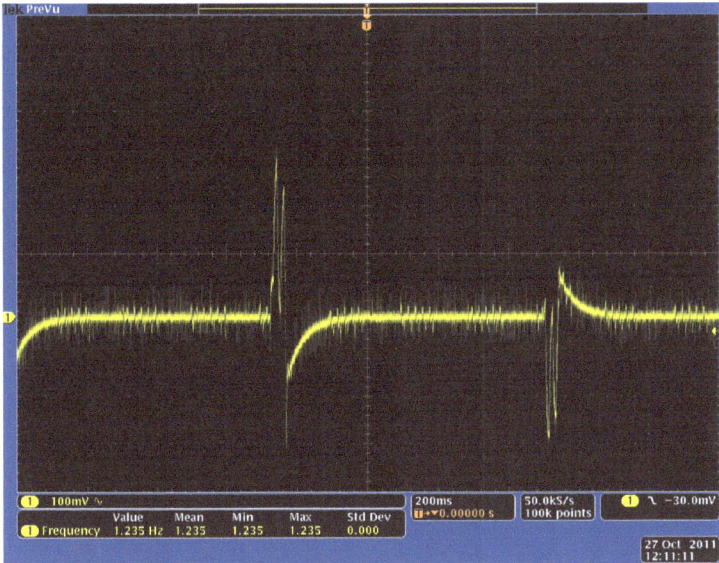

The Same Load Current Step and Operating Conditions, but with Different Oscilloscope Scaling

Figure 2-5

Simulated or Expected Results If Available

As discussed in Chapter 1, one reason for testing is to validate or correlate simulation or mathematical models.

An example is shown in Figure 2-6.

In this case, it is important for us to have the simulated results available so that when we obtain the measurement we can immediately determine its efficacy and if the model and measurement agree.

In the event the measurement and model do not agree this is the most opportune time to determine the reason for the disagreement. The measurement is still set up and you can immediately go into troubleshooting mode, obtaining additional data to sort out the reasons for the differences.

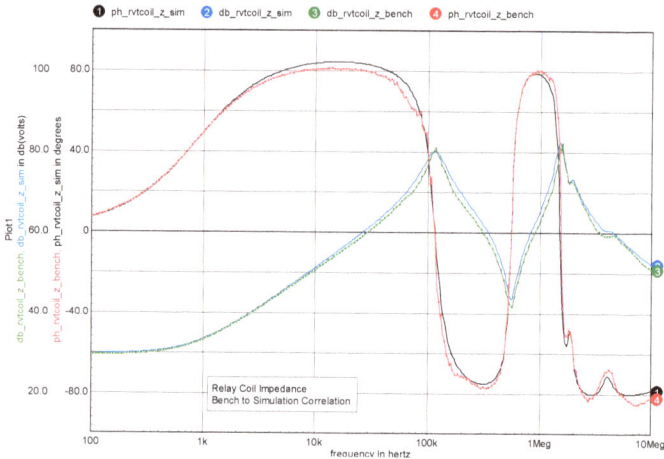

A Comparison between a Measurement of Impedance Phase and Magnitude (Red and Green Traces) and the Expected (Simulation) Result (Black and Blue Traces)
Image courtesy of AEi Systems

Figure 2-6

The Date and Physical Location of the Testing

I recall several instances in recent years where the date and physical location of the testing were not only significant, but fundamental to understanding the root cause of the issue being evaluated.

In one instance, a customer complained that one of the devices they were purchasing consistently overheated. After making measurements and contacting the manufacturer, the issue became obvious.

The unit was designed and tested in Frankfurt, Germany during the winter and the customer was having trouble in Denver in the summer.

While these small bits of information might seem irrelevant, and actually could be, the weather in Frankfurt is quite cold in the winter and the altitude is very close to sea level.

On the other hand, Denver is warm in the summer at an altitude of 5300ft. The root cause of the issue was ultimately determined to be that the natural convection finned heat sinks—intended to provide necessary cooling—are not nearly as effective at 5300ft as they are at sea level.

What appears to be trivial information can provide great insight when troubleshooting such a problem.

Similarly, we could expect to find humidity-related issues to be more prevalent in southeastern states, such as Florida—and more frequently experienced during the summer months.

Operational Test Environment and Conditions

Some circuits are very sensitive to environmental conditions.

It is just good practice to know the conditions and environment in the lab.

In particular, temperature, humidity and altitude have the most significant influence on measurement accuracy. This is somewhat less critical if the equipment is self-calibrating.

If the measurement is made in proximity to the test equipment, it is possible that the test equipment will elevate the temperate surrounding the Device Under Test (DUT) and this information would be worth noting.

All general operating conditions, such as input/output voltages and currents should be recorded for each measurement. This can be very helpful in determining performance anomalies and when correlating simulation models.

The Model of Each Piece of Test Equipment (Including Probes) and Verification they are Calibrated

This is just good practice.

If nothing else, this information provides the assurance that the test equipment fidelity and resolution are compatible with the measurement goals.

In some rare instances, "bugs" or deficiencies are identified in particular models of equipment.

Setup Diagram and/or Photograph

A simple setup diagram or sketch allows a reviewer to understand the measurement and its limitations.

The picture in Figure 2-7 shows a VNA setup to measure a $1m\Omega$ resistor.

The image shows the short 50Ω coaxial connections, the connector mounts to the printed circuit board with the resistor mounted and the inclusion of a common mode coaxial transformer.

This measurement is discussed in detail later in this book, but it is obvious from the picture what connections were made and that they were made considering the necessary fidelity of the measurement.

The connection diagram in Figure 2-8 conveys the same information, but the high fidelity considerations are not communicated nearly as well.

Setup Image for a Low-Impedance Measurement using a VNA Clearly shows the Quality of the Connections

Figure 2-7

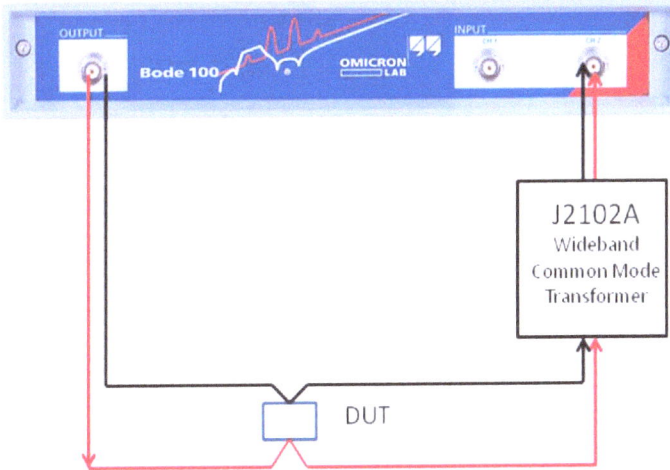

Connection Diagram of the Same Setup as Figure 2-7 Which Does Not Convey the Connection Quality

Figure 2-8

Measurement Annotations and Comments

Most modern test equipment allows the user to annotate measurements on screen, but whether the test equipment can perform this function or not, each trace or measurement should be annotated to indicate what it is a measurement of.

If there is a particular highlight in the measurement, make sure to note it as a comment on the measurement. You may have noticed that while many, if not most, of the figures in this chapter include measurements in the screen capture, most of them are not legible when published.

A similar issue arises in the cases where the test equipment publishes color-coded results and the test report is published and distributed in black and white.

Any Observed Anomalies

Engineers as a community tend to attribute irregular or undesirable events to user error, intermittent probe connections, etc.

These irregular events or anomalies are significant and yet all too often this type of information is ignored. Evidence of such random events can be very helpful in troubleshooting designs and should therefore be well documented.

Summary of the Results and any Follow-On Work

There is no better time to document summary notes and thoughts on further investigations than while the data is still fresh.

Chapter Three
Measurement Fundamentals

IT IS ESSENTIAL to understand the basic theory behind measurement acquisition in order to optimize the measurement fidelity. This chapter explores the process of acquiring data and the impacts of noise and bandwidth on the measurement. The impacts of scaling and the comparative strengths and weaknesses of the three measurement domains: frequency, time and spectrum are also discussed. The fundamental concepts in this chapter are used to maximize measurement fidelity in later sections of this book.

Sensitivity

The 'sensitivity' of a measurement is the smallest detectable change in the signal level. It is generally calculated as a function of the number of bits in the A/D. In theory, this is true, but not all ADCs are implemented equally. The use of dithering and oversampling with filtering can greatly influence the apparent sensitivity. Of course, there are many other impacts that reduces the Effective Number of Bits (ENOB), such as A/D non-linearity and electrical noise. Commonly used oscilloscopes are based on 8-bit A/Ds while some of the newer oscilloscopes use 12-bit A/D converters. Frequency domain instruments and spectrum analyzers often use 24-bits or more.

The minimum detectable signal level, neglecting noise, is based on the full-scale range of the A/D converter. The number of bits is described as follows:

$$Sensitivity = \frac{Full\ Scale}{2^{number\ of\ bits} - 1} \qquad 3.1$$

This sensitivity as a function of the number of bits is shown graphically in Figure 3-1.

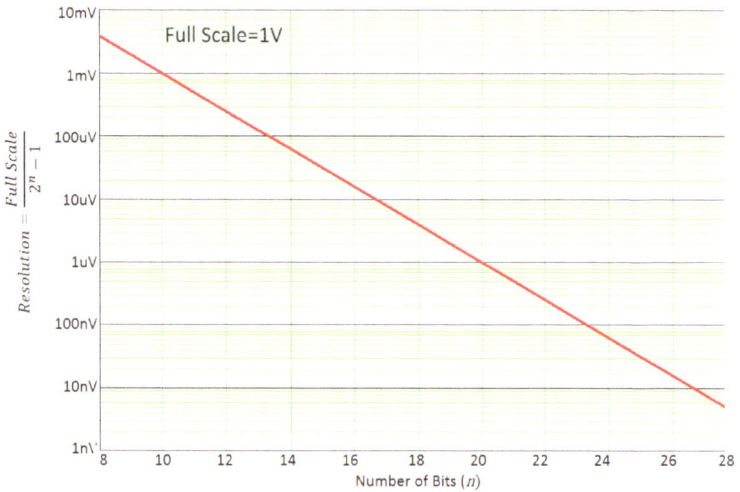

Measurement Sensitivity as a Function of the Number of Bits

Figure 3-1

Therefore, an 8-bit measurement, with a 1V full-scale reference, has a theoretical sensitivity of 3.92mV. The sensitivity of a 12-bit measurement is 240µV and a 24-bit measurement is 60nV.

The relationship demonstrates the significance of the number of bits and how it impacts the resolution of a measurement.

Effects such as circuit noise or input amplifier distortion and ADC clock jitter will degrade this performance.

Noise Floor

One of the most basic limitations of measurement performance is the measurement noise floor. The noise floor of the equipment can generally be determined using the measurement functions within the equipment itself.

For an oscilloscope, you can generally set it up for 4-channel operation with each channel terminated into 50 Ω with no external connections to any of the channels.

The equipment's RMS noise is then measured for each of the four channels. The results of such a test are shown in Figure 3-2.

Note that each channel has a slightly different noise level.

To show this is not specific to a manufacturer or model, the same measurement is repeated in Figure 3-3 using a different manufacturer's make and model of equipment.

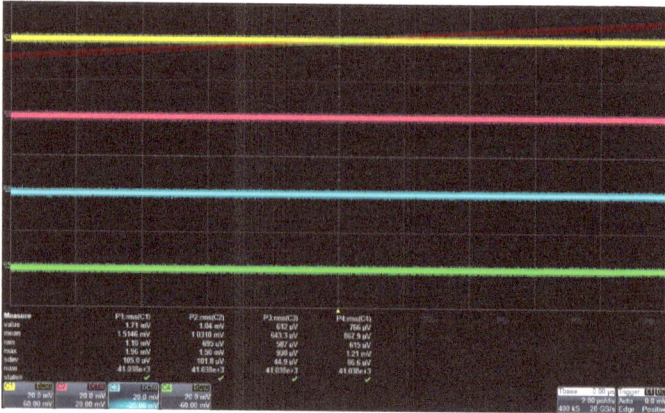

All Four Unconnected Channels set to 50Ω
Termination and 20mV/Div Vertical Resolution

Figure 3-2

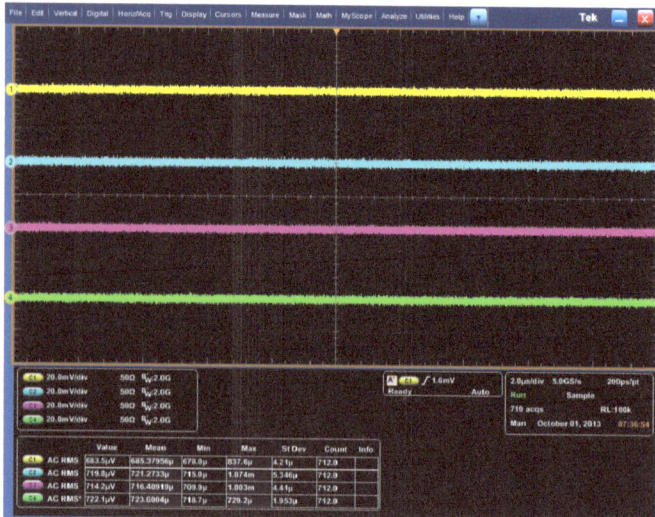

All Four Unconnected Channels set to 50Ω
Termination and 20mV/Div Vertical Resolution

Figure 3-3

Dynamic Range

The dynamic range is the ratio of the maximum measurable signal to the minimum measured signal.

The general calculations for the ADC's dynamic range are also a function of the bit count.

Again neglecting noise, the dynamic range of the ADC measurement is the full scale at the maximum and a Least Significant Bit (LSB) at the minimum, so that the available dynamic range for an 8-bit ADC is:

$$Dynamic\ Range = \frac{Full\ Scale}{1LSB} = 2^8 - 1 \qquad 3.2$$
$$= 255 = 48dB$$

In actuality, the computation of the dynamic range is a bit more complicated.

Using the same oscilloscope setups as in the previous figures, a signal is added to CH1 of each oscilloscope and the amplitude is adjusted until the maximum signal level is reached.

Note that in both cases, the maximum signal level extends beyond the display screen.

In Figure 3-4, the signal level is recorded as 94.03mVrms and represents the full-scale measurement.

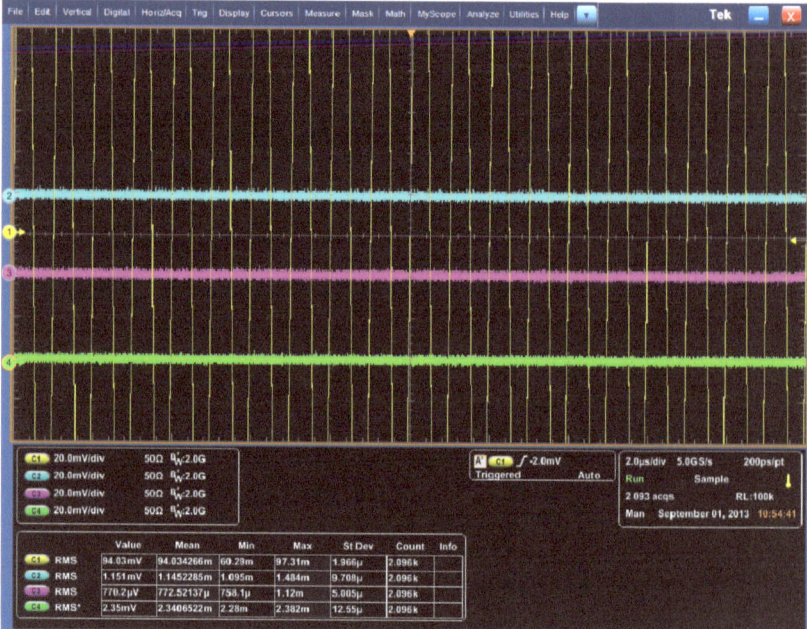

Despite being off the Screen, the Maximum Measurable Signal Amplitude is Recorded

Figure 3-4

We could express the dynamic range as:

$$Measurement\ Dynamic\ Range$$

$$= 20 \cdot log\left[\frac{Max\ Signal}{Noise} + 1\right] \qquad 3.3$$

$$= 20 \cdot log\left[\frac{94.03mV}{712.8uV} + 1\right] = 42.4dB$$

This assumes we could measure both a maximum amplitude signal, as well as a signal at the noise floor of the device (seen in Figure 3).

This is greatly reduced by limiting the signal level to the maximum screen display level. With 10 divisions and 20mV/div, the maximum displayed signal is 200mVpp or 70.71mVrms.

Display Dynamic Range

$$= 20 \cdot log \left(\frac{Max\ Display}{Max\ Noise} + 1\right) \qquad 3.4$$

$$= 20 \cdot log \left[\frac{70.71mV}{712.8uV} + 1\right] = 40dB$$

We can also use this measurement to approximate the ENOB (neglecting distortion effects) by solving Equation 3.2 for the number of bits, n.

$$Measured\ ENOB = log_2 \left[\frac{Max\ Signal}{Noise} + 1\right] \qquad 3.5$$

$$= 6.66\ bits$$

Similarly, we can make the same evaluation for another oscilloscope.

In Figure 3-5, the signal level is recorded as 60.02mVrms.

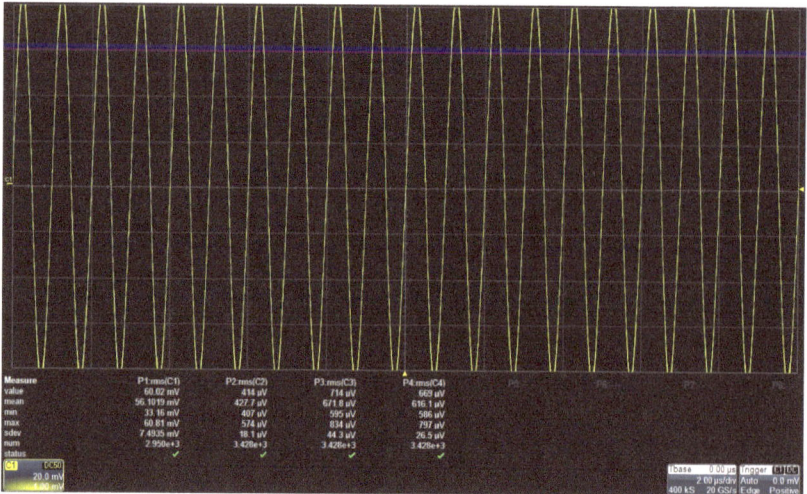

Despite being off the Screen, the Maximum Measurable Signal Amplitude is Measured

Figure 3-5

$$Dynamic\ Range = 20 \cdot log\left[\frac{Max\ Signal}{Noise} + 1\right]$$

$$= 20 \cdot log\left[\frac{60.02mV}{414uV} + 1\right] = 43dB \qquad 3.6$$

Limiting the signal level to the maximum screen display level with 8 divisions and 20mV/div, the maximum displayed signal is 160mVpp or 56.57mVrms.

Display Dynamic Range

$$= 20 \cdot log \left[\frac{Max\ Signal}{Noise} + 1 \right]$$

$$= 20 \cdot log \left[\frac{56.57mV}{414uV} + 1 \right] \qquad 3.7$$

$$= 42.7dB$$

As with the previous ENOB approximation, we can solve using Equation 3.5, resulting in 7.1 bits. Allowing 6dB above the noise floor as a minimum signal, the maximum usable dynamic range is 36.7dB. Finally, the noise measurement of a 12-bit oscilloscope is shown in Figure 3-6.

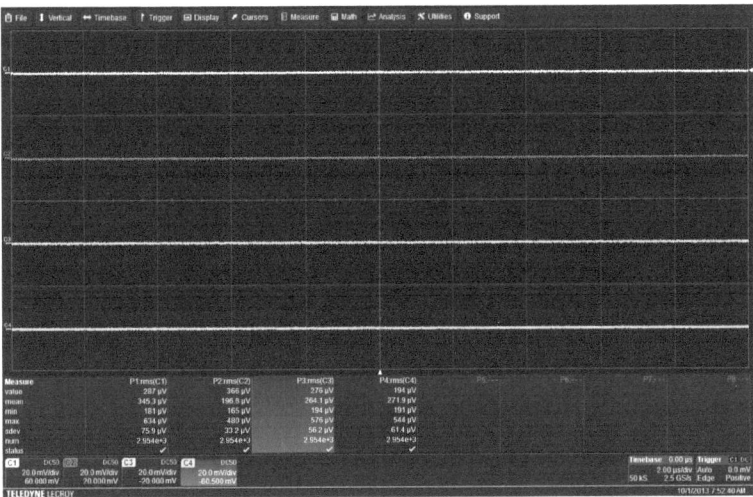

The Noise on an HDO6104 12-Bit Scope—this Scope Shows the Same Maximum Signal (56.57mVrms), but with a Lower Noise Level

Figure 3-6

$$Dynamic\ Range = 20 \cdot log\left[\frac{Max\ Signal}{Noise} + 1\right]$$

$$= 20 \cdot log\left[\frac{60.02mV}{194uV} + 1\right] \qquad 3.8$$

$$= 49.8dB$$

Limiting the maximum signal to the display gives this result:

$$Display\ Dynamic\ Range$$

$$= 20 \cdot log\left[\frac{Max\ Display}{Max\ Noise} + 1\right]$$

$$= 20 \cdot log\left[\frac{56.57mV}{0.194uV} + 1\right] \qquad 3.9$$

$$= 49.3dB$$

As with the previous ENOB approximation, we can solve using Equation 3.5, resulting in 8.2 bits.

Because noise is related to the measurement bandwidth, we cannot directly compare performance of equipment with different measurement bandwidths. A higher measurement bandwidth results in a higher noise level. We can show this by repeating the measurement of Figure 3-2 with the measurement bandwidth reduced to 1GHz and the sampling rate reduced to 5GS/s.

The results, shown in Figure 3-7, show that the noise level on CH1 and CH2 have been significantly reduced, while the noise level on CH3 and CH4 are much less impacted.

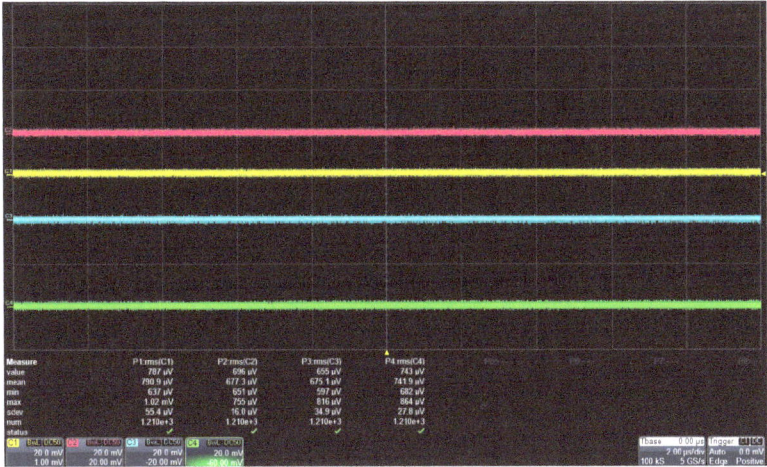

The Noise Measurement is repeated at 5GS/s and with Bandwidth Limited to 1GHz

Figure 3-7

Noise Density

Noise is often measured as noise density, which is the measured noise using a 1Hz measurement bandwidth.

This is often referred to as Resolution Bandwidth (RBW).

The noise amplitude is then a function of the bandwidth of the measurement, such that:

$$Noise = Noise\_Density \cdot \sqrt{RBW} \qquad 3.10$$

That being the case, the narrower the measurement bandwidth, the lower the noise floor.

All domains support RBW reduction to some extent. Frequency and spectrum domain instruments allow adjustments of the measurement bandwidth, often to 1Hz.

Oscilloscopes, due to their wideband nature, cannot do so

in the time domain, though almost all oscilloscopes offer a bandwidth limiting low pass filter to reduce the noise if the higher frequency content isn't required.

Some oscilloscopes allow more sophisticated programmable filters, but in general, the time domain measurement is a wideband instrument and, therefore, has a correspondingly higher noise floor.

There are tradeoffs for a reduced RBW.

First, if the signal jitter is greater than the RBW, the amplitude will not be correct. A reduced RBW also increases the resolution in the spectrum view. In order to discern two closely spaced signals, the resolution bandwidth must be lower than the spacing of the signals.

Another tradeoff for RBW is that the lower the RBW the longer it takes for the sweep.

Many modern oscilloscopes offer spectrum analyzer functionality as an option, most using FFT analysis with a spectrum analyzer user interface. This functionality allows us to demonstrate the relationships between noise and RBW.

Figure 3-8 shows an oscilloscope display using the spectrum analyzer option so that we can display the instrument noise floor, as well as the selected RBW.

The noise level is measured as -64.64dBmV and the actual RBW is 2.2Hz.

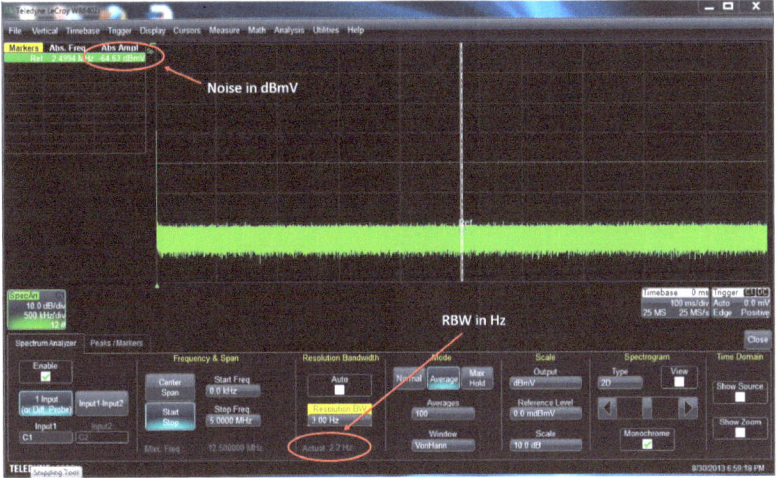

Oscilloscope Using an FFT Spectrum Analyzer Interface to Illustrate the Relationship between Noise and RBW

Figure 3-8

The measurement from Figure 3-8 is repeated for various selected RBW.

Note that the actual RBW is not exactly the same as the selected RBW, and so the actual RBW is used in our noise calculations.

A summary of the results are shown in Table 3.1 with the signal levels converted from dBmV to voltage and then converting to a noise density by dividing the voltage by the square root of the measurement bandwidth.

The result is nearly constant with an average noise density of 401nV/root-Hz.

dBmV	Volts	BW		V/SQRT(BW)
-31.97	2.52E-05	4577.6		3.73E-07
-22.03	7.92E-05	36621.1		4.14E-07
-41.16	8.75E-06	572.2		3.66E-07
-46.04	4.99E-06	143.1		4.17E-07
-57.65	1.31E-06	8.9		4.39E-07
-64.63	5.87E-07	2.2		3.96E-07
		average		4.01E-07

Noise Density related to Noise Voltage and Measurement Bandwidth

Table 3.1

A similar display is shown in Figure 3-9 using SignalVu® to display the average noise level of -83.65dBm using an RBW setting of 10kHz. The maximum measurable signal is also shown. It reveals the dynamic range of the measurement, which is the difference between the maximum signal level and the noise level, as approximately 85dB.

The RBW is a function of ADC oversampling. The apparent bit count increases one bit for each factor of four oversampling. An x4 increase in sampling rate improves the ADC ENOB by 1 bit. This is often referred to as process gain.

Using the same oscilloscopes used for the noise floor at the beginning of this chapter, the maximum measurable signal is determined. The dynamic range is again calculated as the maximum signal level divided by the noise level. In the case of Figure 3-9, a delta marker in the upper right corner directly displays the dynamic range as 84.74dB. Note that this is nearly 35dB better than the equivalent time domain measurement. This is a result of the oversampling and filtering process.

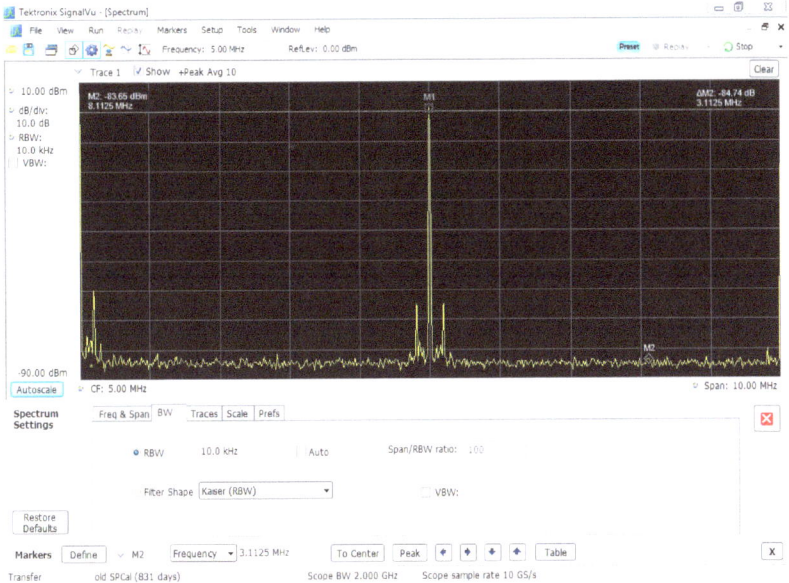

SignalVu® Measurement showing the Average Noise Level and the Maximum Signal Level

Figure 3-9

The measured signal needs to be large enough to rise above the noise.

As a rule, the noise margin should be at least 6dB, and preferably 10dB, meaning that we can measure a signal that is 6dB above the noise floor.

Figure 3-10 shows two signals, close in frequency and set for the maximum dynamic range. The yellow trace is set close to the maximum measurable signal level and the blue trace shows a signal that is 10dB above the noise floor.

The signals are not exactly at the same frequency in order to make it easier to distinguish them.

This measurement confirms the dynamic range of the measurement.

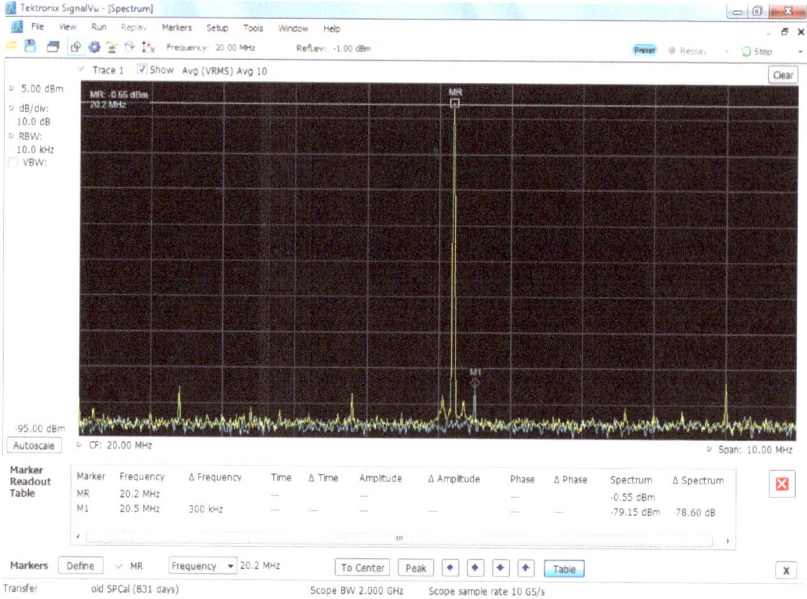

SignalVu® Measurement showing the Dynamic Range with a 10dB Noise Margin

Figure 3-10

Signal Averaging

While it is possible to improve noise by averaging sweeps, this feature must be used with extreme care.

In theory, most noise is white noise, meaning its distribution is random and Gaussian. The averaging of an infinite number of traces should, therefore, average to zero, greatly reducing the measurement noise.

This effect can be demonstrated by repeating the measurement of Figure 3-3 and averaging 256 measurements results as shown in Figure 3-11.

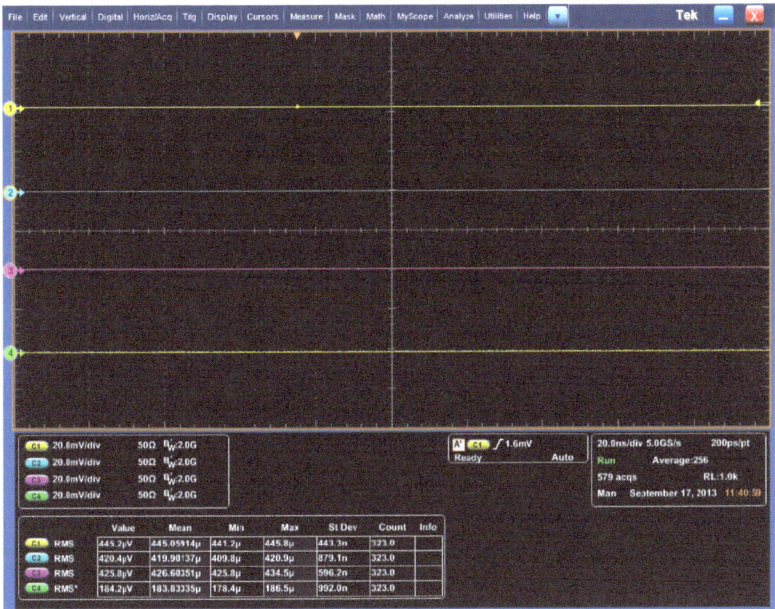

The Same Measurement as the Figure-3 Noise Measurement Repeated with 256 Trace Averages

Figure 3-11

Some channels are shown to have reduced noise, indicating that the noise is white noise while other channels have not significantly changed.

This indicates that the noise is a fixed signal, making the degree of benefit uncertain.

Another issue is that even expected repetitive signals are sometimes not repetitive. For example, consider the step load response of a switching point of load regulator.

While this might be expected to be a repetitive signal, it is not as shown in Figure 3-12.

The positive overshoot resulting from the rapid load reduction is not quite repetitive. The response is somewhat dependent on where in the switching cycle the regulator is when the load was reduced. These two signals are not synchronous and, therefore, the response is not truly uniform.

It is quite easy to lose the signal in the averaging.

This is also often the case when looking at the duty cycle of the switching regulator as seen in Figure 3-13.

Averaging is discussed in more detail later in the individual measurement sections, but for now, be forewarned that if you are not certain about the nature of the signal, you should avoid the use of averaging.

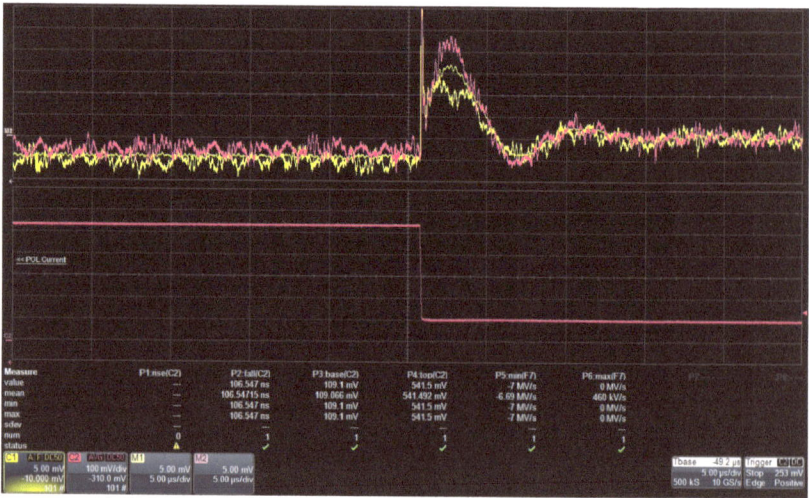

Step Load Response of a POL Regulator showing the Non-Repetitive Characteristics

Figure 3-12

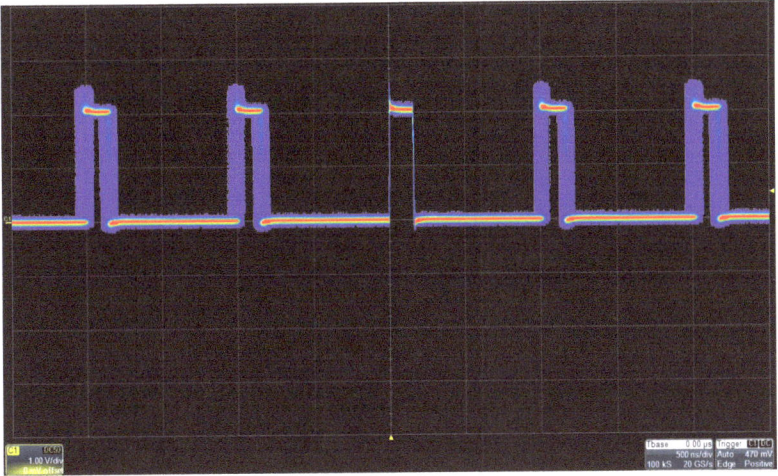

Switch node of a POL regulator showing the non-repetitive characteristics of duty cycle and frequency

Figure 3-13

Scaling

While it is important that the measurement have sufficient dynamic range and noise floor, it is also important to be able to display the results in a meaningful way. The oscilloscope display in Figure 3-13 shows a signal from an AWG delivered via a port splitter to 2 channels of the oscilloscope. One channel is directly connected while the other channel is connected via a 30dB attenuator.

The visually discernible dynamic range, using linear scales, is between 30dB and 40dB, though the oscilloscope measurement functions work beyond our visual range. In some cases the apparent dynamic range can be improved using zoom traces and filters. The display in Figure 3-15 shows the same signals connected though a 60dB attenuator. Note that the attenuated signal is no longer discernible in the upper display window. A second display with increased sensitivity is added, as well as a filtered version of the zoom trace. The oversampling and filter benefits are again illustrated as the dynamic range now exceeds the previous 43dB calculation by nearly 30dB.

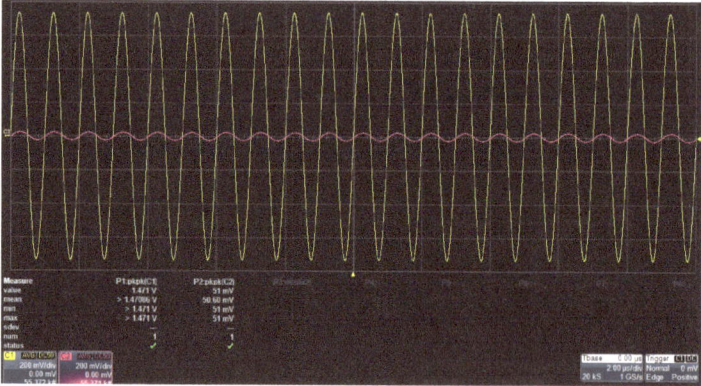

These Two Signals Provide a Visual Representation of a 30dB Dynamic Range on a Linear Oscilloscope Screen

Figure 3-14

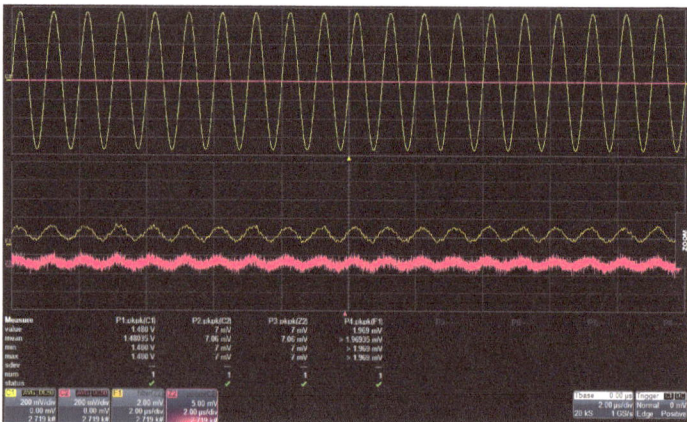

Use Multiple Scales and Filtering to improve the Dynamic Range to 60dB

Figure 3-15

The example in Figure 3-18 shows the open loop gain of an arbitrary op-amp.

The maximum signal level is 85dB and the minimum signal level is -40dB. At the same time, the frequency spans the range of 1Hz to 40MHz or nearly eight orders of magnitude.

Open loop gain plots are generally presented this way in the manufacturer's datasheets. In this case, both axes span large ranges, and so they are both displayed logarithmically.

For comparison, the next two figures clearly illustrate the benefits of the logarithmic scaling.

Attenuators

The equipment internal attenuators allow measuring larger signals, though the noise is proportionally increased.

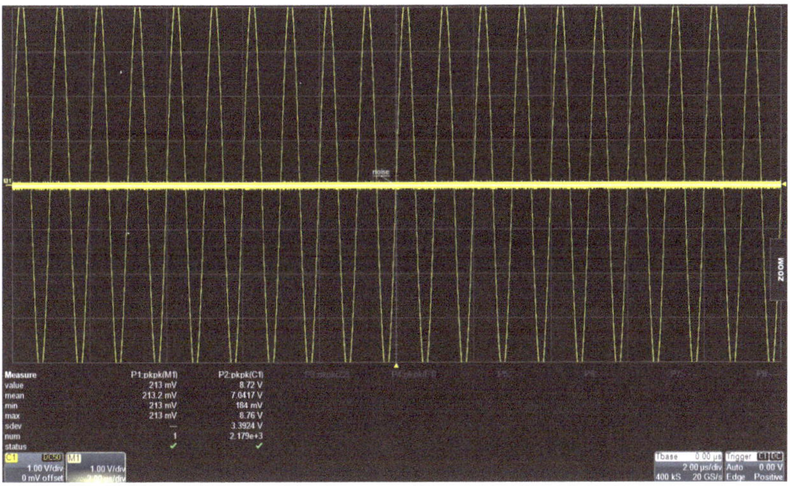

Noise Floor and Maximum Signal at 1V/div

Figure 3-16

$$Dynamic\ Range = 20 \cdot log \left(\frac{Max\ Signal}{Noise}\right) \qquad 3.11$$

$$= 20 \cdot log\ \frac{8.72V}{213mV} = 32dB$$

Comparing this result with the previous result in Equation 3.7 clearly shows the degradation resulting from the addition of the attenuators.

Preamplifiers

In contrast to the impact of the attenuators, it is possible to improve the noise floor and dynamic range with the addition of a low noise preamplifier when measuring signal amplitudes below full scale.

The spectrum analyzer plot shown in Figure 3-17 shows the noise floor without a preamplifier in the upper yellow trace. The marker identifies the noise level as 477nV at 10kHz.

The RBW is set to 10Hz and so the noise density can be determined as 151nV/root-Hz by dividing the 477nV by the square root of the 10Hz RBW.

$$Noise\ Density = \frac{477nV}{\sqrt{10\ Hz}} = 151nV \Big/ \sqrt{Hz} \qquad 3.12$$

The amplifier output noise density should be equal to or better than the noise density of the measurement without the preamplifier.

In this example, the noise density is 151nV/root-Hz and a Picotest J2180A wide band low-noise preamp is added at the input of the measurement instrument. The gain is adjusted in the spectrum analyzer to account for the 20dB external amplifier gain and the result is shown in the blue trace. The amplifier noise density of 2.4nV/root-Hz is much lower than the 151nV/root-Hz noise density.

Therefore, the noise floor is improved by the full 20dB of the preamplifier or approximately 15nV/root-Hz.

The addition of the preamp reduced the apparent noise, by moving the reference level down by the 20dB amplifier gain. The signals are also amplified by 20dB.

Therefore, the addition of the preamplifier increases the sensitivity and improves the dynamic range if the signal is less than full scale using the most sensitive equipment setting.

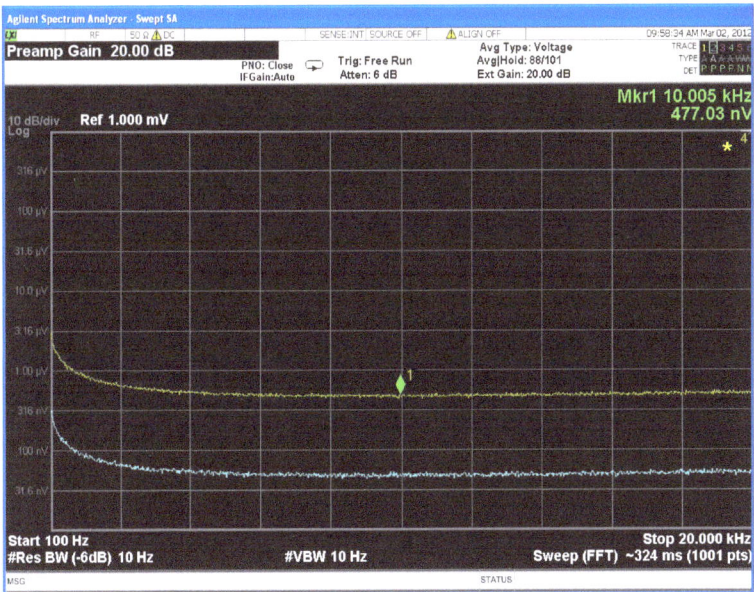

The Addition of a Low-Noise 20dB Preamplifier improved the Noise Floor by 20dB.

Figure 3-17

Linear vs. Log display

In many cases, the dynamic range of the measurement will be much larger than the signals seen so far.

One common example is op-amp open loop gain.

A simulated op-amp model is shown in Figure 3-18.

The maximum gain is 85dB and the minimum displayed gain is -40dB, requiring 125dB dynamic range to display the y axis. Likewise, the measured frequency range spans from 1Hz to 40MHz, equivalent to 152dB dynamic range in the horizontal axis.

Using logarithmic scales for both the gain and frequency axes allows us to display and discern the details despite the large dynamic range requirements.

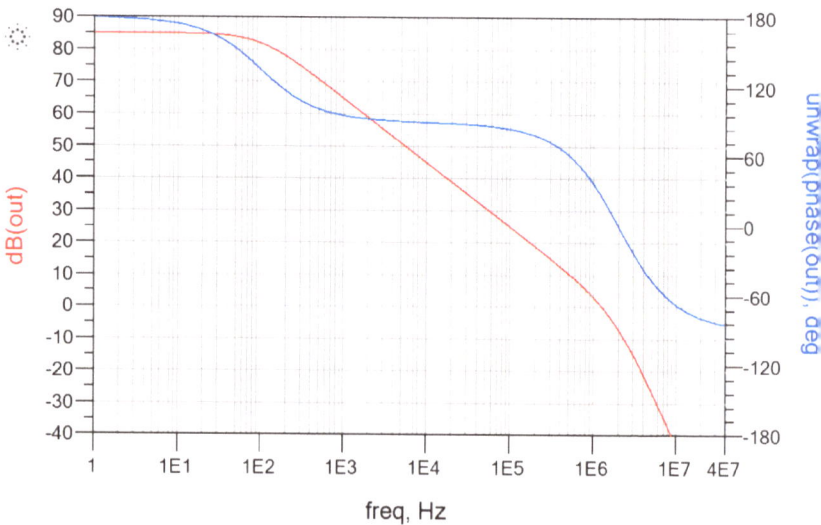

Simulated Op-Amp Open Loop Gain with Logarithmic Gain and Frequency Axes

Figure 3-18

Figure 3-19 shows the same simulation result using linear scales.

In this figure, two markers are placed to show that the low frequency gain and phase shift are represented correctly, though the measurement demonstrates very poor fidelity.

Figure 3-20 again shows the same simulation result, this time using a linear frequency scale and a logarithmic gain scale.

Two markers are placed to show that the low frequency gain and phase shift are both represented correctly, though the measurement lacks the clarity offered by the logarithmic scale.

These examples demonstrate the importance of logarithmic scales when measuring wide dynamic ranges, whether the wide range is in the horizontal or vertical axis.

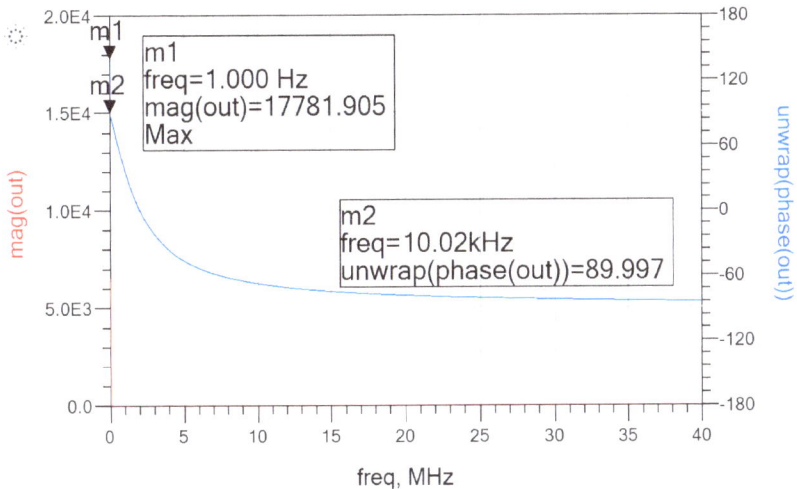

Simulated Op-Amp Open Loop Gain with Linear Frequency and Gain Axes

Figure 3-19

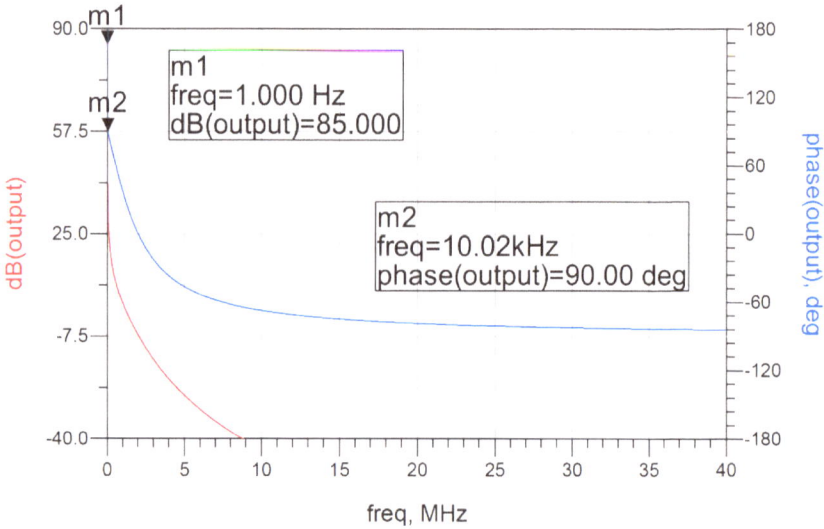

Simulated Op-Amp Open Loop Gain with Linear Frequency Axis and Logarithmic Gain Axis

Figure 3-10

These techniques are applied to individual measurements in later chapters.

Measurement Domains

Frequency Domain

Frequency domain measurements can be either single frequency or frequency sweep measurements.

Frequency domain measurements are either made using a Frequency Response Analyzer or a Vector Network Analyzer (VNA).

One common frequency domain measurement is the gain and phase measurement, such as op-amp open loop gain and phase.

Other gain phase examples include filter transfer functions, control loop Bode plots, common mode rejection and PSRR plots.

A few examples of common frequency domain impedance plots include capacitance and ESR, inductance, DCR and voltage reference output impedance. In high-speed systems, the common power distribution network impedance measurement is also a frequency domain measurement.

The RBW is selectable in frequency response measurements, minimizing noise.

The frequency and amplitude scales can both be independently selected as either linear or logarithmic, thus allowing a very large dynamic range.

The frequency domain instruments generally have higher ADC bit counts allowing very large dynamic range measurements and very high sensitivity.

Gain and Phase

The gain and phase measurement is a vector (real and imaginary) measurement of the ratio of two measured signals.

The results are the magnitude of gain (or attenuation) and phase difference between the two signals. A common example found in most op-amp datasheets is the op-amp open loop gain, as shown in Figure 3-12.

This measurement example illustrates many of the benefits of frequency domain measurements including very high sensitivity, wide dynamic range, low noise and a very broad measurement frequency range.

This measurement is performed with an op-amp output signal level of approximately 100mVrms.

Therefore, the signal level at the op-amp input is reduced by the open loop gain of approximately 85dB at low frequency or 6μVrms.

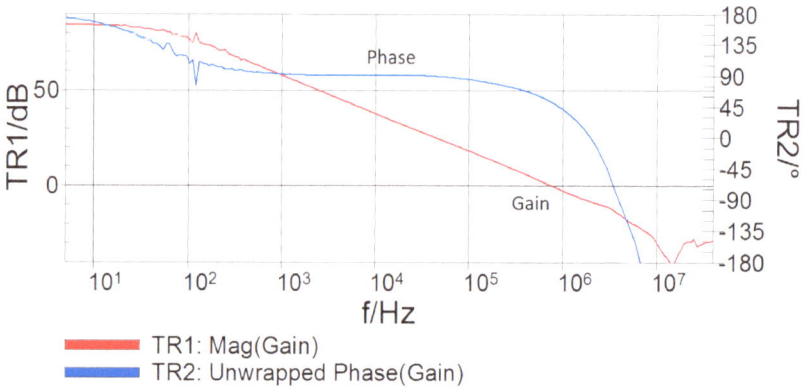

TR1: Mag(Gain)
TR2: Unwrapped Phase(Gain)

Frequency Domain Measurement Example LT1001 Op-Amp Open Loop Gain

Figure 3-21

S-Parameters

Among other tests, RF VNAs perform S parameter measurements, which are helpful in accurately measuring high frequency signals.

One of the fundamental benefits of S-parameter measurements is that S-parameters are generally measured without a probe, but rather use a fixed impedance system. Most VNA connections are compatible with 50Ω, though 75Ω is common in video, cable and satellite TV applications.

S-parameters measure both the transfer function of each port to every other port and to itself. This is accomplished by measuring both the transmitted signal and the reflected signal. If a perfect 50Ω source is connected to a perfect 50Ω load, the transmission is 100% and the reflection is zero.

S-parameter measurements can include one or several ports, and the number of S parameters is the square of the number of ports.

The port labels define the measurement.

For example, if only a single port exists, there is only one S-parameter S11, where S11 is the return loss of port 1. If two ports are used, there are four S-parameters, S11, S21, S12 and S22. S11 and S22 represent the return loss of port 1 and port 2 respectively, while S21 is the voltage appearing at port 2 because of a signal at port 1. S12 is the signal appearing at port 1 resulting from a signal at port 2.

If we establish the source impedance as Z_S and the loads impedance as Z_L then we can define a reflection coefficient as:

$$\Gamma = \frac{reflected}{incident} = \frac{Z_S - Z_L}{Z_S + Z_L}$$

3.13

The return loss, RL, is defined as:

3.14

$$RL = -20 \cdot \log(|\Gamma|)$$

S-parameters include both magnitude and phase terms and as we have shown here can be used to measure transfer functions, as well as return losses. All ports are terminated into the characteristic system impedance (typically 50Ω or 75Ω) using coaxial cables matching this impedance.

Impedance

Since we have defined the return loss as a function of the source and load impedance and the source impedance is the characteristic system impedance, we can convert from return loss to impedance.

$$Z = \frac{1 + \Gamma}{1 - \Gamma} \cdot Z_S$$

3.15

This capability is generally included within the VNA allowing the complex impedance to be displayed directly without the need of adapters, probes, or fixtures. The VNA is also designed to be easily calibrated using a SHORT, OPEN and LOAD calibration for a single port measurement and SHORT, OPEN, LOAD and THRU calibrations for multiple ports.

In addition to measuring transfer functions and impedances, the VNA also provides Group Delay, which is the first derivative of the phase curve. In RF measurements, the Group Delay, generally designated as T_g, is used to measure resonant Q of an LCR network, or other resonator, such as a quartz crystal.

There is a direct relationship between Group Delay and Q:

$$Q = \pi \cdot T_g \cdot frequency \qquad\qquad 3.16$$

This is a convenient method of determining the Q and will be used in later chapters of this book.

Time Domain

Time domain measurements record and decode signal voltage or current as a function of time. The time domain is very helpful for troubleshooting and detecting glitches, decoding logic, establishing timing, jitter and signal integrity metrics, as well as measuring signal edges, overshoots, response times and other characteristics. There are also several limitations to time domain measurements. These include the linear x (time) and y-axes, making it difficult to measure signals having large dynamic ranges. As was shown previously, the wideband nature of the time domain measurement also results in a higher noise floor than other domains. The time domain measurement is also heavily dependent on the sample rate and time base of the instrument to observe high fidelity signals. The measurement in Figure 3-22 illustrates the impact of sample rate on the measurement.

In this measurement, the first leading edge spike, which represents the minimum circuit voltage, is evident at 20MS/s and not visible at 5MS/s.

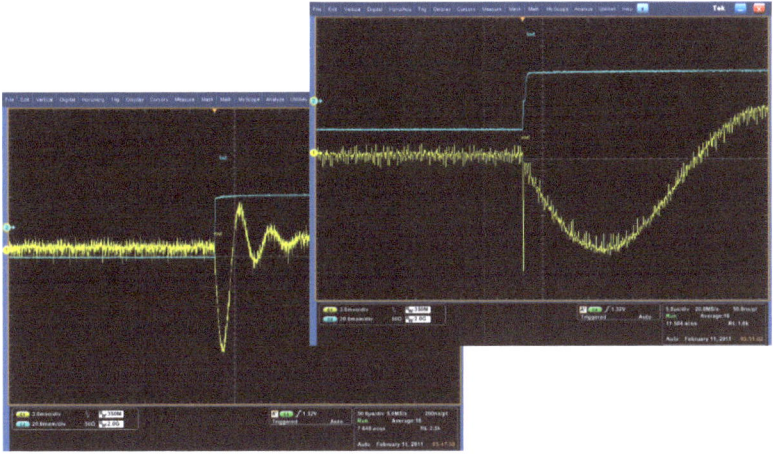

Time Domain Measurement Example of Sample State

Figure 3-22

The measurement in Figure 3-23 also shows a limitation of the linear time scale. In the upper display, the ringing in the voltage waveform is visible, while in the expanded image in the main display a second ringing frequency is evident.

In order to assure no details were missed in the measurement, it is generally helpful to set a long measurement record. Several zoom images with different scales can then be set in order to be sure that significant information is not missed.

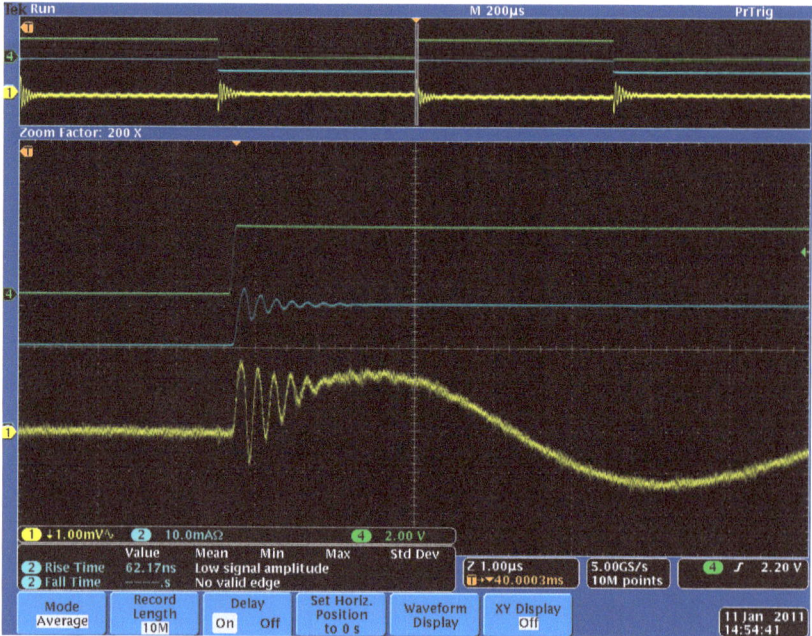

Time Domain Measurement Example of Time Scale

Figure 3-23

While the S-parameter measurement eliminates probe effects and can be easily calibrated over the full measurement range, the oscilloscope is dependent on the characteristics of the probe.

In addition, the oscilloscope lacks the calibration capabilities of S-parameter measurements and often requires external stimulus, which can degrade measurement fidelity.

These issues are examined in further detail in later chapters of this book.

Spectrum Domain

The spectrum domain is similar in some ways to the frequency response measurement.

In the spectrum analyzer, voltage or current signal amplitudes are measured using either FFT or swept methods, allowing distinct signal frequency components to be assessed.

Like frequency domain instruments, the RBW is selectable in frequency response measurements, minimizing noise.

Frequency and amplitude scales can both be independently selected as either linear or logarithmic allowing very large dynamic range measurements in either axis.

The spectrum domain instruments also generally have higher ADC bit counts. This enables very large dynamic range measurements and very high sensitivity.

The spectrum analyzer is often used to assess and locate noise sources, including Electro Magnetic Interference (EMI), clock jitter and power supply ripple signals.

An example shown in Figure 3-24 shows the noise voltage of a linear voltage regulator, a custom low noise regulator, and a voltage reference.

Note that in the case of the linear regulator (an LM317) the device has periodic responses or "spurs" at low frequencies.

Despite these spurs being only 2μV in this measurement, very sensitive electronics, such as high-speed clocks, can be sensitive to these signals.

Additionally, the low frequency nature of these signals can make them difficult to filter.

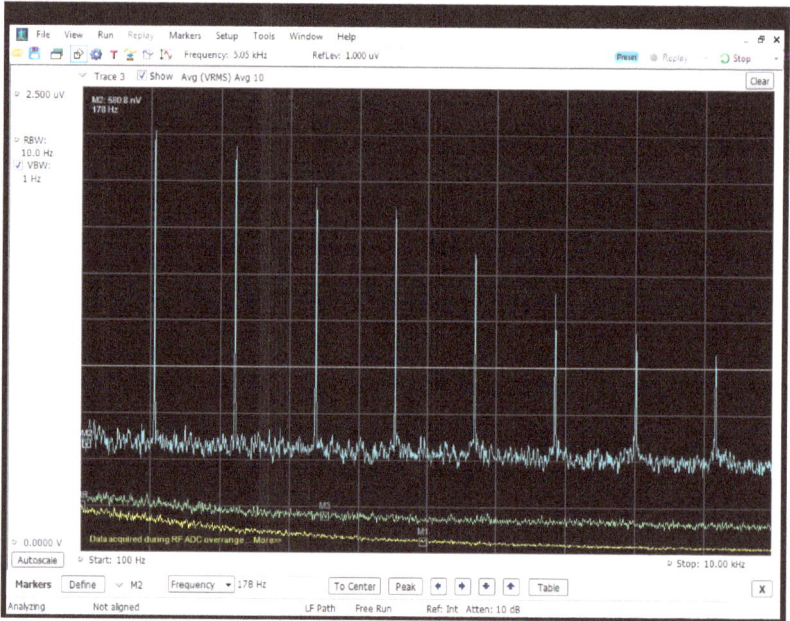

Spectrum Domain Measurement Example

Figure 3-24

Most modern oscilloscopes include spectrum analyzer capability. This spectrum analysis can be used in many applications.

However, they are generally not equivalent to a high-performance, standalone spectrum analyzer in either sensitivity or noise floor.

Additionally, most of these devices offer only linear time scaling. This makes them difficult to use for measuring power supplies, which generally present a very wide dynamic range in the frequency scale. Some of these deficiencies can be improved by adding a low noise preamplifier as discussed previously and as shown in Figure 3-25.

A special class of spectrum analyzers, the signal source analyzer, generally offers superior sensitivity and noise characteristics compared with a spectrum analyzer. The signal source analyzer can accurately measure clock jitter, integrated over a selectable frequency range, as well as identifying spurious responses. The signal source analyzer and spectrum analyzer are both used in later sections of this book.

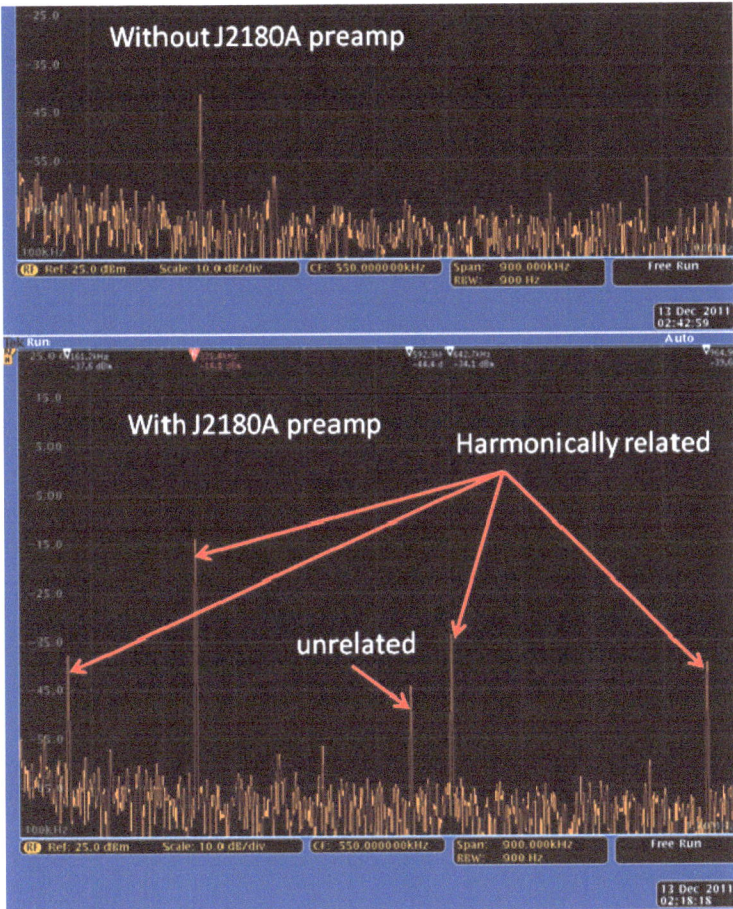

Identifying Noise Sources using a Spectrum Analyzer with Improved Sensitivity via an added Preamplifier

Figure 3-25

Comparing Domains

By showing the same signal in two domains, additional information can be derived.

In the case of Figure 3-26, the output voltage of a voltage reference is shown in response to a step load.

From the upper time domain trace there are significant large signal effects identified by the different ringing frequency associated with each of the two current step levels.

While it appears that each edge rings at a different frequency, the spectrum analyzer plot shows that—in addition to the two ringing frequencies—all of the sum and difference frequencies of the step load repetition rate and the ringing are apparent.

These resulting frequencies can propagate through a system and it is often difficult to isolate the source. This often turns out to be the culprit when unexpected noise signal frequencies show up in the performance measurements.

The example in Figure 3-27 shows the use of three domains to identify an EMI issue.

The EMI signal contains two significant frequency components.

One is identified as the ringing of the switch node.

The other resonance—found using a spectrum analyzer with a near field probe—identifies a resonant printed circuit board plane, which is confirmed by measuring the plane impedance in the frequency domain.

Using the right measurement domain or combination of measurement domains expedites the troubleshooting process and allows circuit optimization with minimal effort.

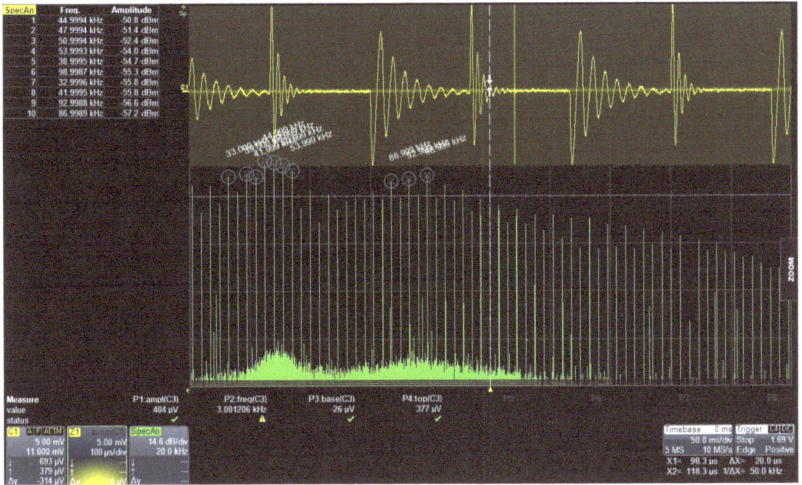

Step Load Response Showing Frequency Comb

Figure 3-26

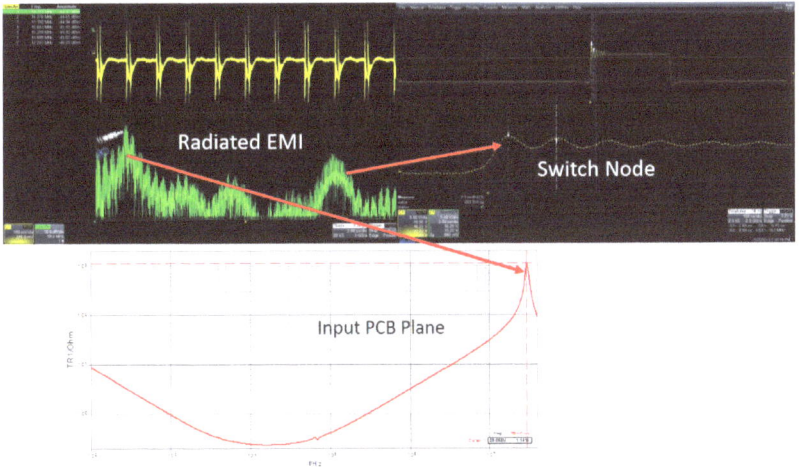

Using Three Domains to Quickly Locate an EMI Problem

Figure 3-27

Tips and Tricks

1. Know your equipment. An easy way to improve the signal to noise of your measurement is simply to use the lowest noise channel for sensitive measurements, especially if you are only using one channel.
2. Keeping the sample rate and measurement bandwidth equivalent is critical when comparing the performance of test equipment from different vendors.
3. Be sure your scope is calibrated. Most modern oscilloscopes include an automatic calibration or signal path compensation calibration.
4. ALWAYS measure the noise floor. The addition of probes and unshielded wires can add significantly to the noise floor.
5. Measure the noise with the DUT turned off. Many labs have significant ambient noise from computer monitors, fluorescent lamps and wireless routers.
6. Try to set the measurement to be close to full scale for the best signal to noise ratio.
7. Oscilloscopes offer much better sensitivity in spectrum analyzer mode. This allows logarithmic scaling in the vertical axis.
8. Signal to noise and sensitivity can often be improved by using an external low-noise preamplifier
9. Look at signals in multiple domains in order to get a clear view of the measurement.

Chapter References

Agilent 5990-5902 *Evaluating DC-DC Converters and PDN with the E5061B LF-RF Network Analyzer* ,
cp.literature.agilent.com/litweb/pdf/5990-5902EN.pdf
Agilent 5968-4506E *New Technologies for Accurate Impedance Measurement* ,
literature.agilent.com/litweb/pdf/5968-4506E.pdf
Agilent 5989-5935 *Ultra-Low Impedance Measurements Using 2-Port Measurements*,
cp.literature.agilent.com/litweb/pdf/5989-5935EN.pdf
Agilent 5990-5578 *Measuring Frequency Response with the Agilent E5061B LF-RF Network Analyzer*,
cp.literature.agilent.com/litweb/pdf/5990-5578EN.pdf
OMICRON—*Lab DC Biased Impedance Measurements Using the Bode 100 and the B-WIC Impedance Adapter*
Steven M. Sandler, Tom Boehler, Charles E. Hymowitz, *Network Analyzer Signal Levels Affect Measurement Results*, Power Electronics Technology January 2011 pages 26-28
Agilent Technologies *Impedance Measurement Handbook*. literature.agilent.com/litweb/pdf/5950-3000.pdf
OMICRON Labs Bode 100 User Manual
Fundamentals of Real Time Spectrum Analysis, Tektronix
The Fundamentals of Spectrum Analysis, Erik Diez Agilent Technologies
http://electronicdesign.com/test-amp-measurement/fundamentals-spectrum-analysis
DPX® Acquisition Technology for Spectrum Analyzers Fundamentals, Tektronix
Agilent 5965-7920E *Spectrum Analyzer Basics*, Agilent
http://www.edn.com/electronics-news/4381122/What-are-S-parameters-anyway-

Steven M. Sandler

Chapter Four

Test Instruments

THIS CHAPTER IS not intended to present a comprehensive list of all of the available power integrity related test instruments, but rather a brief introduction to those that were used in support of this book.

Most of the instruments are shown in Figure 4-1.

The selection of instruments is based on the measurement requirements, as well as some personal choices.

The instruments listed here include the frequency, time and spectrum domains as were discussed in earlier chapters. The most significant characteristics for each instrument type are discussed along with some tips for selecting equipment to meet your own needs.

A summary of the equipment types and their associated measurement applications is shown in Table 4.1.

Many of the Instruments Used in the Creation of this Book are Shown in this Picture of my Lab

Figure 4-1

Equipment Type	Application
Vector Network Analyzers	Stability, impedance (including PDN), PSRR, Reverse transfer, crosstalk, component characterization
Oscilloscopes	Ripple and Noise, troubleshooting, spectrum option adds EMI
Spectrum and Signal Source Analyzers	Ripple and Noise including clock jitter and EMI
Signal Generators	Measure/verify measurement bandwidth, time domain support for injectors, load profiles
TDR/TDT Analyzers	High Freqency VNA applications (including PDN) high frequency S-parameters

Summary of Equipment and Associated Applications

Table 4.1

Steven M. Sandler

Frequency Response Analyzers and Vector Network Analyzers

Frequency Response Analyzers (FRAs) and Vector Network Analyzers (VNAs) fall into the same baseline category of Network Analyzers, but are actually quite different instruments. A Frequency Response Analyzer is generally a low frequency instrument used to measure the transfer functions of control loops and filters.

The FRA generally has a low impedance oscillator and two high impedance inputs, often using differential amplifiers.

The VNA, on the other hand, is generally a high frequency device, sometimes starting at a frequency of 9kHz, with a 50Ω oscillator and two 50Ω inputs all mounted to an RF front panel/groundplane. The VNA is traditionally used to measure scatter parameters or RF devices. The VNA is also capable of measuring reflection, admittance and impedance using a number of different methods, which are discussed in detail in the Measuring Impedance chapter of this book.

FRAs and VNAs are both narrowband measuring instruments, resulting in very high dynamic range and a very low noise floor allowing the measurement of very small signals.

There are two instruments that I am familiar with that are both a FRA and a VNA in a single instrument. These instruments can both measure transfer functions of control loops and filters just like an FRA, but additionally can measure S-parameters, as well as impedance, reflection and admittance. They allow the inputs to be set for high impedance or 50 Ohms and start at much lower frequencies than a typical VNA. They are also very useful for component characterization, as well, and we often employ them for testing related to SPICE modeling. The result is one of the most versatile instruments in the lab, supporting the majority of measurements.

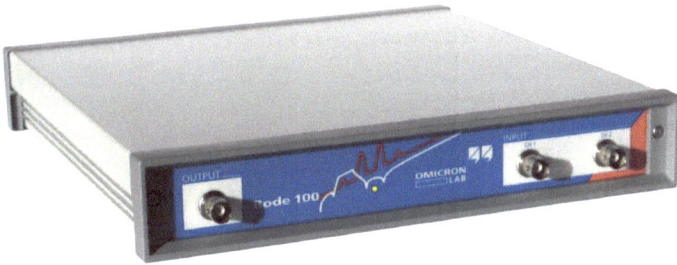

The OMICRON Lab Bode 100 FRA/VNA
Photo Courtesy of OMICRON Lab Reprinted with Permission

Figure 4-2

The OMICRON Lab Bode 100—Key Features

- Highly portable—fits in an average laptop case and weighs only 4.4Lbs
- 8W power consumption—can operate all day on a typical 12V 4500mA-hr CCTV battery
- One of the most cost effective VNAs on the market
- FRA and VNA functionality in one package
- Simple graphical user interface and high resolution output for publishing
- >100 dB dynamic range
- 1Hz-40MHz frequency range
- Low enough frequency to measure PFC loops.
- Assesses stability without access to the control loop
- SOLT calibration
- S-parameters
- Impedance measurements
- Precision 50Ω or 1MΩ selectable inputs
- Can measure impedance below 1mΩ
- Works with all Picotest Injectors
- Includes the Non-invasive Phase Margin Measurement Software

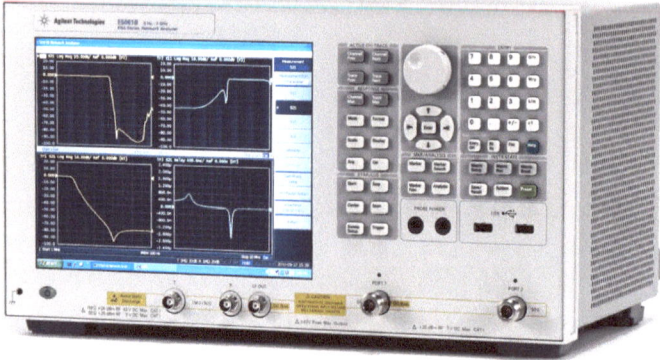

Agilent Technologies E5061B FRA/VNA
Photo Courtesy of Agilent Technologies
Reprinted with Permission

Figure 4-3

Agilent Technologies E5061B—Key Features

- >100 dB dynamic range
- 5Hz-30MHz frequency range (low frequency inputs)
- 5Hz-3GHz frequency range (high frequency inputs)
- FRA and VNA functionality in one package
- Impedance analyzer (optional)
- Can measure impedance below 1mΩ
- SOLT calibration
- S-parameters
- Precision 50Ω or 1MΩ selectable inputs
- Semi-floating inputs on low frequency inputs eliminates low frequency ground loops
- Built in DC Bias $\pm$40V
- Multiple display windows

Oscilloscopes

Several oscilloscopes are used in this book. The selection helps to highlight both the similarities and the differences. The choice of an oscilloscope is based on several figures of merit, which include sample rate, bandwidth, dynamic range and memory depth. Other considerations might include specialty features such as spectrum analyzer functionality, logic decoding and signal integrity functions. Since most oscilloscope probes are specific to a particular brand another significant consideration is the availability of probes to mate with the oscilloscope.

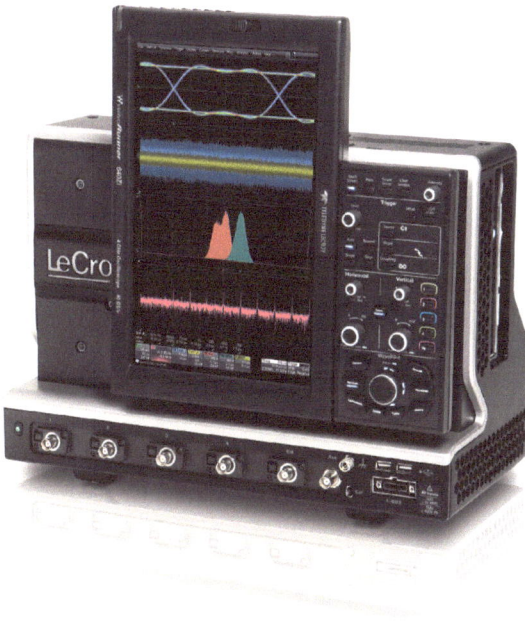

Teledyne Lecroy Waverunner 640Zi
Photo Courtesy of Teledyne Lecroy Reprinted with Permission.

Figure 4-4

Teledyne Lecroy Waverunner 640Zi—Key Features

- A versatile scope in the 400MHz to 4GHz class
- 40GS/s sample rate
- Low noise
- 256 Mpts of analysis memory
- 90° rotating and tilting display
- Small footprint, only 8.1 inches deep
- Two convenient USB ports on the front, two on the side
- Deep toolbox for more measurements, more math, and more power
- Large selection of serial triggers and decoders
- Up to 36 digital channels

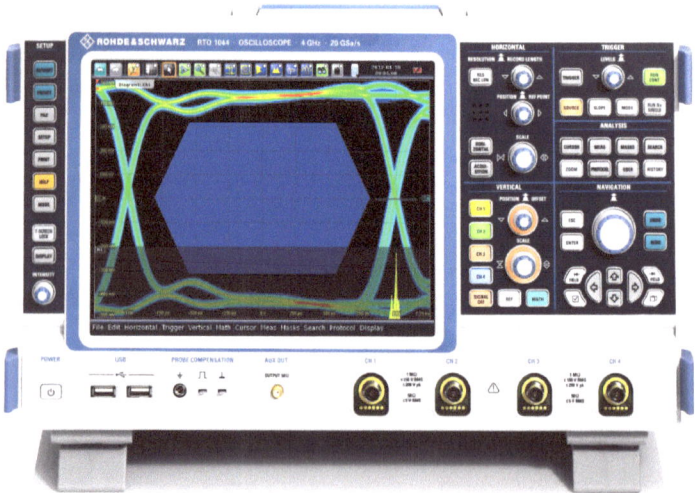

Rohde & Schwarz RTO1044
Photo Courtesy of Rohde & Schwarz Reprinted with Permission

Figure 4-5

Rohde & Schwarz RTO1044—Key Features

- Upgradeable models from 600MHz-4GHz
- Up to 20GSa/s (2 Channel) and 10GSa/s (4 Channel)
- Up to 400Mpts (1 Channel)/100Mpts (4 Channel) sample record length
- Multiple spectrum / RF Signal display
- Fast, flexible colored spectrum with frequency mask capability for use with RF signaling and EMI Debug
- Smart Grid drag and drop user interface for maximum flexibility and signal configuration
- >1,000,000 Waveforms per second in standard operation (no special mode required)
- Superior Reference Clock ±0.002PPM (opt. B4)
- History mode (always on) in both time and frequency (to reduce trouble shooting time)
- Digital trigger for unprecedented trigger stability in noisy environments (WYSIWYG)
- High Dynamic Range (84 dB (typical) signal to noise (1 GHz @ 100 KHz resolution bandwidth)
- True 1mV/Div setting with very low noise to detect weak signals. (1mV/Div—0.08 mV RMS Noise Floor @50Ω)
- Each input can be used as RF and Analog to offer 4 input channels from 500 MHz to 4 Ghz for simultaneous time and frequency measurements
- Many optional analysis packages including Jitter, Power Analysis, OFDM, LTE, etc.

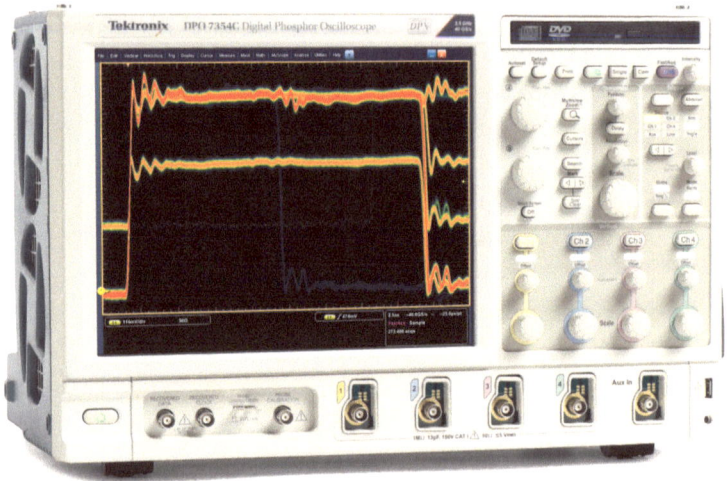

Tektronix DPO7354C
Photo Courtesy of Tektronix Reprinted with Permission

Figure 4-6

Tektronix DPO7354C—Key Features

- Has up to 3.5GHz bandwidth
- Up to 40GS/s for one channel and 10GS/s for 4 channels
- DPX color persistence shows multiple traces overlaid
- 500 Megapoint record length.
- Over than 250,000 waveforms per second capture rate with a special FastFrame™ segmented memory acquisition mode to achieve over 310,000 waveforms per second capture rate
- Runt/Glitch triggers
- Many optional analysis packages

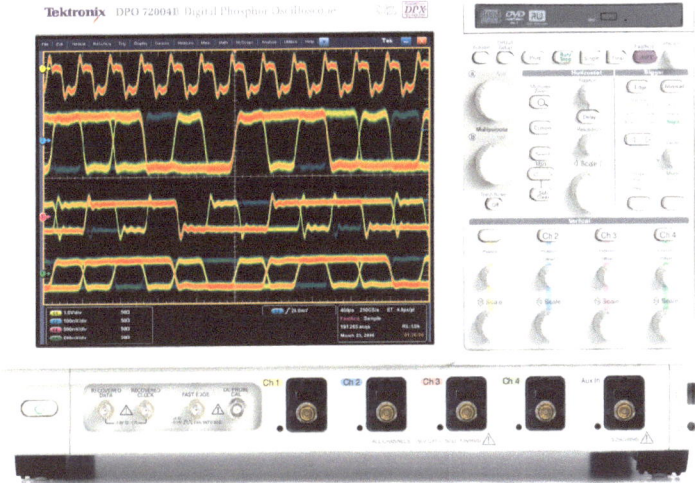

Tektronix DPO72004B
Photo Courtesy of Tektronix Reprinted with Permission

Figure 4-7

Tektronix DPO72004B Key Features

- Up to 33GHz bandwidth
- 100GS/s for 2 channels and 50GS/s for 4 channels
- DPX color persistence shows multiple traces over time
- Up to 500 Megasample record length,
- Fast waveform capture rate with over 300,000 waveforms per second per channel
- Runt/Glitch triggers
- Many optional analysis packages
- Supports the P7600 TriMode probes available to 33GHz

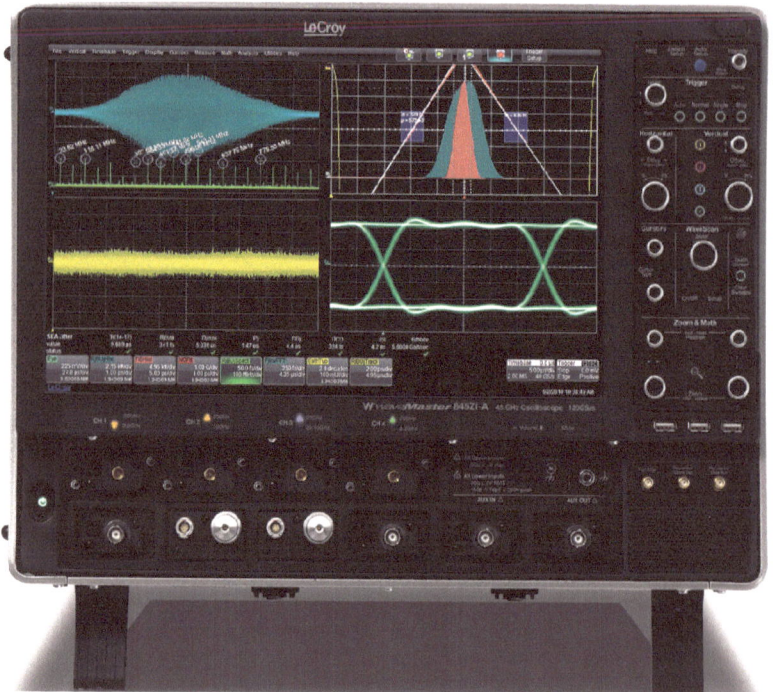

Teledyne Lecroy Waverunner 845Zi
Photo Courtesy of Teledyne Lecroy Reprinted with Permission

Figure 4-8

Waverunner 845Zi—Key Features

- Up to 20 GHz on 4 Channels
- Up to 45 GHz bandwidth at 120GS/s
- 14.1 Gb/s Hardware serial trigger available
- 768 Megapoints of memory
- Crosstalk and vertical noise analysis
- Four simultaneous diagrams and jitter calculations
- Large selection of serial triggers and decoders

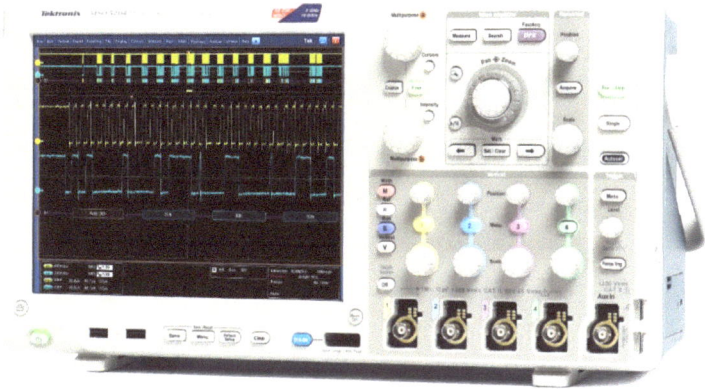

Tektronix MSO5204
Photo Courtesy of Tektronix Reprinted with Permission

Figure 4-9

Tektronix MSO5204—Key Features

- Models with up to 2 GHz bandwidth
- Up to 10 GS/s real-time sample rate on one or two channels and up to 5 GS/s on all four channels
- 250 Megapoints record length
- >11 bits vertical resolution
- Memory acquisition mode with up to 290,000 segments and >310,000 waveforms per second capture rate
- User-selectable bandwidth limit and DSP filters for lower noise and better measurement accuracy
- FastFrame™ segmented memory acquisition mode with up to 290,000 segments and >310,000 waveforms per second capture rate

Teledyne Lecroy HDO6104
Photo Courtesy of Teledyne Lecroy Reprinted with Permission

Figure 4-10

Teledyne Lecroy HDO6104—Key Features

• 12-bit ADC resolution and up to 15 bits with enhanced resolution available
• 1 GHz bandwidth
• Up to 250 Mpts/Ch memory
• Spectrum analyzer Mode
• 16 digital channels with 1.25 GS/s sample rate
• History mode that allows waveform playback
• Clock-data jitter analysis and many other mixed signal analysis options

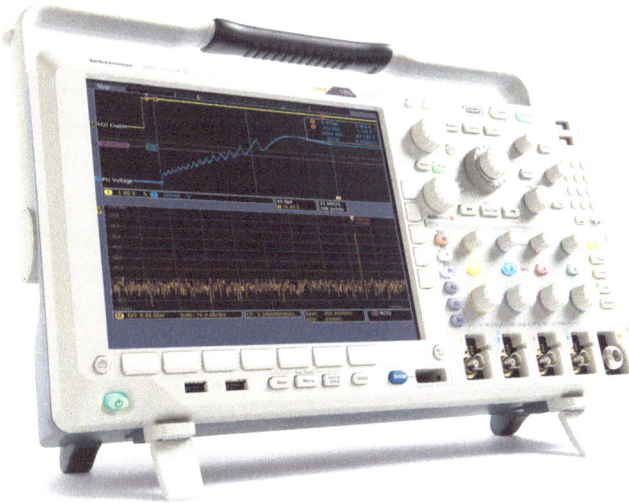

Tektronix MDO4104-6
Photo Courtesy of Tektronix Reprinted with Permission

Figure 4-11

Tektronix MDO4104-6—Key Features

- 4 analog channels and 16 digital channels
- 1GHz Bandwidth
- High-speed acquisition provides 60.6ps fine timing resolution
- 1 spectrum analyzer channel
- Advanced RF analysis software
- Time-correlated analog, digital, and RF signal acquisitions in a single instrument
- Easily navigate time-correlated data from both the time and frequency domains
- Includes automated measurements for channel power, adjacent channel power ratio, and occupied bandwidth

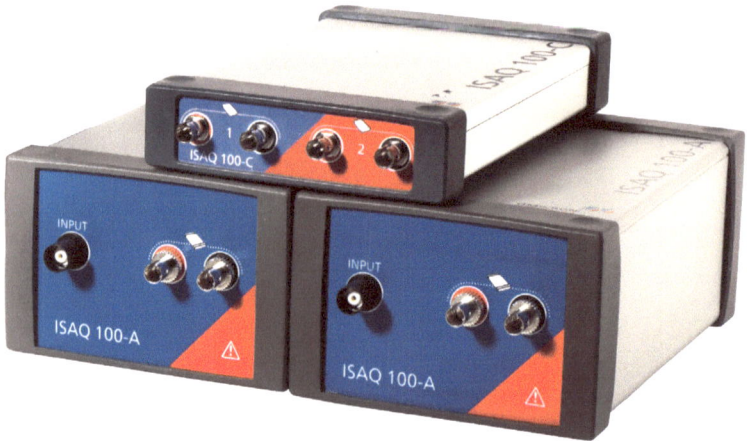

OMICRON Lab ISAQ 100 Fiber Optic Oscilloscope

Figure 4-12

OMICRON Lab ISAQ 100—Key Features

- PC controlled optically isolated data acquisition system
- 18 bit resolution
- 2MSps sampling rate
- Galvanic Isolation > 1MV
- On-desk operation up to 1000V
- Low Radiated Emissions: 30 Hz to 1000 MHz: 2 dBµV/m in 10 m
- Up to 3 km optical data transfer
- 8000 hour battery life
- 2 synchronized optically isolated channels

Spectrum Analyzers

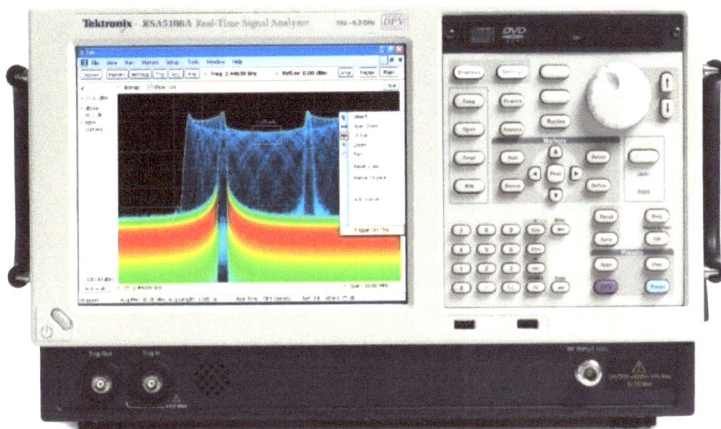

Tektronix RSA5106
Photo Courtesy of Tektronix Reprinted with Permission

Figure 4-13

Tektronix RSA5106—Key Features

- Frequency range from 1Hz to 6.2GHz
- Good for mid-range spectrum analysis
- Reduce time-to-fault and increase design confidence with real time signal processing
- Residual response for frequencies of 3GHz to 6.2GHz is -95dBm
- Image response for frequencies greater than 3GHz to 6.2GHz is less than -70dBc

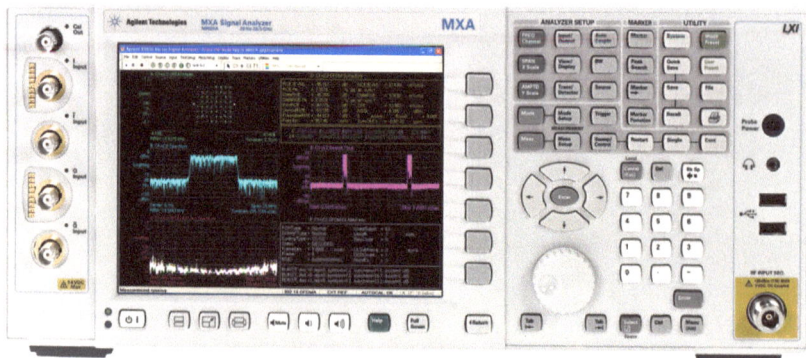

Agilent Technologies N9020A Spectrum Analyzer
Photo Courtesy of Agilent Technologies Reprinted with Permission

Figure 4-14

Agilent Technologies N9020A—Key Features

- Internal preamplifier options up to 26.5GHz
- 25MHz analysis bandwidth
- Basic EMI pre-compliance measurement capabilities available, including CISPR 16-1-1 bandwidths, detectors, amplitude correction factor, band preset, tune & listen at marker, and limit lines
- Supports over 25 measurement applications, covering cellular communication, wireless connectivity, digital video, and general purpose
- Advanced analysis of up to 70 signal formats, software runs inside the MXA
- MATLAB data analysis software for general purpose data analysis, visualization, and measurement automation. MATLAB software for the X-Series

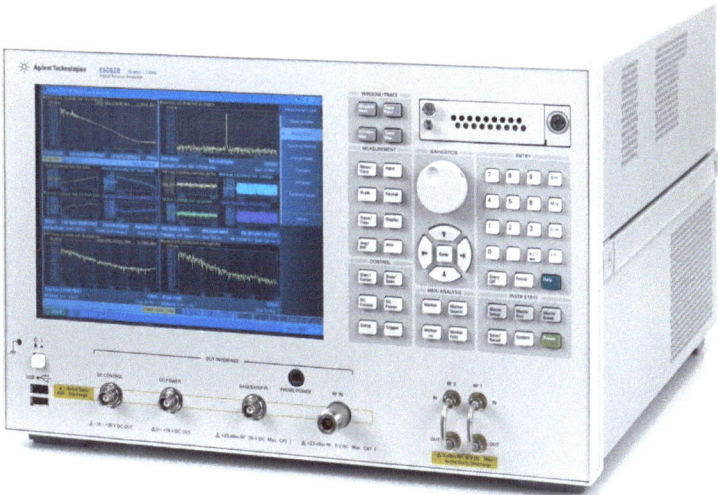

Agilent Technologies E5052B Signal Source Analyzer
Photo Courtesy of Agilent Technologies Reprinted with
Permission

Figure 4-15

Agilent Technologies E5052B—Key Features

- RF Input frequency range: 10MHz to 7GHz
- Analysis offset frequency range: 1Hz to 100MHz
- Phase noise measurements with ultra low noise floor and a cross-correlation method
- Frequency transient capture range up to 80MHz in narrow band, up to 4.8GHz in wide band
- Baseband noise measurement from 1Hz to 100MHz
- Spectrum monitoring up to 15MHz span in a real-time mode

Signal Generators

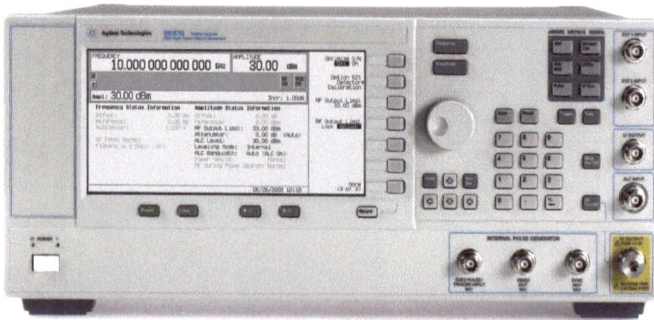

Agilent Technologies E8257D 20GHz Signal Generator
Photo Courtesy of Agilent Technologies Reprinted with
Permission

Figure 4-16

Agilent Technologies E8257D—Key Features

• Broad frequency ranges—250 kHz to 70 GHz

• High output power—typical performance of +23 dBm at 20 GHz, + 17dBm at 40 GHz, + 14 dBm at 67 GHz

• Flexible analog modulation formats—AM, FM, ØM, and pulse

• Dual internal function generators—sine, square, triangular, ramp, and noise waveforms

• Easy frequency extension—up to 325GHz with mm-wave modules

• Narrow pulse modulation—typical 8ns rise/fall times and 20ns pulse width from 10MHz to 67GHz

• Backwards compatible—same form factor and 100% code compatible with previous generations of PSG signal generators

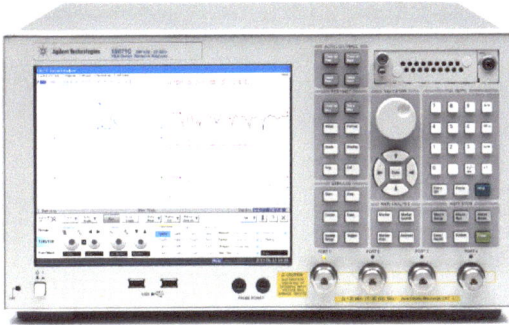

Agilent Technologies E5071C VNA and TDR Analyzer
Photo Courtesy of Agilent Technologies Reprinted with Permission

Figure 4-21

Agilent Technologies E5071C Key Features

- Wide frequency coverage of 9kHz to 20GHz
- Low trace noise measuring below 0.004dBrms at 70kHz IFBW
- Wide dynamic range of over 123dB
- Error correction
- High temperature stability: 0.005dB/°C
- Full calibration available for guaranteeing measurement accuracy.
- ESD robustness

Chapter Five

Probes, Injectors and Interconnects

TEST EQUIPMENT NEEDS to be connected to the DUT before measurements can be made.

Depending on the DUT, connections may include input power, a load, and for some measurements, a stimulus signal (such as for a step load).

This chapter is devoted to the available selection of probes, stimulus injectors, and interconnects, and the impact that probes have on measurement fidelity.

Voltage Probes

One of the most frequently used probes is the voltage probe. Voltage probes are frequently used with oscilloscopes, though in some cases voltage probes may be used with a VNA or spectrum analyzer.

The first considerations are generally the bandwidth or rise time of the probe. The measurement bandwidth is not the same as the probe bandwidth. The measurement bandwidth is altered because of interaction with the circuit being measured. The measurement bandwidth may also be limited by the bandwidth of the instrument to which the probe is connected.

Another consideration is the loading effect of the probe, which can significantly alter the performance of the circuit being measured.

A high capacitance probe connected to the output of a high-speed opamp, for example, might cause the opamp to oscillate.

In addition to such an oscillation preventing us from making a valid measurement, it could also potentially damage the circuit.

Other considerations include the DC and frequency dependent AC voltage rating of the probe and the attenuation, which are somewhat related.

A higher voltage probe generally also has greater attenuation.

Greater attenuation leads to reduced SNR.

Yet another consideration is probe connectivity, meaning whether the probe is specifically keyed to a particular brand or family of instruments or whether it has a universal connection that allows it to be used with all instruments.

There are four basic classifications of voltage probes: Passive, Active, Active Differential and Low Impedance (sometimes referred to as *transmission line probes*).

Each classification has advantages and disadvantages as summarized in Table 5.1.

	Advantages	Disadvantages
Passive probes	Lower cost High voltages available	Higher capacitance Limited bandwidth/rise time Often keyed to a particular brand
Active probes	High bandwidth/rise time Very low capacitance	Higher cost Low max. allowed voltage Often keyed to a particular brand
Differential probes	Measurements can float No low frequency ground loops High bandwidth/rise time	Higher cost Low max. differential voltage Often keyed to a particular brand
Low Impedance probes	High bandwidth/rise time Very low capacitance Low cost Standard 50Ω connection	DC resistance as low as 50Ω

Voltage Probe Comparison Matrix

Table 5.1

Probe Circuit Interaction

The probe bandwidth and rise time are related and the derivation can be found in Chapter 13, *Measuring Switching Edges.*

$$Probe\ rise\ time = \frac{0.34966}{BW} \qquad 5.1$$

The voltage probe loading is represented by a shunt resistance and/or a shunt capacitance along with some series resistance as shown in Figure 5-1. The series resistance can be estimated from the bandwidth and the probe capacitance as:

$$R_{series} = \frac{1}{2\pi \cdot BW \cdot C_{series}} \qquad 5.2$$

The series resistance can also be measured using the techniques described in the *Measuring Impedance* chapter. A simplified schematic showing the probe loading is shown in Figure 5-1.

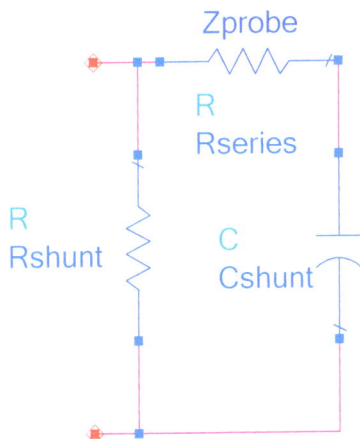

Typical Simplified Probe Model

Figure 5-1

From a DC standpoint, the circuit is loaded by Rshunt. At higher frequencies, it is loaded by C_{shunt}. The bandwidth is determined by the combination of R_{series} and C_{shunt}.

The probe interacts with the source impedance of the signal being measured, as well as the impedance of the connections from the probe to the DUT. The circuit impedance can be represented by any combination of resistance, inductance and capacitance, while the probe connections are generally represented by resistance and inductance. The probe, circuit impedance, and probe connections are shown in Figure 5-2.

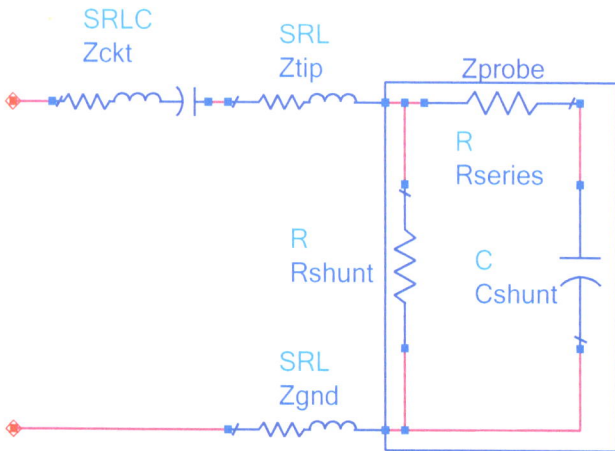

Typical Probe along with the Circuit Impedance and Probe Connections

Figure 5-2

It may now be obvious that the probe loading impedance forms a divider with the circuit impedance and probe interconnects such that:

$$V_{probe} = \frac{Zprobe}{Zprobe + Ztip + Zgnd + Zckt} \qquad 5.3$$

Note that each of these impedance terms is represented by a complex impedance comprised of resistive, inductive and capacitive terms.

A simulation model of a typical 500MHz passive probe loading along with the probe tip and ground interconnects is shown in Figure 5-3. For the sake of simplicity, the probe is shown as a unity gain device (though these are typically 10X probes) so that the results are scaled accordingly. The probe and connections are simulated using a fast step edge (100ps) to the probe.

The probe connections are simulated using approximately ¼ inch leads, representing the exposed probe tip and a ground spring clip. Using a typical wire inductance of 30nH/inch, each connection is represented by a 7.5nH inductance. The circuit is also simulated with a 3.5 inch ground clip, represented as 100nH.

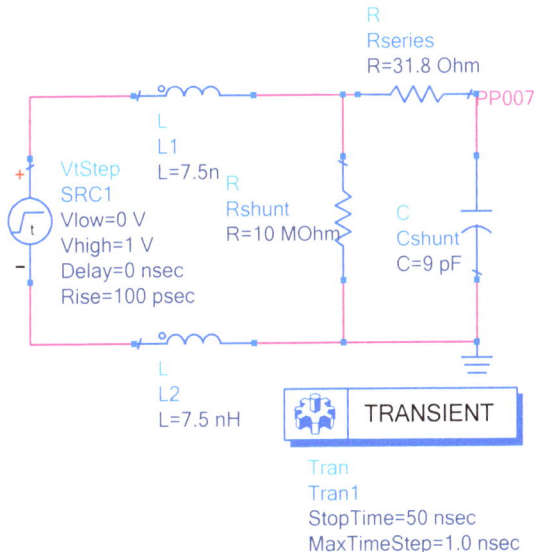

A 9pF Probe Connected using ¼ inch Leads

Figure 5-3

The simulated response for the short leads and the longer ground clip are shown in Figure 5-4. In both cases, the probe shows significant overshoot. This figure shows the simulated AC response for both ground connections.

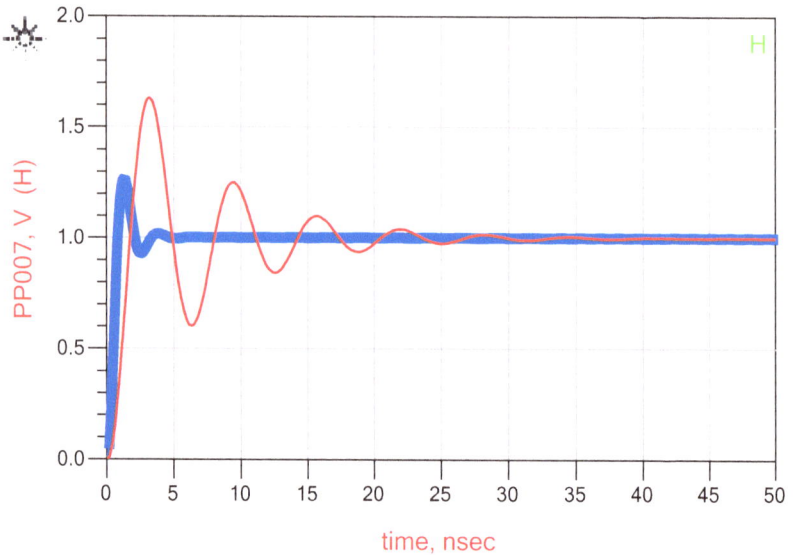

Simulated Transient Response to a Fast Step using ¼ inch leads (Blue) and a 3.5 inch Ground Clip (Red)

Figure 5-4

Figure 5-5 shows the response is not flat, with peaking in both cases. The -3dB bandwidth also changes in the two simulated cases. With the 3.5 inch ground clip, the peaking is very significant at 10dB and the 500MHz bandwidth is reduced to approximately 166MHz.

With the shorter connection, the peaking is closer to 3dB and the bandwidth is close to the nominal 500MHz.

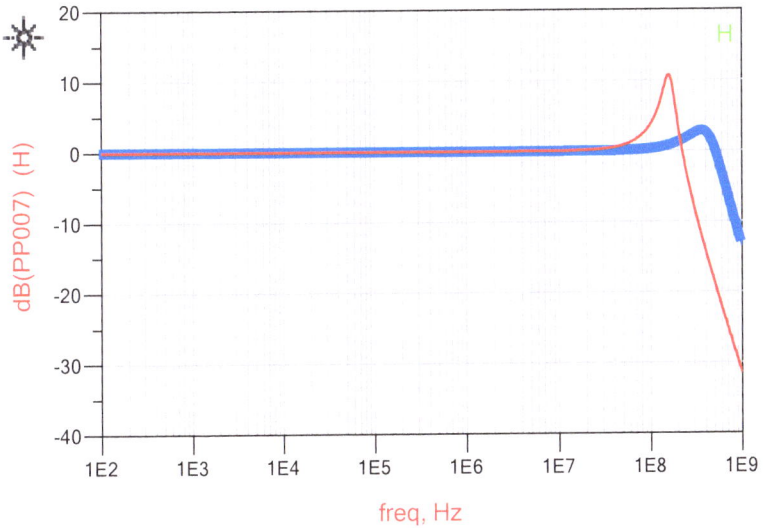

Simulated AC Response using ¼ inch leads (Blue) and a 3.5 inch Ground Clip (Red)

Figure 5-5

Flattening the Probe Response

One way to flatten the response is to add an additional series resistance to provide damping. The series damping resistance can be approximated from the total series inductance and the probe shunt capacitance. The total series inductance can be approximated from the ringing frequency and the probe capacitance.

$$L = \frac{1}{[2\pi \cdot 166MHz]^2 \cdot 9pF} = 102nH \qquad 5.4$$

And the series damping resistance to flatten the response can be calculated as:

$$R_{damping} = \frac{0.85}{2\pi \cdot 166MHz \cdot 9pF} = 91\Omega \qquad 5.5$$

The simulated AC response is shown in Figure 5-6 with the 3.5 inch ground clip with and without a 91Ω series resistor.

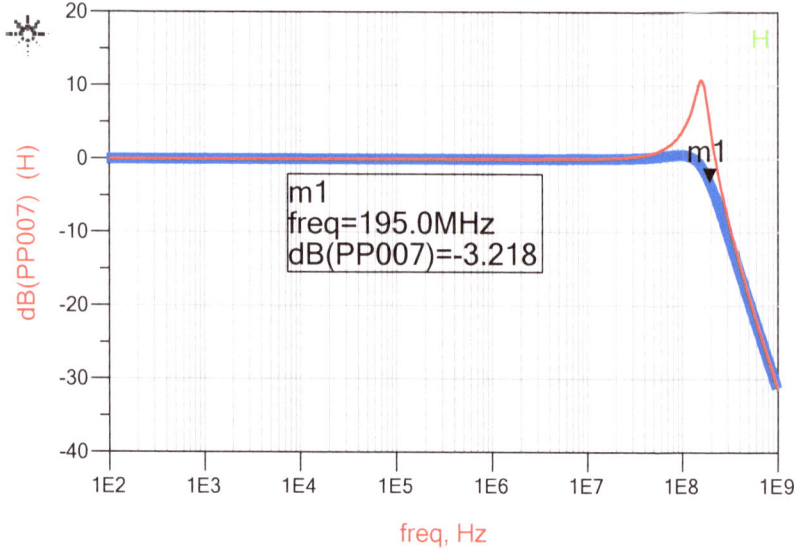

Simulated AC Response with (Blue) and without (Red) the addition of a 91Ω Damping Resistor to Improve Flatness

Figure 5-6

The addition of the damping resistor flattens the response and eliminates the ringing; however, the bandwidth is still significantly reduced due to the inductance of the probe connections.

Another way to minimize the ringing is to limit the allowable rise time measurement. The simulation results using the ground clip are shown for rise times of 10ps, 15ns and 25ns as shown in Figure 5-7.

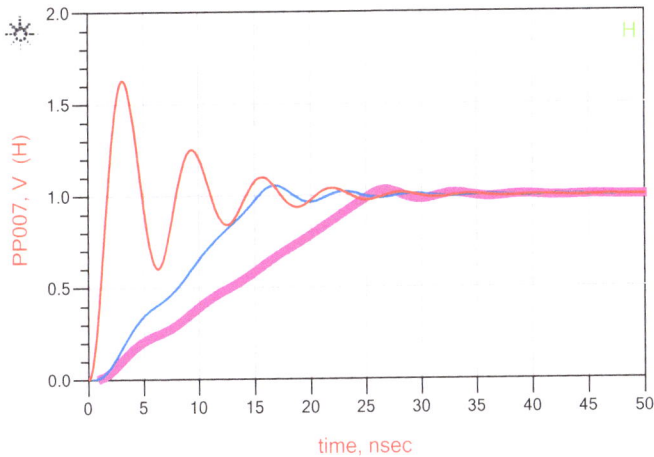

Simulated Transient Response with Ground Clip for 10ps, 15ns and 25ns Rise Time Signals

Figure 5-7

The fastest rise time that can be measured without significant overshoot or ringing can be determined as:

$$Minimum\ t_{rise} = \frac{4}{f_{ring}} = \frac{4}{166MHz} = 25ns \qquad 5.6$$

Confirming Measurements

A fast edge pulse with an attenuator is used to present a 50Ω impedance 200pS impulse to a PP007 probe with a 3.5 inch ground clip. The measured probe responses with and without a 100Ω series damping resistor are shown in Figure 5-8. The ringing frequency is measured to be 169MHz. The probe was also measured using a rise time of 5ns and a rise time of 20ns as shown in Figure 5-9. Note the slight distortion in the rising edge with the 20ns rise time. Eliminating this distortion would require further limiting of the rise time.

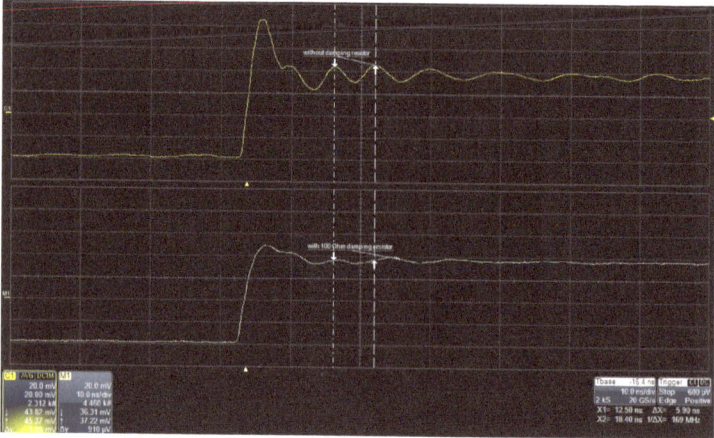

PP007 Response using 3.5 inch Ground Clip without Damping Resistor (Top) and with 100Ω Damping Resistor (Bottom)

Figure 5-8

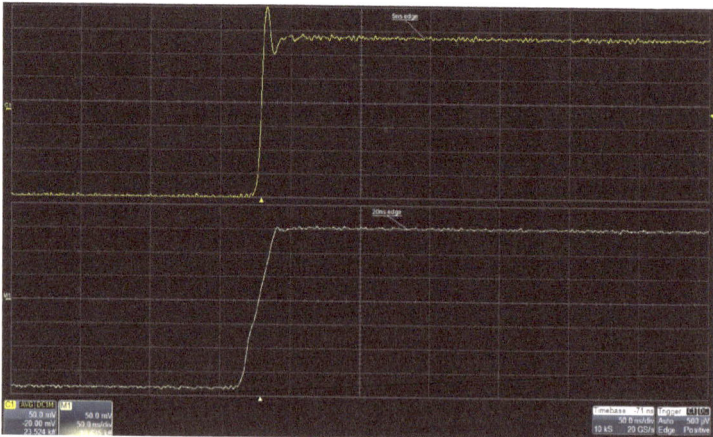

PP007 Response using an Impulse Rise Time of 5ns (Top) and 20ns (Bottom)

Figure 5-9

Selecting a Voltage Probe

In order minimize the interaction between the probe and the circuit, the probe's loading impedance should be greater than sum of the circuit impedance and the probe connection impedance up to at least the minimum acceptable measurement bandwidth.

$$Zprobe \geq Ztip + Zgnd + ZcktforFreq \qquad 5.7$$
$$\leq desiredbandwidth$$

If $Ztip+Zgnd+Zck$ is purely inductive, and exactly equal to the capacitive impedance of the probe, the circuit will be resonant at the minimum acceptable bandwidth.

The peaking that occurs at this resonance can then be eliminated using a series damping resistor, as determined above, and the desired minimum bandwidth will be achieved. In the case that $Ztip+Zgnd+Zck$ is purely resistive, there would be no peaking and the -3dB bandwidth would be at the minimum acceptable bandwidth. In either case, the minimum bandwidth is achieved.

The loading impedance versus frequency for several probes is shown in Figure 5-10. While a 9pF probe capacitance provides a high impedance at low frequency, it falls to 500Ω at approximately 35MHz. A 3.9pF probe falls to 500Ω by about 80MHz.

Low capacitance active probes provide much higher impedance, minimizing interaction between the probe and the circuit.

Two low impedance probes present significantly more loading at low frequencies, but above a few hundred MHz, the low impedance probes present lower loading than the active probes.

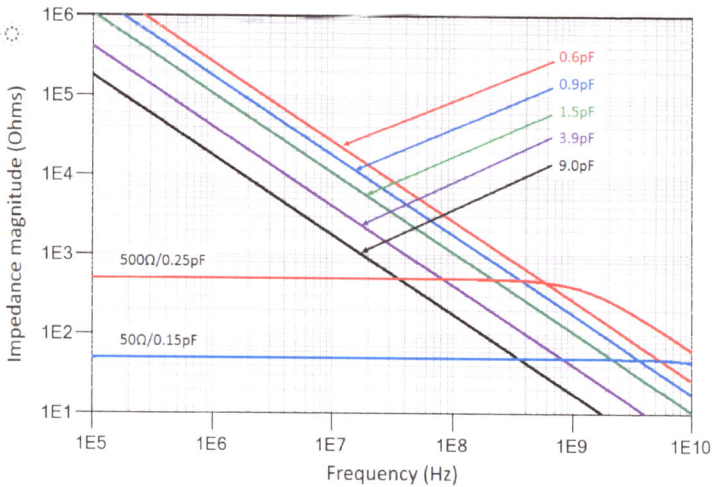

Probe Loading Impedance for Several Probe Resistances and Capacitances

Figure 5-10

In order to select the probe loading that complies with Equation 5.7, it is important to determine the circuit impedance and to estimate the tip and ground impedance. If the circuit impedance is not known, it should be measured using the techniques presented in the *Measuring Impedance* chapter.

Passive Probes

The Teledyne Lecroy PP007 500MHz 10pF probe, the Tektronix 1GHz TPP1000 3.9pF probe and the Tektronix TPP0850 800MHz 1.8pF high voltage probe are used for the measurements in this book. The Teledyne Lecroy probe is used for the Lecroy oscilloscopes, though the PP007 uses a generic BNC connector allowing it to mate with all standard equipment. The Tektronix TPP1000 offers the lowest capacitance and highest bandwidth for a passive 10X probe. This probe is specifically keyed to the TekVPI^{TM} system and is designed for use with MSO/DPO4000B and MSO/DPO5000 and MDO/DPO7000 series oscilloscopes.

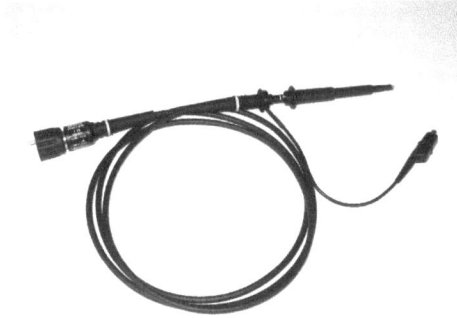

Teledyne Lecroy PP007 500MHz Passive 10X Probe

Figure 5-11

Tektronix TPP1000 1GHz 10X Passive Probe is the only 1GHz passive probe available from any manufacturer. The probe loading is 3.9pF, which is significantly lower than typical 10X passive probes.

Tektronix TPP1000 1GHz 10X Passive Probe

Figure 5-12

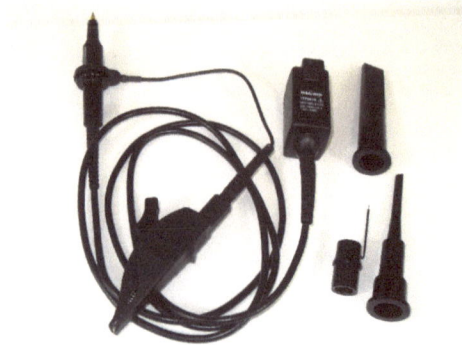

Tektronix TPP0850 1kV 50X Probe with Impressive 500ps Rise Time and 1.8pF Capacitive Loading

Figure 5-13

Active Probes

Teledyne Lecroy offers a wide variety of active probes up to 33GHz. The ZS2500 2.5GHz, 0.9pF probe and the ZS4000 4GHz, 0.6pF active probes, which are compatible with the Waverunner 6Zi series oscilloscopes, are used in this book.

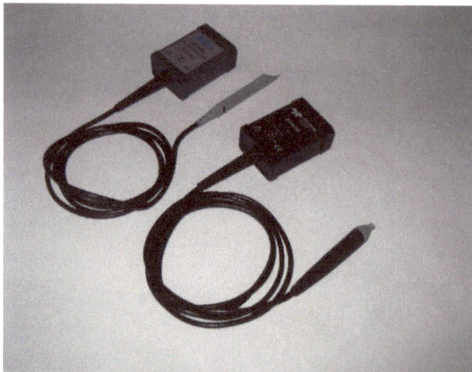

ZS4000 (ZS2500 is similar)

Figure 5-14

Key Features for ZS2500 (ZS4000)

- Easy to Use, Integrated with Oscilloscope
- High Signal Fidelity to 2.5 GHz (4GHz) Bandwidth
- Low Load—High Input Impedance/Low Capacitance
- $\pm$ 8V Input Dynamic Range
- 1MΩ Input Resistance
- 0.9pF (0.6pF) Input Capacitance
- $\pm$ 12 V Offset Range

Tektronix also offers many high performance active probes including the P7500 series TriMode probe, which is available up to 20GHz and the P7600 series TriMode probe, which is available up to 33GHz.

Differential Probes

The Teledyne Lecroy ZD1500 has a differential range of 18Vpp with a differential offset of $\pm$8V and a common mode range of $\pm$10V. The differential input capacitance is 1pF and the system noise is 4mVrms.

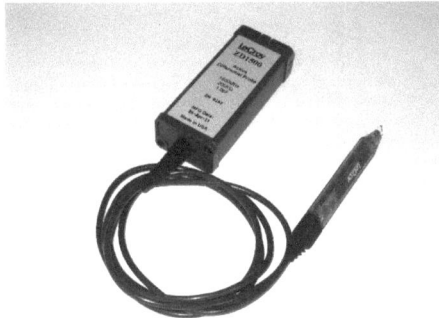

Teledyne Lecroy ZD1500 1.5GHz Differential Probe

Figure 5-15

117

Specialty Probes
Low Impedance Probes

Low capacitance probes from Teledyne Lecroy include the PP065 1GHz, 2pF, 5kΩ (100X) probe and the PP066 7.5GHz, 0.25pF, 500Ω (10X/20X) probe. Low capacitance probes from Tektronix include the P6158 3GHz, 1.5pF, 1kΩ (20X) probe and the Tektronix P6150 3GHz/9GHz, 0.15pF, 50Ω/500Ω (1X/10X) probe.

Agilent, Rohde Schwarz and Tektronix also offer similar transmission line probes like the Teledyne LeCroy PP066. Some are available as a 1X 50Ω probe.

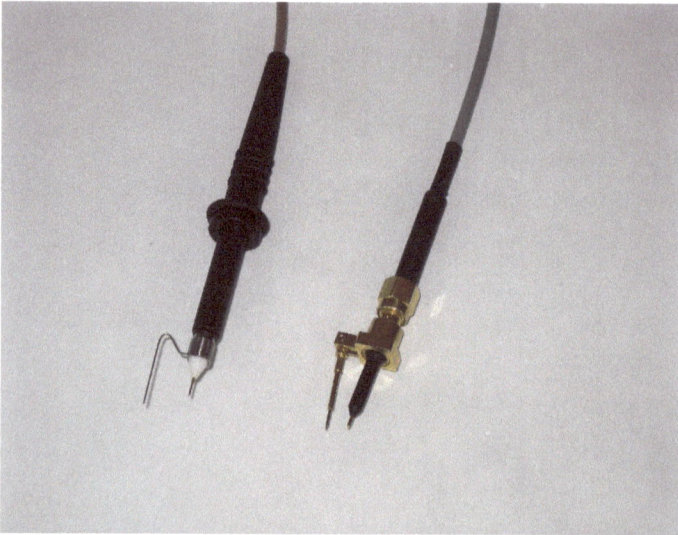

Transmission Line Probes used in this Book include the Teledyne Lecroy PP066 10X 7.5GHz 500Ω Probe (Left) and PP065 100X 1GHz 5kΩ (Right)

Figure 5-16

Multiport Probes

Low impedance two-port probes are available from Picotest, supporting the single port impedance measurement, the two-port shunt through impedance measurement, and the non-invasive stability measurement.

A 2-Port Probe used to Measure PDN Impedance for an FPGA

Figure 5-17

Port Splitter

While not presently offered as a product, several three-element, two-port, resistive splitters were manufactured for this book, allowing the same measurement to be placed on two instruments simultaneously.

A picture and schematic of the splitter are shown in Figure 5-18.

In some cases, the splitters are used to allow different instrument types to be connected, such as an oscilloscope and a spectrum analyzer, but more frequently the splitter is used to connect the same signal to multiple oscilloscopes or spectrum analyzers (for simultaneous, comparative phase noise measurements).

The forward and reflection performance of the splitters are measured using am Agilent E5071C VNA and the results indicated good performance up to approximately 5GHz as seen in Figure 5-18.

The forward and reverse S-parameters for the splitter are shown in Figure 5-19 confirming a usable bandwidth to greater than 4GHz.

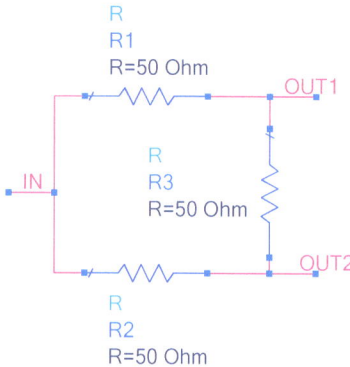

Picture and Schematic of the Three-Resistor, Two-Port Splitters Manufactured for this Book

Figure 5-18

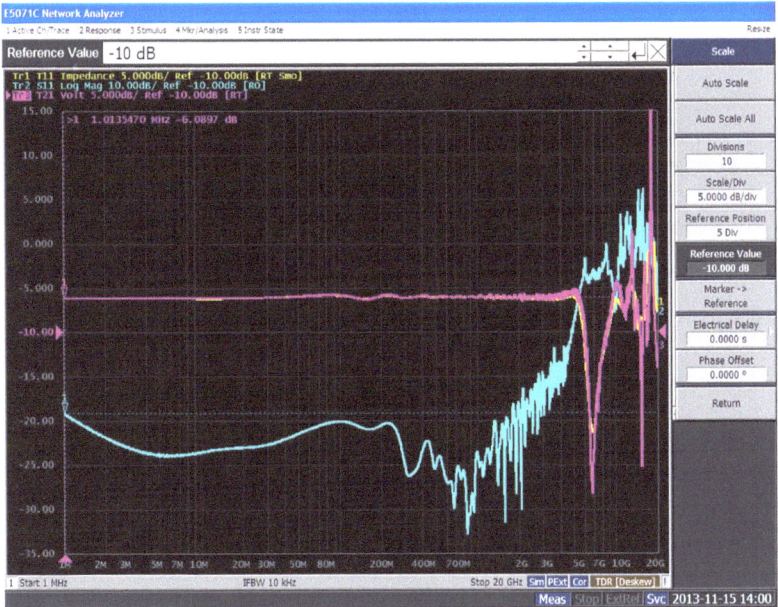

S-Parameter Measurements for Three-Resistor, Two-Port Splitters. Note that at 4GHz, the Return Loss is Greater than 16dB

Figure 5-19

Performance Probes

High performance oscilloscope probes are available from Teledyne Lecroy and Tektronix.

The Tektronix P7500 and P7600 series probes use the TekConnect® interface for use with the Tektronix DPO/MSO/DSA70000 series oscilloscopes. These high performance probes are tri-mode, supporting single ended, differential and common mode measurements.

A wide assortment of accessories is available for solder-in and standard probing to 33GHz. The Tektronix P7300 series offers single ended and differential measurements up to

12.5GHz for solder in and hand probing.

Teledyne Lecroy offers differential probes with solder in and probing tips to 25GHz for use with the WaveMaster 8Zi and LabMaster 9Zi and 10Zi series oscilloscopes.

Direct Coaxial Cables

Many high-speed measurements are made using a 50Ω coaxial cable to connect the DUT directly to a 50Ω instrument port; however, not all coaxial cables are created equal.

Two different probes are shown here in Figure 5-20.

A 200pS edge is measured using two different coaxial cables.

Note that one cable is close at 226pS while the other (equal length) cable shows 567pS. The odd shape of the slower edge is often related to dribble effect, which is discussed in detail in the *Measuring Edges* chapter.

For measurements below a few GHz, standard RG58 or RG174 cable (the author prefers double shielded) is adequate.

For optimum performance, a high quality 18GHz or more coaxial cable should be used. The high performance cables often use silver plated stranded and twisted center conductors and foil layers to reduce the high frequency losses, which are responsible for this dribble effect.

The slower cable in this measurement is a non-descript unbranded cable.

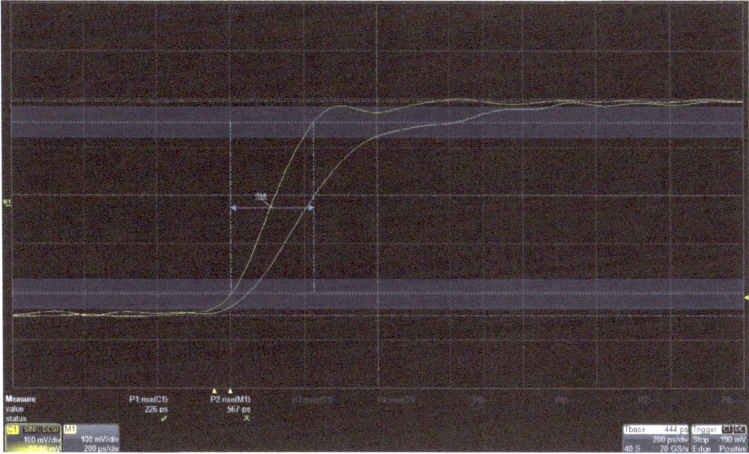

Rise Time for Two Equal- Length 50Ω Coaxial Cables

Figure 5-20

WARNING! Many higher frequency devices use 2.92mm connectors (SMK) often referred to as K connectors. The SMK connector looks very similar to a SMA connector and the two can even mate. If the male connector is SMA, it can potentially damage the SMK female connector. SMK connectors are precision air dielectric connectors while SMA connectors are not. It is worth purchasing high quality cables and connectors in order to protect the sensitive instrument connectors. Replacing the instrument connectors can be very expensive.

Current Probes

Current probes are occasionally helpful for measuring impedance and for EMI troubleshooting. Many current technology POL and switching regulators exhibit faster edge speeds than typical current probes can measure. The insertion impedance of a current probe can also significantly slow the speed of the switches, especially if you are measuring the current at the switch's source lead.

	Advantages	Disadvantages
Passive probes	Lower cost	No DC operation High minimum frequency Droop for low frequency pulse
Hall probes	Operate down to DC	Requires periodic degaussing Often larger with higher insertion impedance
Rogowski probes	Lowest cost No saturation effects Very small—fits in small spaces Low frequency operation Works with all equipment Zero insertion impedance	No DC operation Highest output noise

Current Probe Comparison Matrix
Table 5.2

The current probes used in this book include the PEM CWT15 Rogowski coil, 30Amps, 100Hz-10MHz, the Tektronix P6022passive probe, 6Amps, 935Hz-120MHz and the Teledyne Lecroy CP031 Hall current probe, 30Amps, DC-100MHz.

The three probe heads are shown in Figure 5-21.

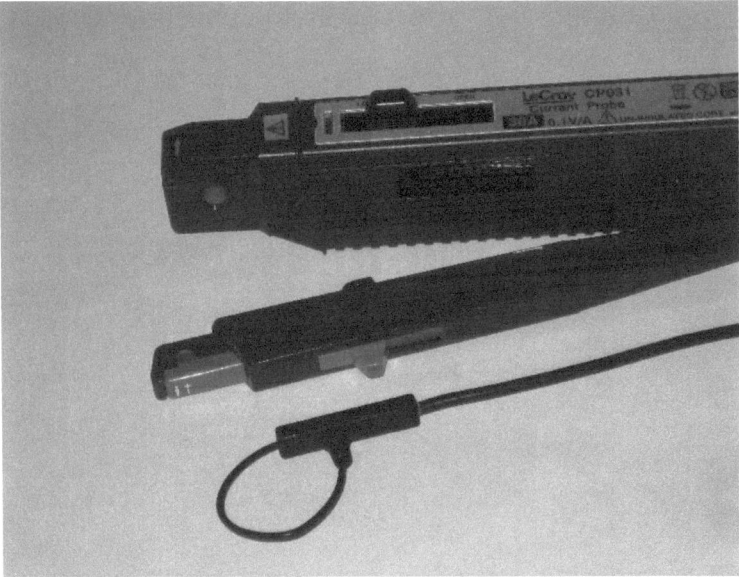

Teledyne Lecroy CP031 (Top) Tektronix P6022 (Middle) and PEM CWT-15 (Bottom)

Figure 5-21

Insertion Impedance

While the voltage probe loading impedance is capacitive, the insertion impedance of a current probe is inductive. The clamp-on current probe may require the addition of a wire loop to clamp around.

This loop adds additional inductance.

The insertion impedance of the CP031 probe is measured including the wire loop inductance, as well as with the loop nulled. The results shown in Figure 5-22 indicate an inductance of approximately 23nH, including the minimally sized wire loop.

With the wire loop nulled, the probe is approximately

1.5nH, so the majority of the insertion loss is not from the probe but from the wire loop required for the probe to clamp around.

For this reason, it is always advisable to measure on an existing wire or trace than adding a wire loop.

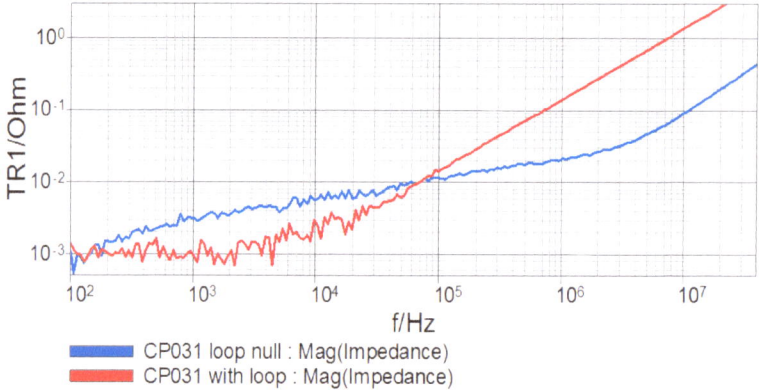

CP031 Current Probe Insertion Impedance with the Wire Loop Included (Red) and with the Wire Loop Nulled (Blue)

Figure 5-22

The three current probes simultaneously measured the Teledyne Lecroy DCS015 Deskew Calibration Source.

The Teledyne Lecroy DCS015 Deskew Calibration Source is used to deskew accurately the voltage and current signals to account for the mismatched time delays of the voltage and current probes.

The results of the three measurements are shown in Figure 5-23. The probes all indicated an approximate 22nS rise time with little or no overshoot.

A hint of peaking is observable in the CWT15 probe, as well as in the CP031 probe, but all three probes responded well.

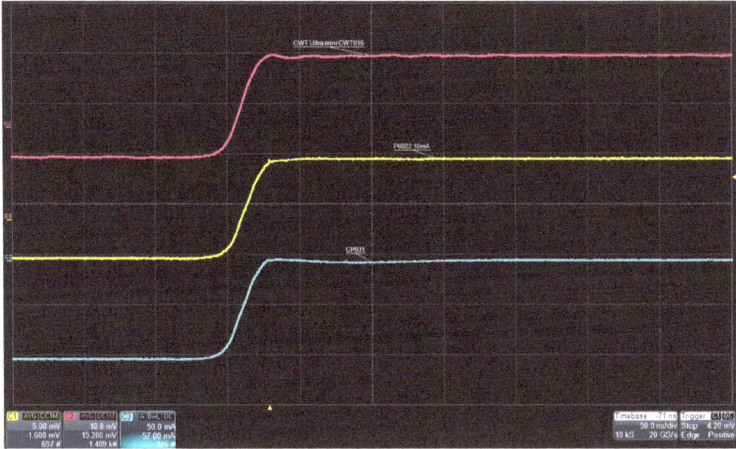

25nS Current Step CWT15 (Red) P6022 (Yellow) and CP031 (Blue)

Figure 5-23

Improving Sensitivity

Clamp-on probes are designed to clamp around a wire, forming a single turn. The sensitivity of the Hall and passive clamp on probes can be increased by placing more than one turn of wire in the head.

The sensitivity is increased in proportion to the number of turns. The additional turns will also increase the insertion impedance of the probe.

Differential or Common Mode Current

Current probes can also measure common mode current or differential current by placing two current carrying wires in the probe head with current flowing in opposing directions.

Other probes the author has used and can recommend include the Tektronix TCP0030A, which is similar in performance to the Teledyne Lecroy CP031 but interfaces with the Tektronix TekVPI interface. Tektronix also offers the

CT1/2/6, which are small passive AC current transformers.

These current transformers are not clamp-on but rather have a through hole allowing a wire to be passed through and soldered into the circuit. The CT1/2/6 probes are not designed to tolerate much DC current without saturating but these probes offer bandwidths up to 2GHz.

Near-Field Probes

Near-field probes are helpful when it is necessary to troubleshoot EMI issues. E-field probes are sensitive to voltage-induced signals while H-field probes are sensitive to current induced signals.

The probes often come in sets, with different sizes of E-field and H-field probes. The H-field probes are open loops. The larger the loop, the greater the probe sensitivity and the lower the selectivity.

The E-field probes include stubs and spheres.

The larger sphere has the highest sensitivity, while the stub has lower sensitivity and greater selectivity. Starting with larger probes you can get in the vicinity of an issue while then reducing to smaller probes to help you pinpoint the source.

The addition of a preamplifier can improve the sensitivity of the probes. In this book, the author uses the Beehive Electronics101A EMC probe, shown in Figure 5-24, which support a frequency range up to approximately 5GHz.

The probes are best used with a spectrum analyzer or the spectrum analyzer feature of an oscilloscope.

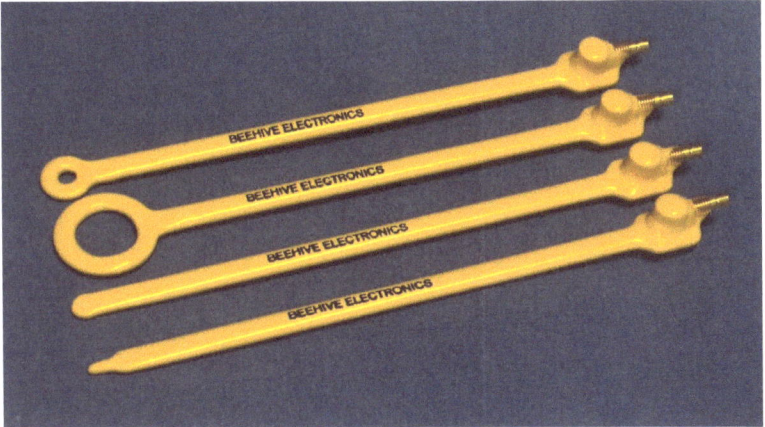

Beehive Electronics 101A EMC Probe Set
Photo Courtesy Beehive Electronics Reprinted with Permission

Figure 5-24

EMI compliance testing measures far-field signals. There is little relationship between the far-field and near-field measurements so the goal is not to use near-field probes to measure EMI.

The near-field probes are used to help identify and locate troublesome signals and to identify characteristics of signals that may help identify these sources in far-field testing.

Signal Injectors

Many measurements require external stimulus to be provided to the DUT. The stimulus can be provided in many different ways depending on the measurement being made.

Picotest manufactures a complete line of high fidelity signal injectors for this purpose as seen in Figure 5-25.

Picotest Signal Injectors

Figure 5-25

Injection Transformers

An injection transformer is a precision wideband transformer that is generally used to connect an AC signal from a VNA to the DUT, often for measuring the control loop stability.

A high quality injection transformer offers very wide bandwidth and flat response. The three injection transformers used in this book are all CE certified for 600V category II, allowing them to be used in most applications.

The Picotest J2100A transformer offers the lowest frequency range of 1Hz-10MHz allowing the measurement of PFC circuits, as well as most switching and linear regulators.

The Picotest J2101A offers a range of 10Hz-40MHz and the OMICRON Lab BWIT is between these two ranges.

In most cases, the transformer is connected across an injection resistor for the purposes of control loop testing. The low frequency bandwidth of the transformer is impacted by the value of this injection resistor.

The lower -3dB point occurs at the frequency at which the impedance of the transformer is equal to the parallel resistance across the transformer, or the injector termination resistance.

For this reason, the transformer must present high inductance in order to allow operation at low frequency.

All of these injection transformers use a core that is specially annealed to obtain the highest inductance. It is also clear from this reasoning that the lower the injection resistance, the lower the inductance required for a given frequency of operation.

Using a 4.99Ω injection resistor offers good low frequency performance and is generally small enough that it can be permanently installed in the circuit without any significant impact.

In some cases, injection transformers can be used to isolate the stimulus source ground from the DUT ground to eliminate ground loops or improve the measurement noise floor. These cases are addressed in the individual measurement sections within this book.

Solid State Voltage Injectors

The injection transformer does not always provide a wide enough bandwidth. The Picotest J2110A solid-state injector offers a bandwidth of DC to approximately 100MHz.

The solid-state injector can provide lower noise measurements at low frequency. It also has sufficient bandwidth to measure the loop gain of many operational amplifiers.

The J2110A is used in several measurements in this book.

Solid-State Current Injectors

Solid-state current injectors are generally used to generate small signal, high-speed step loads or current profiles. The current injector is also useful for generating a signal for non-invasive PSRR testing, non-invasive stability testing, and for measuring impedance.

The current injectors offer a bandwidth from DC to more than 10MHz. The Picotest J2111A allows step loads of 100s of uA to 10s of mA with a typical rise time of 20nS. The Picotest J2112A can produce steps up to 1A peak with a rise time of 10nS for testing POL regulators and other high current, low voltage sources.

The connection between the current injector and the DUT must be very low inductance and a good solution is to use special low inductance coaxial cable.

Both the Picotest J2111A and Picotest J2112A current injectors are used in this book.

Specialty Coaxial Cables

The connection between the DUT and a current injector (or electronic load) and between an external power supply and the DUT is critical. Temp-Flex® low inductance coaxial cable is used for many such connections in this book.

Temp-Flex offers an inductance of approximately 1nH/inch with a characteristic impedance of approximately 9Ω, so it is quite capacitive.

DC Bias/Blocker

The Picotest J2130A is a resistor-capacitor bias/DC-blocking device, useful for biasing semiconductor junctions, capacitors and low current voltage references.

The capacitor allows a low frequency bandwidth of

approximately 100Hz. The Picotest J2130A is also useful for AC coupling the DUT to a 50Ω instrument port. The Picotest J2130A is included in many test setups within this book.

Attenuators

Attenuators are useful to reduce the signal level of the VNA for measuring Bode plots of low power control loops and to verify the noise floor and dynamic range for sensitive measurements.

The Picotest J2140A attenuators are cascadable 10dB, 20dB and 40dB 50Ω attenuators allowing up to 70dB of attenuation. In some cases, two J2140A attenuators are used to achieve more than 70dB attenuation.

Line Injector

The Picotest J2120A is designed to be between the bench power supply and the DUT input power, allowing small signal modulation of the input source for PSRR testing.

The Picotest J2120A is used in the *Measuring PSRR* chapter of this book.

Preamplifier

The Picotest J2180A preamplifier is a very low noise (2.4nV/root-Hz) 0.1Hz to 100MHz device with a high input impedance and a 25Ω output impedance. This allows high impedance probes to be connected to 50Ω equipment. A50Ω thru terminator can also be used to provide a 50Ω input impedance if desired.

The Picotest J2180A preamplifier is useful for low impedance measurements, near-field measurements, and ripple and noise measurements. It can also be used to extend the dynamic range of a measurement or to improve the SNR of a measurement.

Common Mode Transformers

The Picotest J2102A common mode transformer is a coaxial wound transformer designed to maintain a 50Ω characteristic impedance while eliminating the low frequency ground loops that can occur in low frequency measurements.

Impedance Fixtures

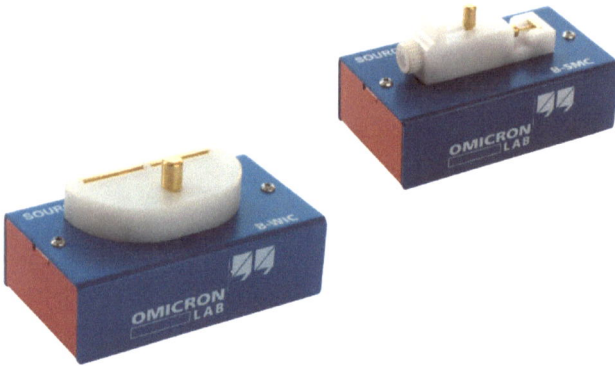

OMICRON Lab B-SMC (Surface Mount) and B-WIC (Through Hole) Impedance Adapters
Photo Courtesy of OMICRON Lab, Reprinted with Permission

Figure 5-26

The OMICRON Lab B-WIC and B-SMC impedance adapters, seen in Figure 5-26 are designed for precise impedance measurements of passive components with the Bode 100.

The impedance adapters operate over the full range of 1Hz-40MHz and support OPEN-SHORT-LOAD calibration. The impedance range for measurement is 0.02Ω-600kΩ.

The B-SMC is used to measure the insertion loss of the current probes, but is also ideal for measuring inductors and capacitors, including DCR, ESR, ESL, capacitance, inductance and self-resonance.

The B-WIC is used to measure the insertion loss of the current transformers and the capacitance of the voltage probes for this book.

Other Connections

Banana Leads

Most power supplies and electronic loads are furnished with banana plugs. Unfortunately, the banana plug is a very poor choice for many of these connections as the wire pair presents a relatively high inductance of approximately 60nH per inch.

It also lacks shielding.

A banana to BNC adapter is often helpful, allowing the use of coaxial cable for such connections, though the loop area of the adapter still introduces undesirable inductance.

Chapter References

1. Q. A. Kerns , F. A. Kirsten and C. N. Winningstad *Counting note: Pulse response of Coaxial Cables*, 1956 *lss.fnal.gov/archive/other/lbl-cc-2-1b.pdf*

2. *ABC of Probes* http://www.tek.com/document/primer/abcs-probes

3. S.M. Sandler, C.E. Hymowitz, T. Boehler, *Making the Connection Key to Accurate Test Data*, Power Electronics Technology, June 2012 pp 30-32 *https://www.picotest.com/blog/?p=1029*

4. S.M. Sandler, *Essential Test Adapters for your Network/Impedance Analyzer* https://www.picotest.com/blog/?p=950

5. S.M. Sandler C.E. Hymowitz, *Selecting Test Signal Injectors, June 2011,* Defense Electronics

6. https://www.picotest.com/blog/?p=539

7. S.M. Sandler C.E. Hymowitz, *Signal Injection Transformers—What Makes Them Special* https://www.picotest.com/blog/?p=340

8. *Introduction to the Solid State Injector* https://www.picotest.com/blog/?p=864

9. *New Injector Supports Testing POL's* https://www.picotest.com/blog/?p=1037#more-1037

Chapter 6
The Distributed System

MODERN PRODUCTS REQUIRE multiple power supply voltages to operate.

In most cases, these voltages are generated by multiple switching and linear regulators. The voltages are provided to circuits within the system through a Power Distribution Network (PDN).

A PDN includes the PCB traces and planes, decoupling capacitors, and in many cases, ferrite beads. These beads are used to help isolate and attenuate high frequency noise signals from particularly sensitive circuits.

In high power devices, such as FPGAs and CPUs which have large and dynamic operating currents, it is the role of the voltage regulator module (VRM) combined with the PDN, to insure that the voltages to the devices remain within acceptable limits.

Power supply noise can degrade the performance of many types of circuits, including ADCs, RF low noise amplifiers, precision clocks and sensitive instrumentation circuits. In these low power circuits, the choice of the voltage regulators and the design of the PDN are significant contributors to the system performance.

In this chapter, we explore the various noise sources and the pathways that distribute them throughout a system.

A typical representation of a simple distributed system is shown in Figure 6-1.

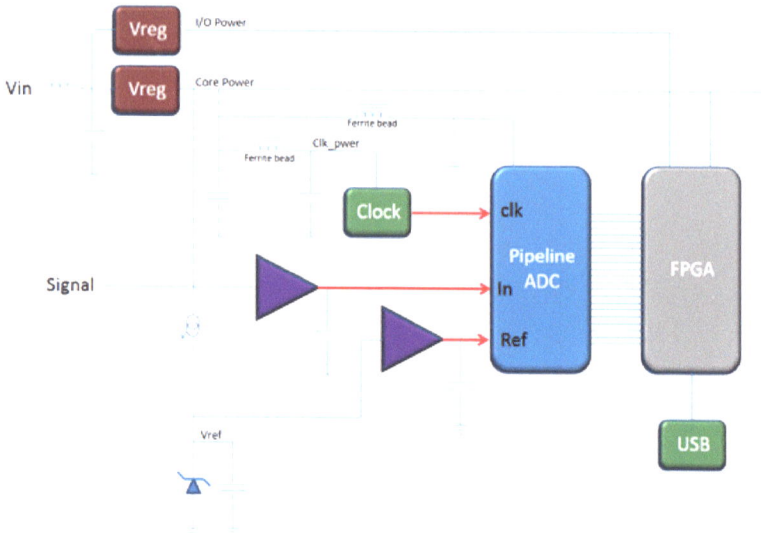

Block Diagram of a Typical Distributed Power System

Figure 6-1

Noise Paths within a Voltage Regulator

There are four noise paths associated with a single voltage regulator. Three of these appear as output noise—internally generated noise, power supply rejection ratio (PSRR), and output impedance, which interacts with the output load current variations.

A fourth noise path appears as input current noise but actually results from load current variations. This is known as Reverse Transfer. This input current interacts with the distribution impedance at the regulator's input, creating a noise voltage. This noise voltage can then propagate through the system. Each of these paths is discussed briefly here and the details of each path are found in their respective chapters.

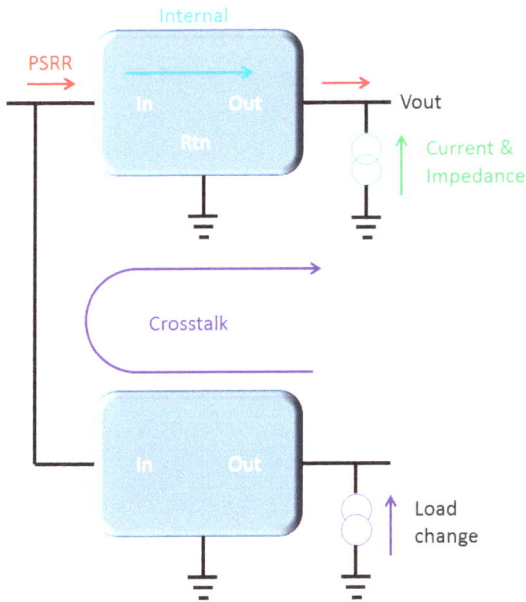

The Four Major Noise Paths through a Single Regulator and Two Regulators

Figure 6-2

In a similar way, these same noise paths exist in many other common electronic devices, including op-amps and voltage references.

For example, a signal on the supply voltage of an op-amp

Steven M. Sandler

results in an output signal due to the PSRR. Most op-amp datasheets include frequency dependent curves for the PSRR. Variations in the op-amp output current are transmitted through its output transistors to the power supply lines as reverse transfer and crosstalk.

The op-amp also has internally generated noise and finite output impedance completing the same four noise paths.

These individual noise contributions are additive with random terms being added using the square root of the sum of the squares method and non-random terms being directly summed. Internal noise, excluding spurious responses can be considered random, while dynamic loading is generally not random. The total noise is the sum of the responses as shown in Equation 6.1.

$$Total\ noise = \sqrt{\sum (random\ noise^2)} \\ + non\ random\ noise \qquad 6.1$$

Internal Noise

All devices generate noise internally.

In linear regulators, this noise is primarily comprised of flicker noise (referred to as 1/f or pink noise), thermal noise (referred to as Johnson or white noise) and shot noise.

It is also common to see small spurious outputs, which are fixed frequency signals (and harmonics) that appear as very small oscillations (blue trace) as shown in Figure 6-3.

In this case, we can see a 2.5µV spurious response at approximately 3.6kHz and a second harmonic of 1µV at 7.2kHz.

The low frequency flicker noise can also be seen at the lower frequencies and the thermal noise at the higher frequency end of the plot.

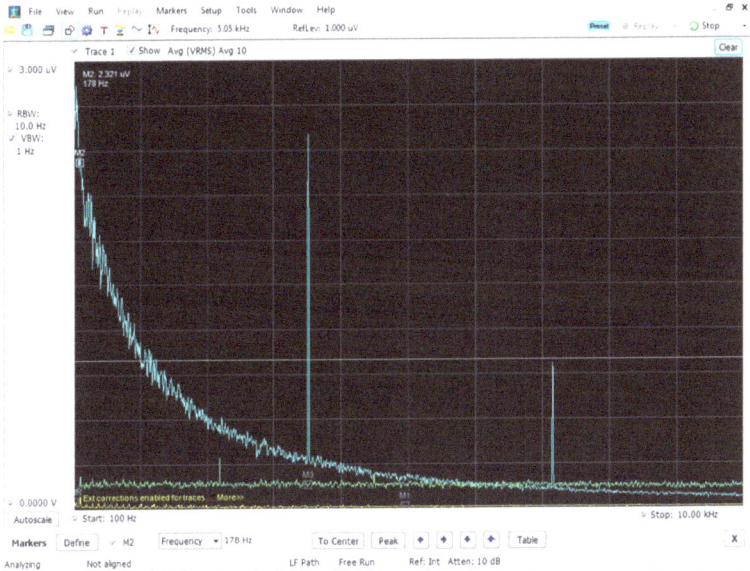

The Blue Trace is a Monolithic LDO while the Green Trace is a Custom Designed Low Noise Linear Regulator and the Barely Visible Yellow Trace is the Measurement Noise Floor

Figure 6-3

In the case of a switching regulator, internal noise is produced as ripple due to the switching action of the regulator.

There are also high frequency components in the output voltage as shown in Figure 6-4.

Note the high frequency content on the leading edge, as well as the switching frequency ripple.

The red trace is the load current.

Internal noise can also be generated in a switching regulator as a result of switching frequency and/or duty cycle jitter.

A severe case of both is shown in Figure 6-5.

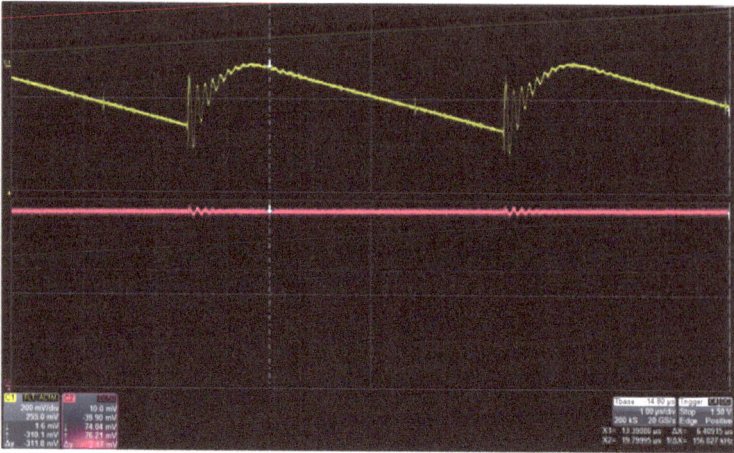

Switching Power Supply Ripple Voltage (Yellow Trace)

Figure 6-4

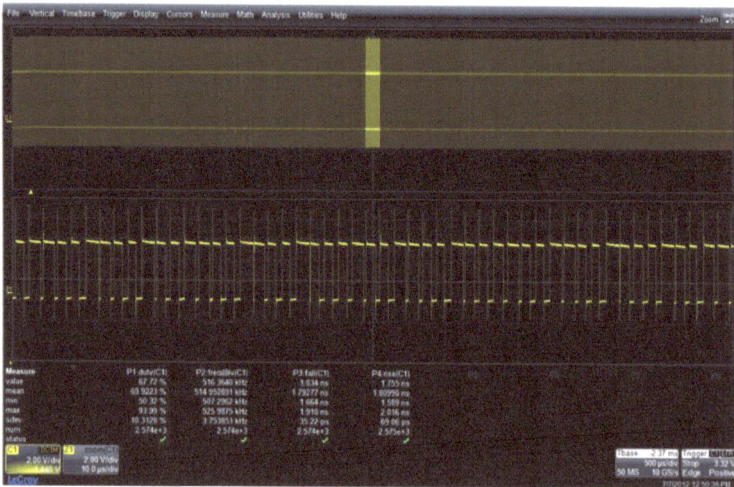

A Severe Case of Duty Cycle and Switching Frequency Jitter

Figure 6-5

Low frequency noise can also be generated in a switching regulator from the "power saver", Pulse Frequency Modulation (PFM) and burst modes incorporated into some switching regulators in order to improve low-power operating efficiency.

Operation in these modes can result in low frequency ripple with much larger ripple amplitude than in the PWM operating mode as shown in Figure 6-6.

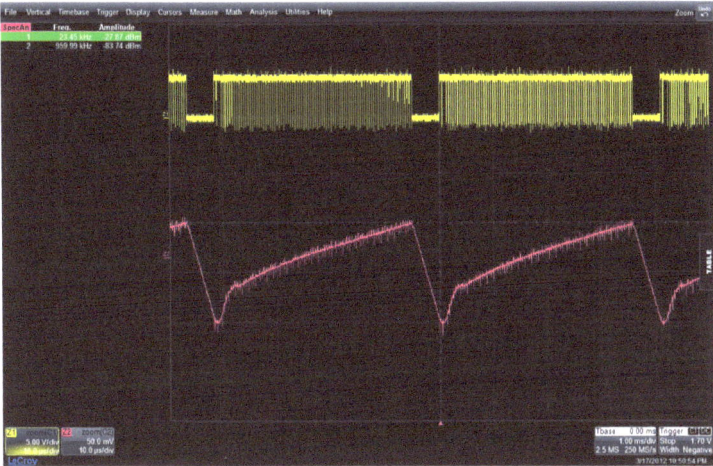

Large Low Frequency Voltage Ripple Resulting from
"Power Saver" Mode

Figure 6-6

Power Supply Rejection Ratio (PSRR)

The power supply rejection ratio is a measure of the ability of a device to attenuate a signal presented at its input. In the case of the linear regulator the PSRR is simply the ratio of the input signal level and the resulting output signal.

This is generally measured using a vector network analyzer

and presented as a rejection measure defined as $\Delta V_{out}/\Delta V_{in}$.

Typical PSRR plots for a LM317 voltage regulator and a custom designed high performance regulator are shown in Figure 6-7.

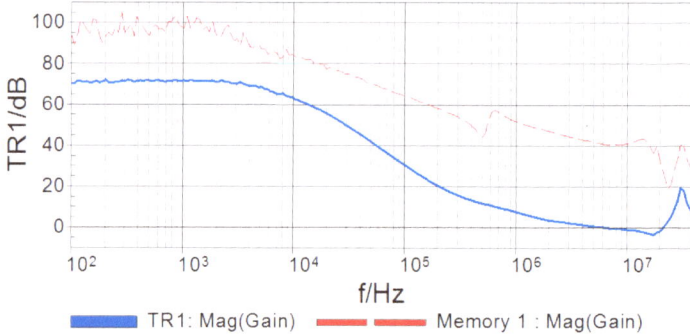

Typical PSRR for a LM317 Linear Regulator (Blue Trace) and a Custom-Made Low Noise, High-PSRR Regulator (Dashed Red Trace)

Figure 6-7

In some cases, such as with low jitter clocks, it is more direct to assess PSRR using clock jitter in place of the regulator's output voltage.

In the example shown in Figure 6-8, a 250kHz 50mVpp signal is added to the regulator input voltage and the resulting phase noise and jitter are measured, shown here as -76.7dB with respect to the carrier signal.

This measurement is repeated at each frequency of interest.

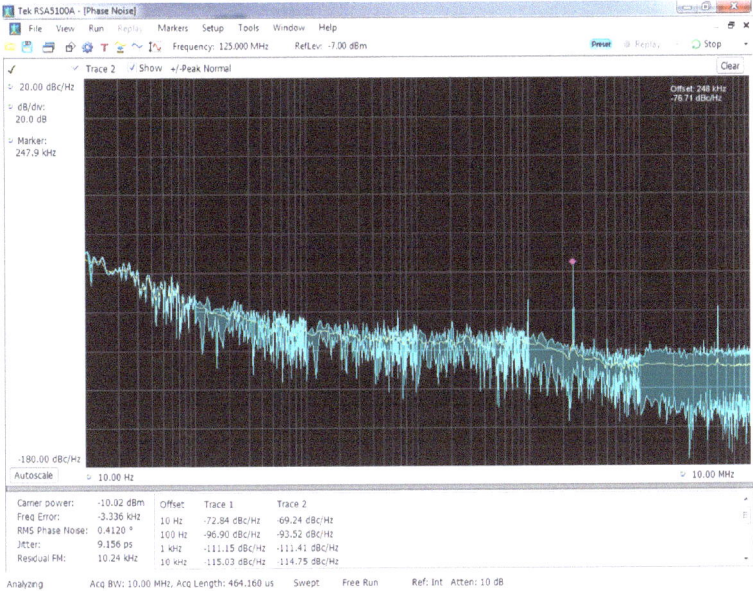

Clock Jitter Resulting from a 50mVpp Sine Modulation at 250kHz

Figure 6-8

Output Impedance

Load current variations are converted to voltages due to the finite regulator output impedance.

In the example shown in Figure 6-9, we can see the load current in the bottom trace along with the regulator output voltage in the top trace.

The ripple and high frequency noise contributions are also seen in the output voltage.

145

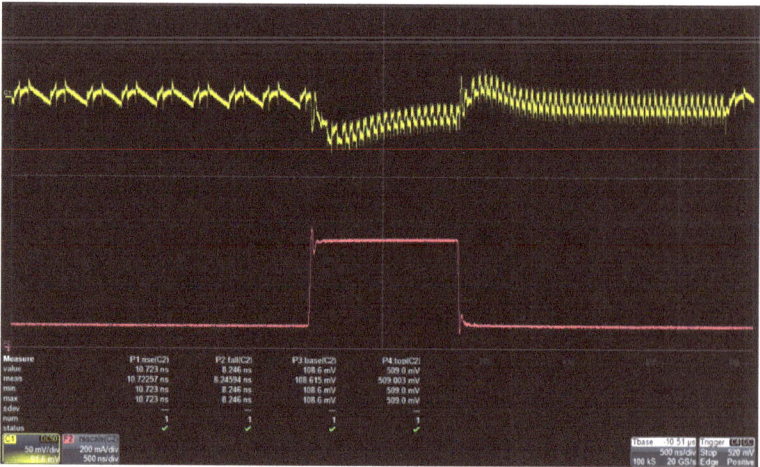

Voltage Change Due to Load Change and Regulator Output Impedance

Figure 6-9

Reverse Transfer and Crosstalk

Linear regulators, as well as voltage references and op-amps, reflect the load current directly to the input, at least at lower frequencies below the control loop bandwidth.

At higher frequencies, decoupling and parasitic elements further attenuate the current at the input.

The result is that whatever current is applied to the regulator's output also flows through the regulator's input, where combined with the finite impedance connecting the regulator and its power source, it becomes a voltage signal.

This voltage signal appears at all devices connected to this node.

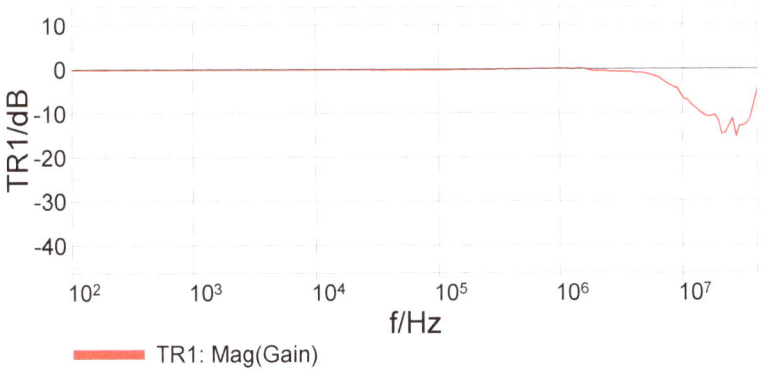

LM317 Reverse Transfer with no Input or Output Capacitors

Figure 6-19

Shunt regulators offer significant reverse transfer isolation and, therefore, reflect much less current at the input. An example of a shunt regulator reverse transfer is shown in Figure 6-11 indicating that for this regulator the resulting input current is nearly 50dB lower than the change in output current.

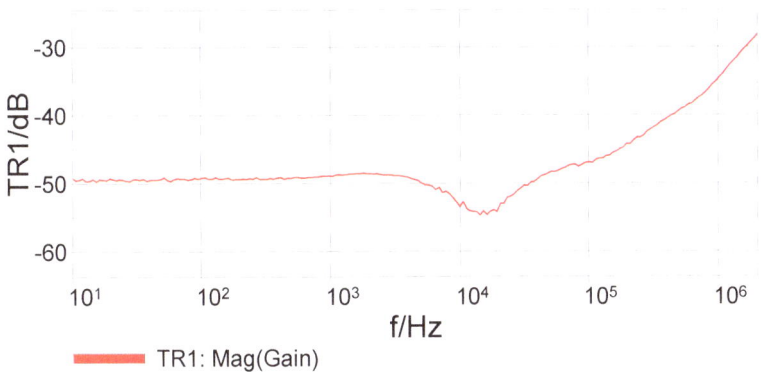

Shunt Regulator Reverse Transfer

Figure 6-11

Switching regulators are very efficient.

The output power and input power are nearly the same magnitude, especially given POLs and VRMs, which are generally more than 90% efficient.

$$Pin = Pout \qquad\qquad 6.2$$

$$Vin \cdot Iin = Vout \cdot Iout \qquad\qquad 6.3$$

$$Iin = \frac{Vout}{Vin} \cdot Iout \qquad\qquad 6.4$$

The low frequency reverse transfer of a switching regulator is the ratio of its output voltage to its input voltage, at least for frequencies below the control loop bandwidth and also below the corner frequencies of input and output filters.

We have now seen that the majority of the noise is related in some way to impedance.

The internal regulator noise, regulator output impedance, PDN impedance, and reverse transfer combined with input source impedance have all been shown to be significant contributors to the noise.

It is for this reason that one of the largest chapters in the book is Chapter 7, which is dedicated to measuring impedance.

The load current variations are another significant term, though in most cases there is little we can do about these variations as they are a function of the system we are powering.

Control Loop Stability

Another significant contribution to the noise is the result of the PDN impedance (including the decoupling capacitors) interacting with the control loop of the voltage regulator in a negative way.

In a simple case, we can consider a voltage regulator

connected to a single decoupling capacitor. The decoupling capacitor interacts with the voltage regulator control loop and can degrade the stability margins of the regulator.

All of the paths we have discussed; PSRR, output impedance, reverse transfer and internal noise are impacted by the stability of the control loop.

Poor stability margins are seen as degraded noise performance near the frequency, or frequencies, at which the stability is poor.

Impact on Output Impedance

Poor control loop stability is revealed as a peaking of the impedance.

The regulator's output impedance plots shown in Figure 6-12 have the same bandwidth, but very different phase margins.

The blue trace has a phase margin of 62 degrees while the dashed red trace has a phase margin of 24 degrees.

The poor phase margin results in a much higher peak output impedance near the frequency of instability, which is 100kHz in this example.

The sharp impedance drop at 450kHz is due to the resonance of the PDN.

A single voltage regulator can have poor control loop stability at more than one frequency, in which case the degradation would be evident near each of these frequencies.

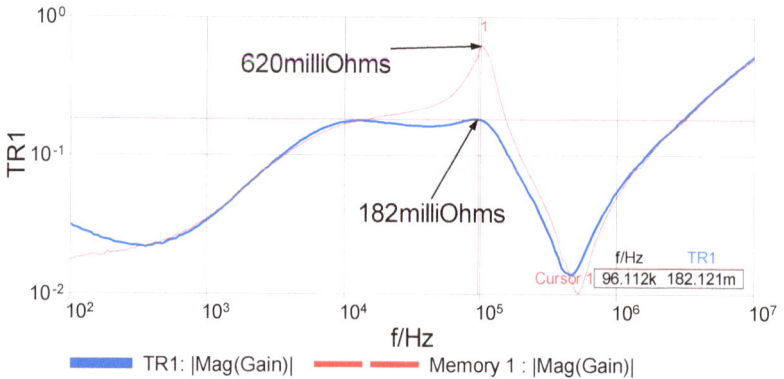

Regulator Output Impedance with Phase Margin of 62 Degrees (Blue Trace) and 24 Degrees (Dashed Red Trace)

Figure 6-12

Impact on Noise

The regulator's output noise is related to the output impedance and the output impedance is impacted by stability, so that the output noise is indirectly impacted by the control loop stability.

This relationship is evident from the measurement plots shown in Figure 6-13.

These plots are included in a voltage reference datasheet and a simple analysis illustrates the relationship between control loop stability, impedance, and noise.

The impedance plots on the right show slight peaking with the 100µF capacitor (green) no peaking with 10µF (purple) and significant peaking with 2.7µF (red).

We already showed that the peaking is related to poor stability margins, so let's consider the noise. The plots on the left show that the significant impedance peaking with the 2.7µF capacitor (red) results in significant noise peaking.

The 10µF capacitor (green) shows no noise peaking and the

100µF capacitor (purple) results in some noise peaking.

The degree of peaking seems much greater in the plots on the left, but this is an illusion due to the left plot using a linear scale and the right plot using a log scale. Upon careful inspection you can see that the impedance peak for the 2.7µF capacitor is approximately 6dB, increasing from 600mΩ peaking to 1.2Ω corresponding with a nearly 6dB increase in noise from approximately $50nV/\sqrt{Hz}$ to $100nV/\sqrt{Hz}$.

Similarly with the 100µF capacitor the impedance peaking is approximately 3dB, from approximately 0.4Ω peaking to 0.6Ω corresponding with the noise increase from approximately $50nV/\sqrt{Hz}$ to $70nV/\sqrt{Hz}$.

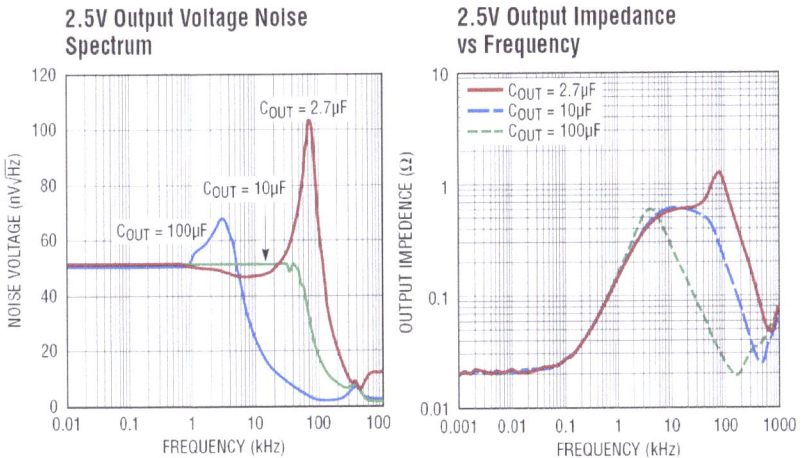

Output Impedance and Corresponding Output Noise for Various Capacitors

Figure 6-13

Impact on PSRR

The PSRR measurement plots in Figure 6-14 also show the significant degradation in PSRR resulting from poor control loop stability.

In this case, the PSRR at 95 kHz is nearly 50dB without a destabilizing output capacitor and the performance is degraded nearly 20dB by the addition of a destabilizing load capacitor.

The PSRR peak at approximately 1.5MHz is due to the series resonance of the load capacitor.

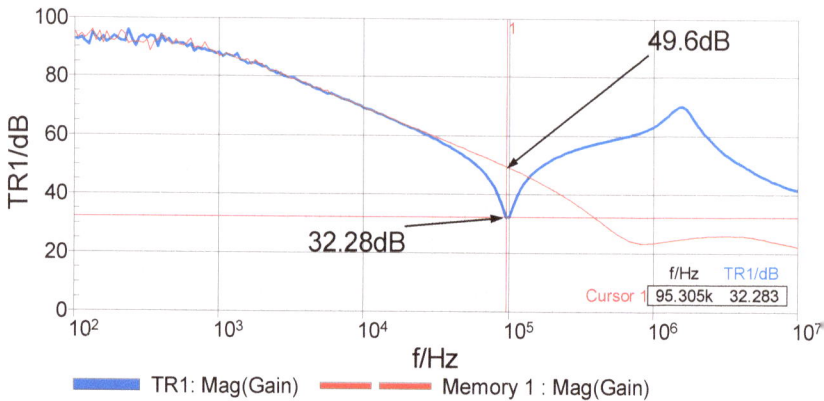

PSRR with Poor Stability (Blue Trace) and Good Stability (Red Trace)

Figure 6-14

Impact on Reverse Transfer

The reverse transfer is also impacted in a similar way.

In this case, the input current is amplified as a result of the poor stability margin.

The example in Figure 6-15 shows the reverse transfer is 0dB for low frequencies. Near the frequency of poor stability, the peak response is a function of the phase margin of the control loop.

With a phase margin of 13 degrees the reverse transfer is increased by 15dB or a nearly six-fold increase in the input current noise generated due to the poor stability.

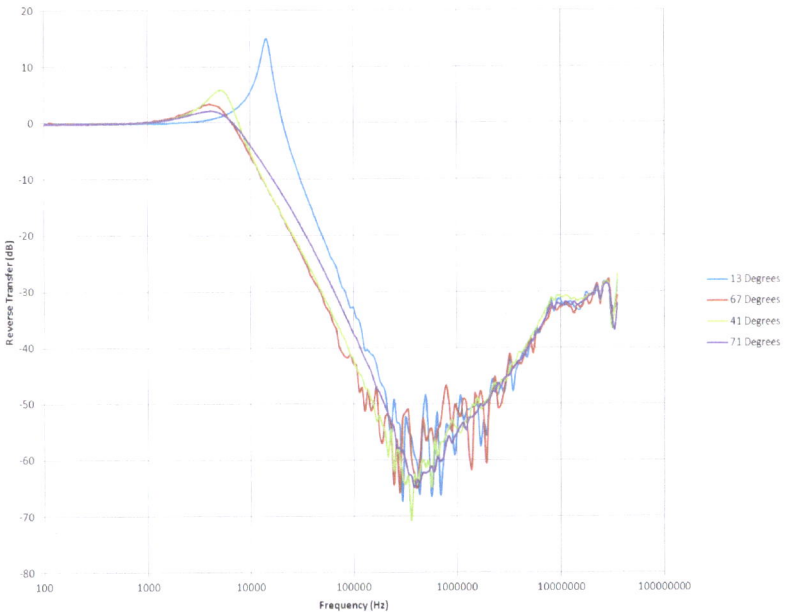

Reverse Transfer with Several Phase Margins

Figure 6-15

The negative impact of poor control loop stability on all of the noise paths is clear, and it is essential to verify all control loops have good margins in order to achieve optimum system performance.

Many military and high reliability guidelines require the PM and GM to be greater than 45 degrees and 10dB, respectively, over the lifetime of the circuit's operation.

How Poor Stability Propagates Through the System

The reverse transfer reflects the variations in the load current to the regulator input current. Since the impedance at the regulator input is finite, this current is transformed to a noise voltage at the input of the voltage regulator. This noise voltage is presented to all other devices connected to the same node. The addition of a second regulator adds a reverse transfer/crosstalk path as seen in Figure 6-2. In the case of crosstalk, load changes at the 1st regulator imposes a noise signal at the common input voltage connection to the voltage regulators where it can then flow through the PSRR of the 2nd regulator.

The peaking that results in each of the regulator's noise paths is representative of a 2nd order system response resulting in a decaying ring in response to an impulse as shown in Figure 6-16.

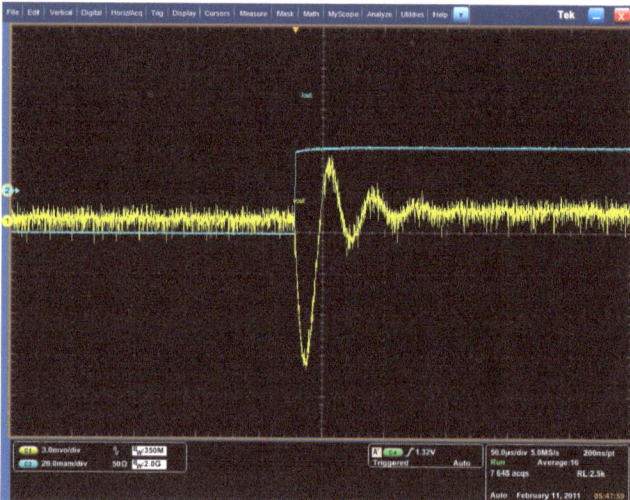

Impulse Response of a Single Peak

Figure 6-16

154

While such decayed responses seem benign, the harmonic content of the signal is quite rich, with the frequency range primarily limited by the edge speed of the impulse and the duty cycle of the impulse signal. An example of such a ring, and the associated harmonic content, is shown in Figure 6-17. In this example, we can see the operating current's impact on the ringing frequency and the very rich harmonic content associated with the ringing.

The separation between the harmonic spurs is the repetition rate of the impulse. This may seem counterintuitive, but the lower the impulse frequency the closer together these spurs will be. The result is a very large number of noise signals over a large frequency range. These signals then travel through the system "looking for" resonant response peaks that coincide with one of these harmonic noise spurs. Poor stability is the most common cause of system level noise problems.

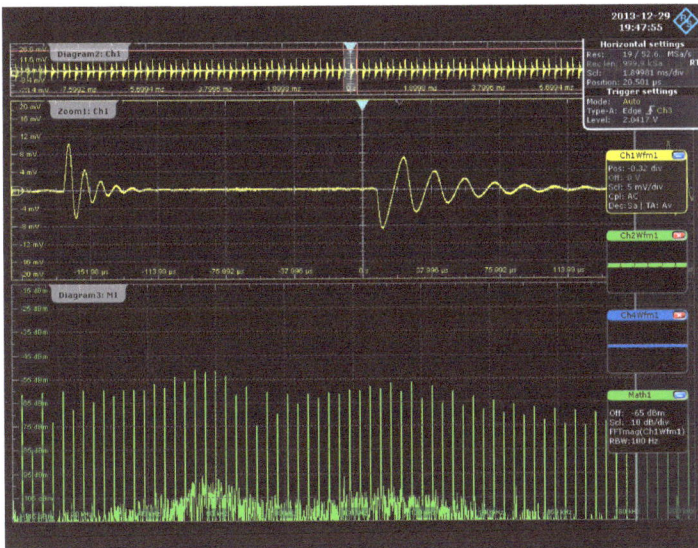

Spectral Response of the Decaying Ring

Figure 6-17

If the impulse response repetition frequency is equal to the frequency of the peak in the noise path, than the noise signal will be much larger as shown in Figure 6-18. In this case, the signal amplitude increases rather than decaying with a number of cycles required to reach full amplitude. The full amplitude response for a sine wave current signal is simply the product of the impulse signal and the peak noise path response. In the case of a square wave impulse, the peak response can be 27% greater than a sine wave due to the fundamental Fourier component. In Figure 6-18, the upper trace (yellow) is the output voltage, the middle trace (blue) is the input current showing the reverse transfer and the lower trace (green) is the load current step.

$$Noisepk\_fundamental = \frac{4}{\pi} \cdot Noise\ impulse \qquad 6.5$$

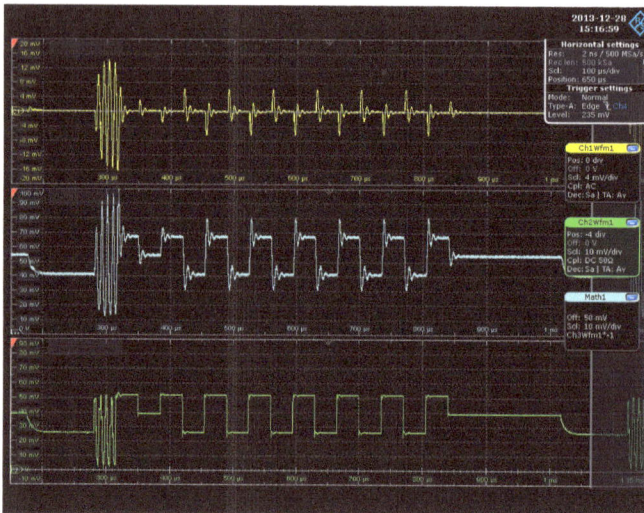

The Impulse Response for the Cases where the Repetition Rate is Equal to and Much Lower than the Frequency of the Peak in the Noise Path

Figure 6-18

Adding the PDNs

So far we have considered the noise generated by the regulator paths and local stability.

The PDN impedance at the regulator's output connects the regulator to the load circuits through printed circuit board traces or planes and frequently includes ferrite beads, series resistors and decoupling capacitors. The result may or may not be "clean" voltages at the load circuits.

In high speed circuits, the PDN impedance is often responsible for the majority of the output noise.

The image in Figure 6-19 shows the output voltage of a high-speed CMOS gate with a 10MHz signal in the upper trace.

The lower trace is the voltage measured from the logic gate Vcc to the logic gate ground, directly on the pins of the device.

The peak-to-peak noise at the logic gate supply pins is approximately 600mV and is caused by the shoot thru current of the logic gate during its switching transitions, as well as the PDN impedance of the decoupling capacitor and its interconnections to the regulator.

In Figure 6-19, the upper trace (yellow) is the logic gate output used to trigger and the lower trace (blue) is the voltage at the device pins.

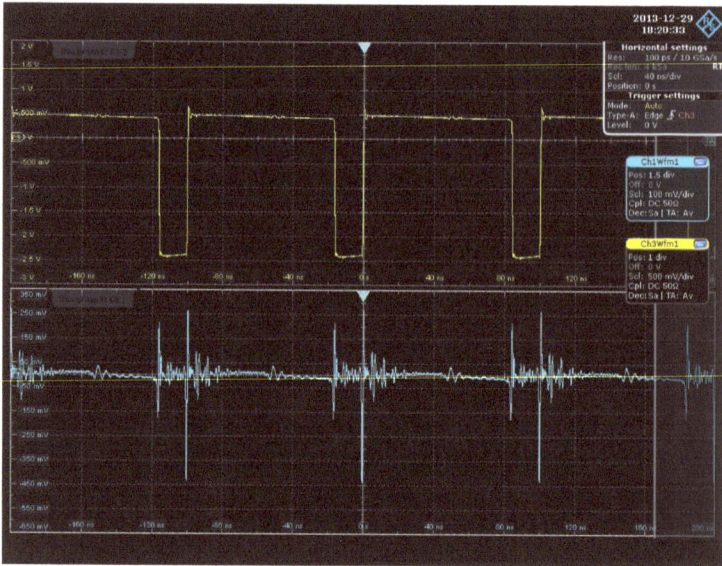

Noise Seen across a Logic Gate's Supply Pins

Figure 6-19

The voltage across the regulator's decoupling capacitor located directly next to this logic gate is 110mV peak-to-peak as shown in Figure 6-20.

This tells us that the majority of the voltage drop is due to the PDN connecting the regulator to the logic gate pins. In this circuit, the PDN is represented by two very short PCB traces and a series connected 1Ω chip resistor (for monitoring the gate shoot through current).

This connects a single 10nF decoupling capacitor to the logic gate supply pins in this example.

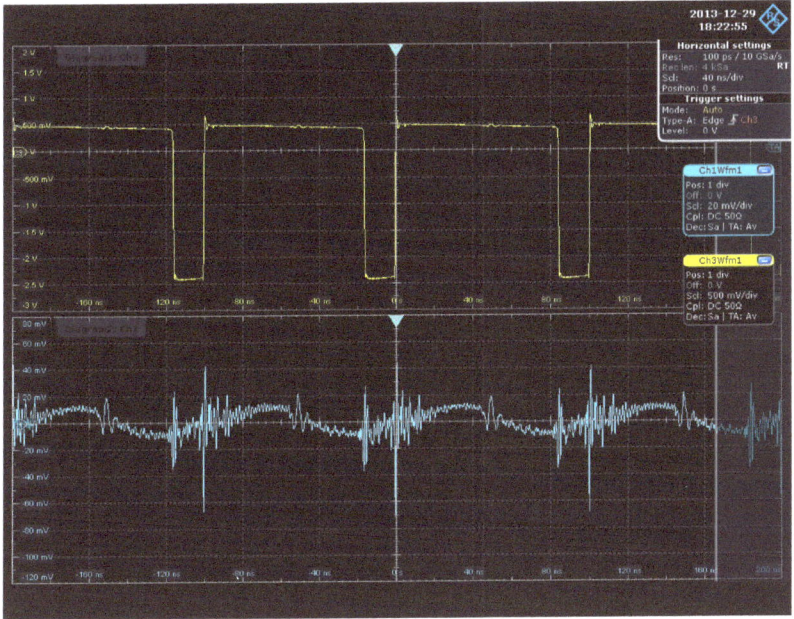

Voltage Seen at the Local Decoupling Capacitor Adjacent to the Logic Gate

Figure 6-20

In this chapter, we have developed a fundamental understanding of the propagation paths through the distributed system.

The remaining chapters are dedicated to the equipment and measurement techniques required to perform measurements of each of these noise paths, as well as some helpful troubleshooting techniques.

As we have shown, the performance of the system is significantly influenced by the loading (i.e. PDN impedances). It is generally best to make measurements in the system so that the impact of the PDN is also represented in the measurement.

Chapter References

1. Ron Mancini, *Opamps for Everyone*, Design Reference Texas Instrument SLOD006A, Sept. 2001

2. Steven M. Sandler, *Target Impedance Based Solutions for PDN May Not Provide Realistic Assessment*,
   EDN, May 2013
   http://www.edn.com/design/test-and-measurement/4413192/4/Target-impedance-based-solutions-for-PDN-may-not-provide-a-realistic-assessment

Chapter 7

Measuring Impedance

IMPEDANCE IS LIKELY the most generally applicable measurement in power electronics, as well as in instrumentation and high speed circuits. Impedance is used to characterize individual component characteristics, such as semiconductor junction parameters, capacitor equivalent series resistance (ESR), capacitor equivalent series inductance (ESL), inductor and transformer mutual and leakage inductance, and self-resonant frequency (SRF) to name just a few.

Impedance is also used to characterize power distribution network (PDN) performance, closed loop power supply performance, system stability, battery performance and many other characteristics.

Selecting a Measurement Method

There are several different techniques that can be used to measure impedance. Each method has pros and cons, though they are generally selected based on the impedance range of the measurement to be made, the required measurement bandwidth and the operating voltage of the device or circuit being measured.

	Impedance Range[1]
Single Port	0.5Ω to $2.5k\Omega$
Two Port Shunt	$250u\Omega$ to 25Ω
Two Port Series	25Ω to $1M\Omega$
Voltage and Current	$1m\Omega$-$2k\Omega$
Impedance Adapter	0.1Ω to $400k\Omega$

[1]Approximate range, based on proper calibration of measurement

Impedance Range of the Various Measurement Methods

Table 7.1

Single Port Measurements

Introduction

The single port measurement uses the controlled port impedance, Zo and the reflection coefficient of the measurement, Γ in order to determine the impedance of the device being tested.

The reflection coefficient is related to the impedance of the VNA, typically 50Ω and the impedance being measured as shown in Equation 7.1.

$$\Gamma = \frac{Z_{DUT} - Z_o}{Z_{DUT} + Z_o} \qquad\qquad 7.1$$

The sensitivity of the measurement can be seen in Figure 7-1, showing the reflection coefficient as a function of the impedance being measured. The measurement is most accurate at Zo and is severely limited at impedances below approximately 1Ω and above approximately $3k\Omega$. At these impedance extremes, the accuracy of the measurement is limited by the resolution and noise floor of the instrument being used for the measurement.

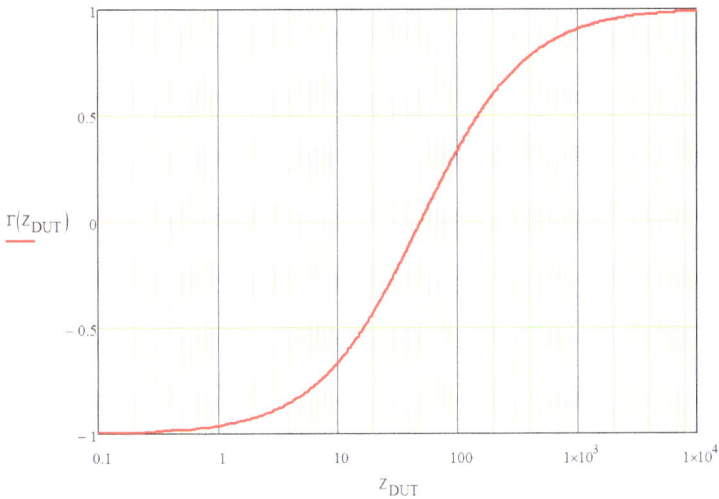

Reflection Coefficient as a Function of the Impedance of the Device Being Measured

Figure 7-1

Pros

- The frequency range of the measurement is limited only by the accuracy of the calibration and the bandwidth of the VNA.

- The measurement can be made at zero volts and also biased, allowing measurements in the powered off state as well as the powered on state.
- The impedance range is quite usable for many passive devices, such as capacitors, inductors and semiconductors as well as low power circuits such as voltage references, regulators and other voltage feedback devices.
- The measurement can be AC coupled using a DC blocker in order to eliminate the DC loading of the DUT by the equipment impedance Zo.
- Very simple measurement requiring only one cable.

Cons

- The measurement is more sensitive to calibration than other impedance measurements, requiring at least 3 calibrations; OPEN, LOAD and SHORT. Some devices also allow electrical length compensation to remove the measurement cable influence without the need of additional calibration.
- Cannot generally measure DCR of inductors or transformers or ESR of capacitors, since they are often below the measurement range.

Tips and Tricks

1. The physical construction of the calibrator must precisely match the construction of the DUT. For example, if the DUT is a 0805 surface mount chip device, the calibrator should also be constructed using similar 0805 devices.
2. I often use a 20MΩ resistor for the OPEN calibrator position. This resistance is high enough not to influence the measurement, but does replicate the solder and

capacitance that the DUT package might contribute.

3. It is nearly impossible to create a zero impedance "SHORT". I often use a low impedance planar current sense resistor in the same size package as the DUT. This is not perfect, but adequate in most applications.

4. At high frequencies even short traces and wires can contribute significant inductance, so be sure to make the measurements at the point of interest. For example if you want to measure the ESL of a capacitor, be sure you measure <u>at the capacitor</u> as even short traces or leads will dominate the measurement.

5. <u>Always</u> measure a known device when creating a new setup to validate both the setup and the calibration.

6. Use a very low level oscillator signal. A large signal will not produce the correct result. You can verify the signal level by reducing it and checking whether the results change. As a rule, the lower the operating current of the device the more sensitive it is to signal amplitude.

Device Setup

The Agilent E5061B provides the impedance measurement access through selection of the MEASUREMENT menu.

The impedance measurement requires the option 05 for this analyzer. The OMICRON Lab Bode 100 allows selection of this measurement from the Measurement pull down menu. These are both shown in Figure 7-2.

Consult your particular equipment manual for additional setting information for other VNA devices.

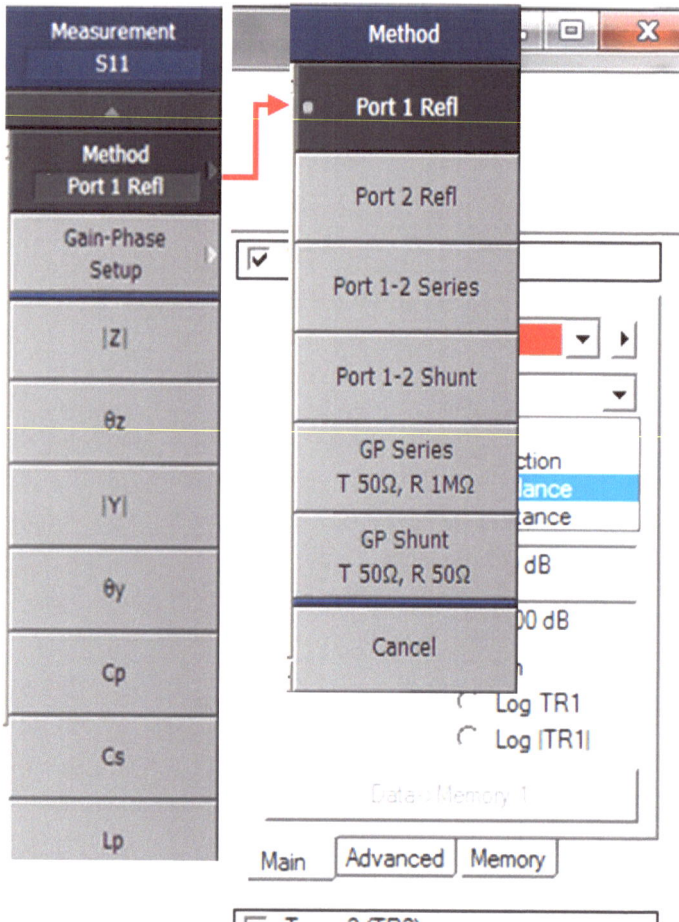

Selection of the Single Port Impedance Measurement is Shown for the Agilent E5061B (Left) and the OMICRON Lab Bode 100 (Right)

Figure 7-2

Calibrating the Single Port Measurement

You should first adjust the measurement settings, such as the frequency range, signal level and resolution bandwidth. <u>If any of these settings are changed or anything changes in the setup, recalibration is required</u>.

The minimum single port impedance calibration typically consists of OPEN, SHORT and LOAD calibration sweeps.

The OMICRON Lab Bode 100 calibration is performed using the calibration menu as shown in Figure 7-3. Note that the display indicates the state of the calibration. The Bode 100 indicates the calibration has been performed by stating it is "performed" and also by changing the color of the state from RED to GREEN. The additional calibrations shown are THRU, used for gain measurements, delay time through the SHORT and the exact resistance value for the LOAD calibrator.

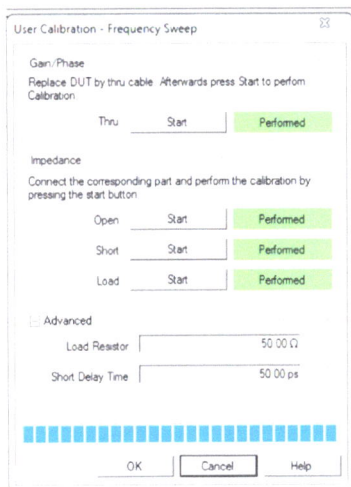

Calibration Screen of OMICRON Lab Bode 100 Showing the OPEN, SHORT and LOAD Calibration

Figure 7-3

The Agilent E5061B calibration is performed using the Calibrate menu in the RESPONSE block as shown in Figure 7-4.

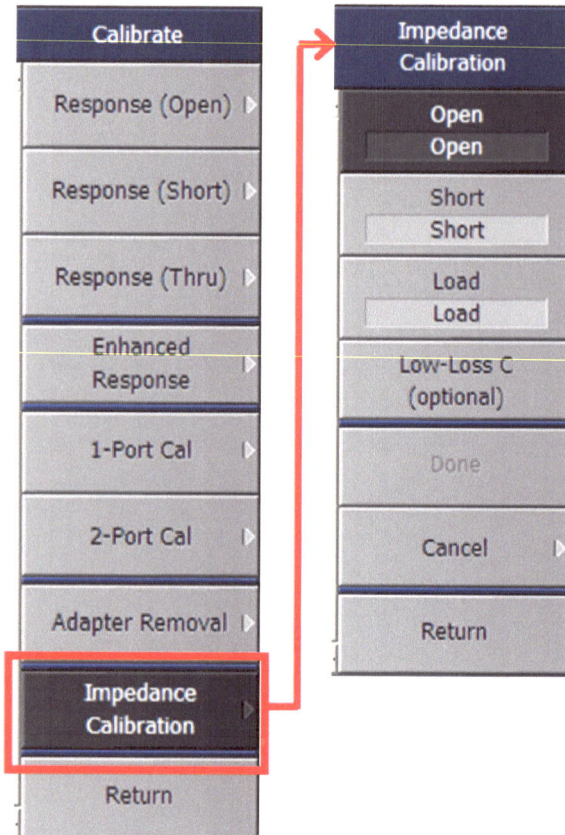

Calibrate	Impedance Calibration
Response (Open) ▷	Open / **Open**
Response (Short) ▷	Short / Short
Response (Thru) ▷	Load / Load
Enhanced Response ▷	Low-Loss C (optional)
1-Port Cal ▷	Done
2-Port Cal ▷	Cancel ▷
Adapter Removal ▷	Return
Impedance Calibration ▷	
Return	

Calibration Screen of the E5061B Showing the OPEN, SHORT and LOAD Calibration

Figure 7-4

The general procedure is to connect the OPEN calibrator and perform the OPEN calibration. Then connect the SHORT calibrator and repeat the calibration for the SHORT and finally connect the LOAD and repeat using the LOAD calibration. The order of the calibration is not significant.

For other VNA's consult the user manual or help functions to determine the procedure for your particular instrument.

Although I know I am repeating myself, the calibration should be performed each and every time changes are made to the equipment setup, such as changing the frequency range or resolution bandwidth of the measurement.

DC Coupled vs AC Coupled

When measuring low power devices, especially in circuit, it is frequently the case that the circuit cannot tolerate the 50Ω impedance of the test equipment. For example, measuring the output impedance of a voltage reference can be performed using a single port impedance measurement; however, the equipment impedance would alter the operating bias current of the voltage reference or overload it. In this case a blocking capacitor can be included in the measurement to isolate the DC impact of the equipment impedance. Be sure the calibration is performed with this blocking capacitor installed if it is being used in the measurement.

Examples

1. 1Ω Chip Resistor

A PCB that includes three BNC connectors, one connected to a 50Ω precision SMD resistor in a 0805 package, one connected to a short (short trace) and one connected to an open is used to perform the OPEN, SHORT and LOAD calibration.

An identical calibrator board is used to mount the 1Ω resistor to be tested with a matching BNC. The matching BNC assures that the measurement is made using the same BNC interconnecting impedance in all three calibrations as well as the device measurement. The setup is shown in Figure 7-5

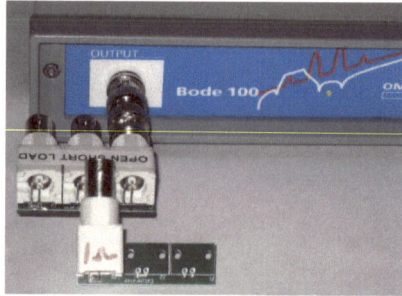

*Setup Showing the OPEN SHORT LOAD Calibrator
and the Matching DUT Board with a 1Ω Resistor
Mounted on the DUT PCB*

Figure 7-5

After performing the calibration the DUT PCB is connected to the VNA using the same barrel adapter that was used to attach the calibrator. The resulting impedance measurement is shown in Figure 7-6. Note that while the 1Ω measurement is close to the expected lower impedance limit, the measurement is well within the 1% resistor tolerance.

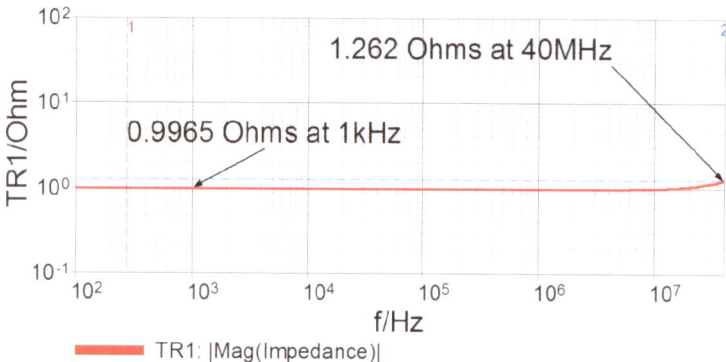

*The Result of Measuring a 1Ω Resistor after
Calibration*

Figure 7-6

The impedance at 40MHz shows that the impedance is just slightly inductive. Assuming a simple series equivalent circuit model consisting of the resistor and a series inductance, we can calculate the inductance, as shown in Equations 7.2 and 7.3 since we know the resistance, frequency and impedance of the measurement.

$$Z = \sqrt{0.9965^2 + (2\pi \cdot 40MHz \cdot L)^2} = 1.262 \qquad 7.2$$

$$L = 3.08\ nH \qquad 7.3$$

The resistor measurement is very close to the 1Ω nominal value and well within the 1% specified tolerance while the inductance of the resistor is approximately 3nH. There can be other small inductance terms, such as the solder joint that attaches the resistors to the PCB. We expect those to be very similar in the calibration PCB and the DUT PCB so this effect is theoretically calibrated out of the measurement. This is one reason we try to assure that the physical connections of the DUT and calibrator are as close to identical as possible.

VNA's can generally display the measurement results in many ways. For example, the impedance can be shown as a series L-R at a single frequency, 10MHz as shown in Figure 7-7. Note that the inductance displayed is very close to the 3nH calculated above.

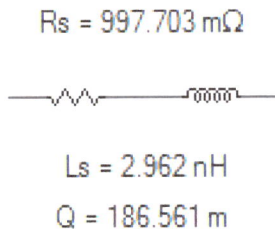

Rs = 997.703 mΩ

—/\/\/———͡‸͡‸͡‸———

Ls = 2.962 nH

Q = 186.561 m

Displaying the 1Ω resistor as a series L-R at 10MHz

Figure 7-7

Steven M. Sandler

2. Voltage Reference

This example demonstrates the measurement of the output impedance of a voltage reference.

An LT1009 shunt reference is biased at 1mA and the impedance is measured with and without a 0.1μF ceramic output capacitor. A schematic of the setup is shown in Figure 7-8.

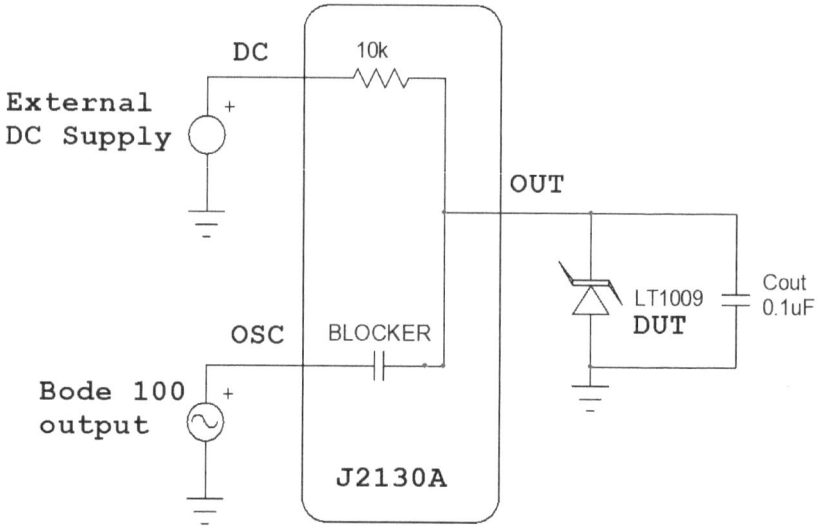

Schematic of LT1009 1mA Output Impedance Measurement Circuit

Figure 7-8

A picture of the device setup is shown in Figure 7-9.

A Picotest J2130A DC blocker, which includes a 10kΩ resistor between the DC+ input and the output allows low current biasing.

The figure shows that the calibration is performed with the DC blocker installed and power applied.

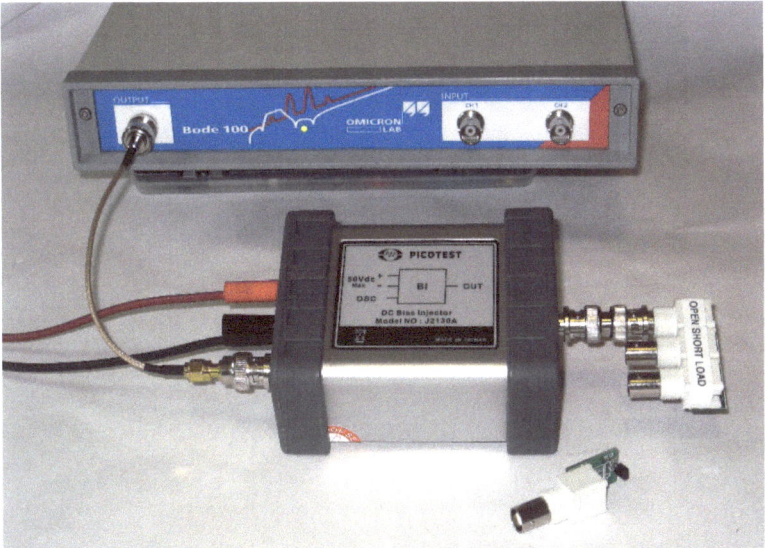

Setup for Measuring LT1009 Shunt Reference—Note the LT1009 PCB in the Foreground

Figure 7-9

After calibration the LT1009 PCB with a similar BNC connector mounting is attached to the barrel connector and the external power supply is adjusted for 1mA bias current.

The output impedance trace is shown in Figure 7-10.

The solid blue trace is the impedance without an output capacitor while the dashed (red) trace is with the addition of a 0.1μF output capacitor in parallel with the LT1009.

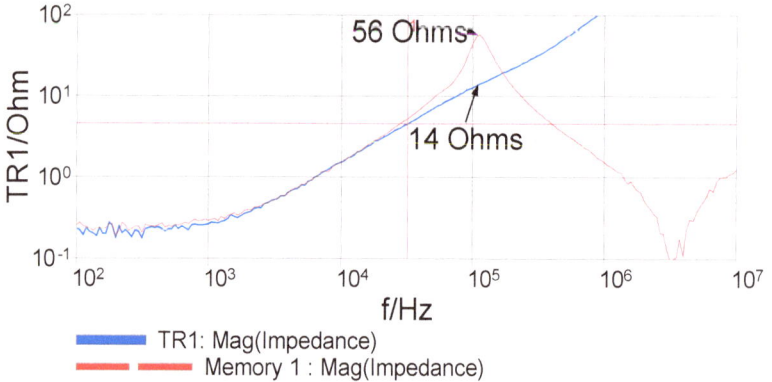

*Output Impedance of the LT1009 at 1mA with (Red)
and without (Blue) 0.1μF Output Capacitor*

Figure 7-10

The results in Figure 7-11 also shows the low frequency resistance of approximately 0.2Ω, which is well below the 1Ω recommended range, making this particular measurement characteristic uncertain.

The degraded stability with the addition of the output capacitor is also evident in that the impedance at 100kHz increased from 14Ω without the capacitor to 56Ω with the capacitor.

At frequencies above 200kHz the addition of the output capacitor reduces the output impedance of the voltage reference circuit.

3. Measuring a Small-Signal Transistor

Using the same measurement setup as the previous example the LT1009 is replaced with a silicon 2N3904 silicon transistor. The bias current is again set to 1mA and the impedance is measured as in the previous example.

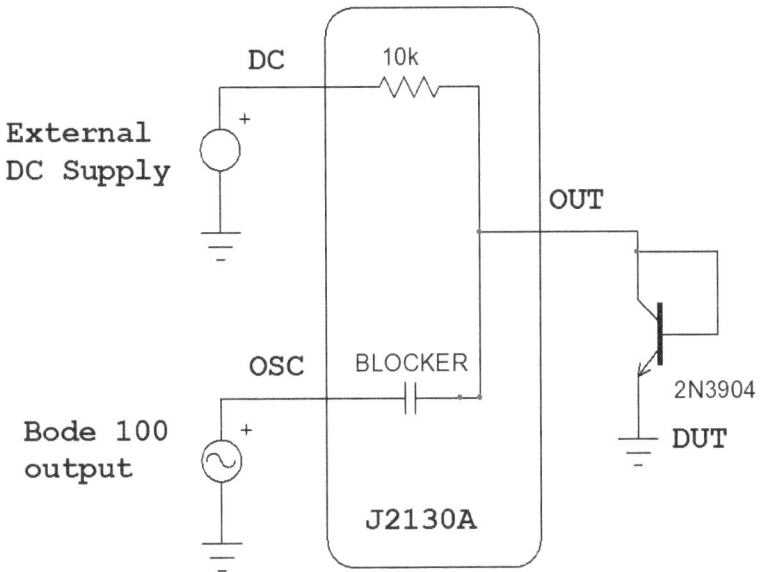

Schematic of NPN Transistor at 1mA Output Impedance Measurement Circuit

Figure 7-11

In this example the impedance is measured with two signal amplitudes, -23dBm and -13dBm in order to show the impact of a signal that is too large.

With the signal source set to -23dBm the device measurement is 27Ω and with the signal source at -13dBm the device measurement is 54Ω.

The results are shown using an Agilent E5061B analyzer in Figure 7-12 and the OMICRON Lab Bode 100 in Figure 7-13 in order to show that this phenomenon is not due to the analyzer.

Figures 7-12 and 7-13 demonstrate the need for small signal levels.

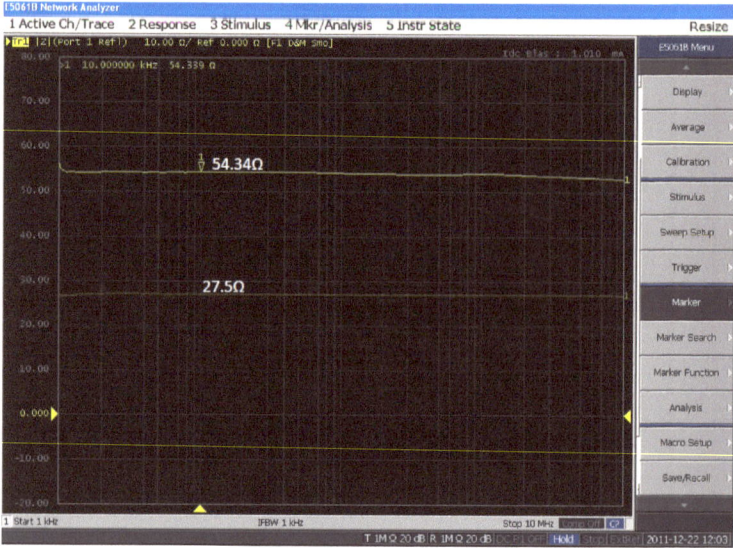

The 2N3904 Junction Impedance at 1mA using the
Agilent E5061B with -13dBm and -23dBm signals

Figure 7-12

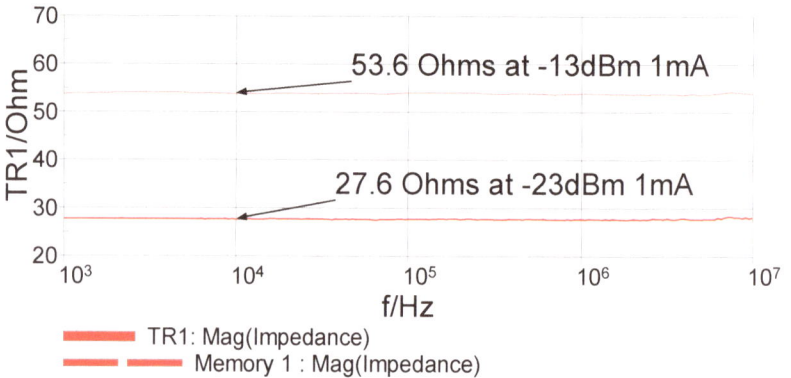

The 2N3904 junction impedance at 1mA using the
OMICRON Lab Bode 100 with -13dBm and -23dBm

Figure 7-13

Note that both analyzers produce similar results and both results are correct, though the 27Ω result is a small signal measurement and the 54Ω is a large signal measurement.

This is another reason we always measure a known quantity.

A reasonable starting point is to limit the signal level to 50mVpp maximum across the junction, corresponding with approximately -22dBm for a junction impedance of 50 Ωs and -27dBm for a junction at near zero bias current.

4. SPICE Model of a Diode

The single port impedance measurement can also be used to determine the SPICE parameters for a semiconductor junction. The impedance measurements are performed in both the forward and reverse biased directions in order to determine the forward SPICE parameters, N and IS and the junction capacitance parameters CJO, Vj and M.

A picture of the single port setup is shown in Figure 7-14.

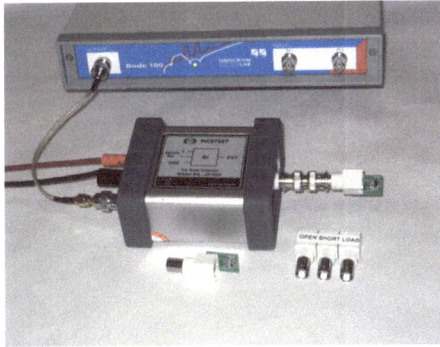

Setup for Measuring Diode SPICE Parameters using a DC Bias Injector

Figure 7-14

The SPICE equation for a forward biased silicon diode junction is shown in Equation 7.4.

$$VD = \ln\left(\frac{Ij + IS}{IS}\right) \cdot N \cdot VT + Ij \cdot RS \qquad 7.4$$

In Equation 7.4, Ij is the junction current, RS is the series resistance of the junction, IS is the junction saturation current, N is the emission coefficient and VT is the thermal voltage of the junction, approximately 26mV at room temperature.

Assuming that Ij is much greater than IS, and the RS term is very small compared with the junction voltage, Equation 7.4 reduces to Equation 7.5.

$$VD = \ln\left(\frac{Ij}{IS}\right) \cdot N \cdot VT \qquad 7.5$$

The junction impedance is obtained from the derivative of the voltage with respect to current, shown in Equation 7.6.

$$ZD = N \cdot \frac{VT}{Ij} + RS \qquad 7.6$$

And assuming that RS is small compared with the junction resistance and the approximate value of VT at room temperature is 26mV, the equation reduces to Equation 7.7.

$$ZD = \frac{0.026}{Ij} \cdot N \qquad 7.7$$

The emission coefficient, N, can therefore be determined from the junction impedance at a known junction current as in Equation 7.8.

$$N = \frac{Ij}{0.026} \cdot ZD \qquad 7.8$$

With N and Ij known, a measurement of the junction voltage allows the determination of IS. It is difficult to measure the RS term using a steady state measurement as the operating junction current heats the junction altering the result. The RS term is generally determined from a pulsed current measurement or from the component datasheet. The resistance is set to adjust the junction voltage to the correct value at a relatively high operating current

The SPICE equation for the junction capacitance is shown in Equation 7.9.

$$Cj = Cjo \cdot \left(1 + \frac{V}{Vj}\right)^{-M} \qquad 7.9$$

After setting the external and connected DC power supply to zero Volts and performing the OPEN, LOAD and SHORT calibration, the external power supply is adjusted to achieve a desired junction current. I generally use a current of 1mA. The forward impedance measurement is shown in Figure 7-15.

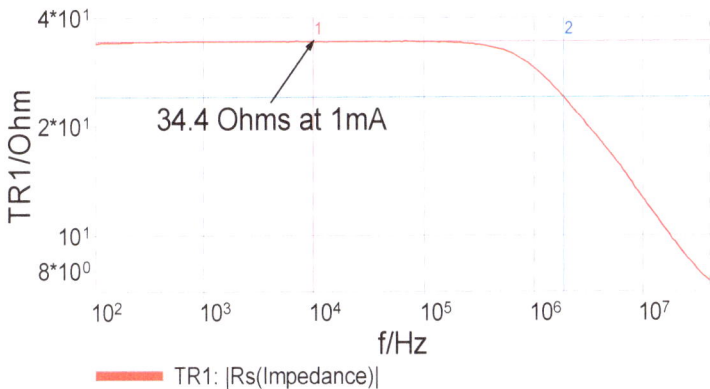

Forward-Biased MUR110 Diode Impedance at a Junction Current of 1mA

Figure 7-15

From this junction resistance N can be calculated using Equation 7.8 as 1.323.

The forward voltage of the diode is measured using a DC voltmeter using the same setup. As the AC sweep signal can impact the reading, the VNA modulation should be turned off or removed for the DC measurement. In this example the diode junction voltage, VD, is 0.472V.

With both N and Ij known, the equation can be solved for the missing term, IS. Substituting 0.472V for V_D and 1.323 for N, Equation 7.10 can be solved for IS as follows:

$$IS = I\,j \cdot e^{-\frac{VD}{N \cdot VT}} = 1.09832 \cdot 10^{-9} A \qquad 7.10$$

Again using the same measurement setup, but reversing the polarity of the external power supply the diode is reverse biased. In this condition we can measure the junction capacitance.

The junction capacitance contains three variables and so the capacitance is measured at three voltages. In this example the capacitance is measured at 0.1V, 3V and 20V and the results are presented in Table 7.2.

The 3V measurement is shown in Figure 7-16.

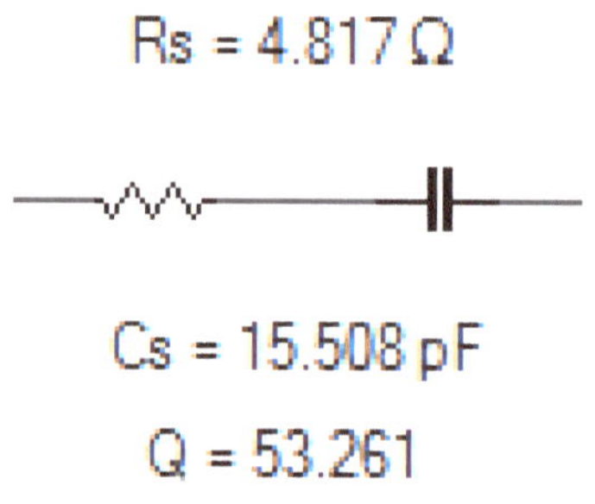

Rs = 4.817 Ω

Cs = 15.508 pF

Q = 53.261

Junction Capacitance Measurement at 3V Reverse Bias

Figure 7-16

Junction voltage	Junction capacitance
0.1V	31.6pF
3.0V	15.5pF
20.0V	8pF

Measured Junction Capacitance at Three Bias Voltages

Table 7.2

With three variables and three data points, the simultaneous equations can be solved using your preferred math solver. We have shown the Mathcad Minerr solution in Figure 7-17. However, EXCEL can also be used as can other math packages.

Given

$$31.6 \cdot 10^{-12} = CJO \cdot \left(1 + \frac{0.1}{Vj}\right)^{-M}$$

$$15.5 \cdot 10^{-12} = CJO \cdot \left(1 + \frac{3}{Vj}\right)^{-M}$$

$$8 \cdot 10^{-12} = CJO \cdot \left(1 + \frac{20}{Vj}\right)^{-M}$$

$$\text{Minerr}(CJO, Vj, M) = \begin{pmatrix} 3.437 \times 10^{-11} \\ 0.391 \\ 0.369 \end{pmatrix}$$

Mathcad MINERR Function is used to Solve for the Three Unknowns Simultaneously

Figure 7-17

The three capacitance values can be verified using Equation 7.9 and are in excellent agreement with the desired results as shown in Table 7.3.

Junction voltage	Measured capacitance	Calculated capacitance
0.1V	31.6pF	31.6pF
3.0V	15.5pF	15.49pF
20.0V	8pF	7.99pF

Measured Junction Capacitance vs Calculated Junction Capacitance

Table 7.3

The extracted SPICE model and model statement are shown in Figure 7-18, which also shows the forward voltage to be in good agreement with the measured value.

Note that this is a small signal model and does not include the breakdown parameters or the internal series resistance, RS.

This SPICE model simulation is performed and the result is shown in Figure 7-19.

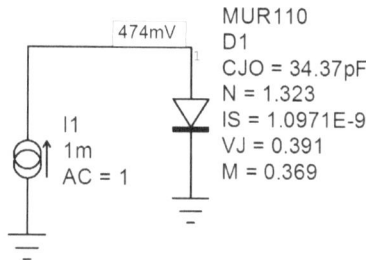

474mV

MUR110
D1
CJO = 34.37pF
N = 1.323
IS = 1.0971E-9
VJ = 0.391
M = 0.369

I1
1m
AC = 1

Extracted MUR110 SPICE Model

Figure 7-18

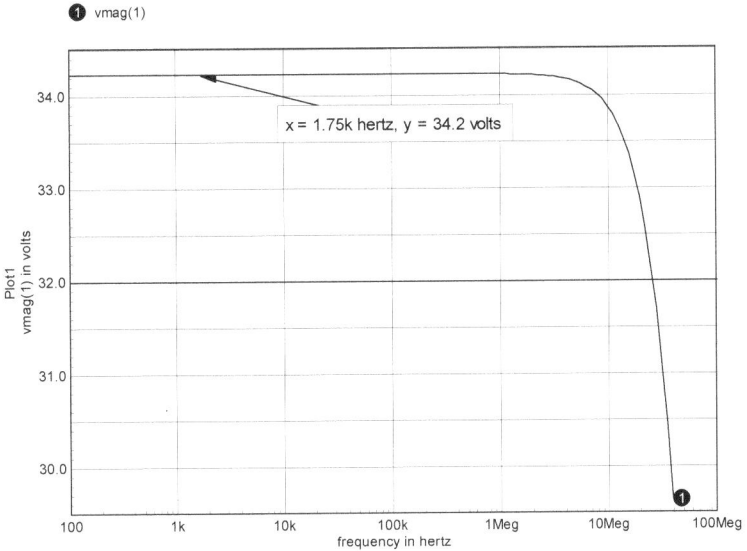

Simulated Diode Junction Impedance at 1mA

Figure 7-19

Two Port Measurements

Introduction

There are two configurations for the two-port impedance measurement: the series-thru and the shunt-thru.

In both cases, the measurement is a gain measurement. In the case of the two-port shunt-thru measurement the DUT is placed in shunt with the thru connection of port 1 and port 2.

In the case of the series-thru measurement the DUT is placed in series between port 1 and port 2.

Both configurations are shown in Figure 7-20.

The shunt-thru method is used much more frequently in power electronics than the series-thru measurement.

PORT 1   **PORT 2**

RS 50   V2   DUT   V3

RL 50

Series Thru

PORT 1   **PORT 2**

RS 50   DUT

RL 50

Shunt Thru

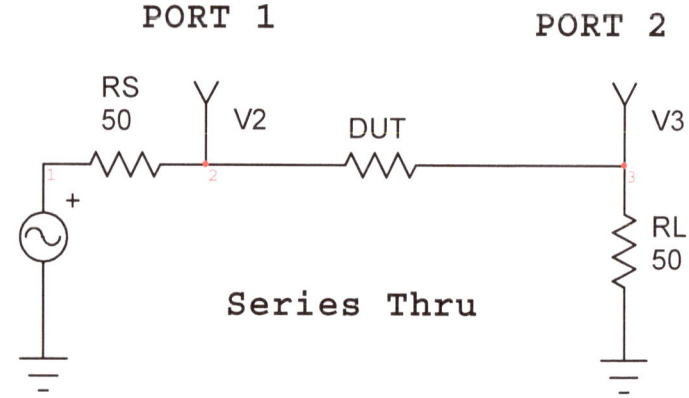

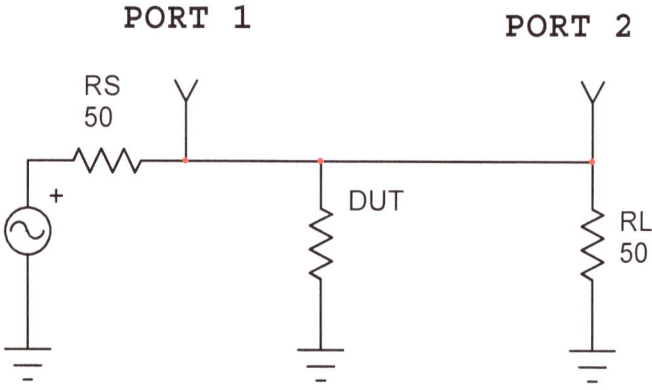

Configuration for the Two-Port Series and Two-Port Shunt-Thru Impedance Measurements

Figure 7-20

In the case of the series thru impedance measurement, the DUT is placed between the two VNA ports.

A voltage divider is formed by port 2 in the bottom of the divider and the source plus the DUT in the top of the divider.

184

The gain, S21, can be evaluated as:

$$S21 = \frac{2 \cdot RL}{RL + RS + Z_{DUT}} \qquad 7.11$$

Solving for the DUT impedance as a function of the series S21 measurement results in:

$$Z_{DUT}(S21) = \frac{2 \cdot RL}{S21} - RS - RL \qquad 7.12$$

In the case of the shunt thru measurement the DUT is in parallel with the port 2 impedance, therefore the gain S21 from port 1 to port 2 is calculated as:

$$S21 = \frac{\dfrac{2 \cdot RL \cdot Z_{DUT}}{RL + Z_{DUT}}}{RS + \dfrac{RL \cdot Z_{DUT}}{RL + Z_{DUT}}} \qquad 7.13$$

Solving for the impedance of the DUT using the shunt measurement method:

$$Z_{DUT}(S21) = \frac{-S21 \cdot RS \cdot RL}{S21 \cdot RS + S21 \cdot RL - 2 \cdot RL} \qquad 7.14$$

The factor of two in both S21 equations accounts for the fact that each port presents the same impedance, and so the internal voltage source E is divided by 2.

   This can be seen by connecting port 1 directly to port 2, representing a gain of 1.

$$Vport2 = E \frac{RS}{RS + RL} = 1 \qquad 7.15$$

And solving for E:

$$E = 1 \cdot \frac{RS + RL}{RS} \qquad 7.16$$

And it is clear that if RS and RL are equal then E is equal to 2.

The relationships between Z_DUT and S21 in both the series and shunt measurement are shown graphically in Figure 7-21. In the series thru measurement the sensitivity decreases rapidly as S21 approaches 0, so that this measurement is most accurate for high impedance measurements and has a realistic lower limit of approximately 25Ω or 30Ω.

We can also see that the sensitivity of S21 in the shunt case decreases rapidly as S21 approaches 1 making this measurement most accurate for low impedances. The shunt thru measurement has a realistic maximum limit of 25Ω or 30Ω maximum.

A common approximation is that the impedance is S21*25Ω and this rule of thumb is also plotted. You can see in this graph the approximation is valid only for low values of resistance. A 5% error results from this rule of thumb at 1.25Ω.

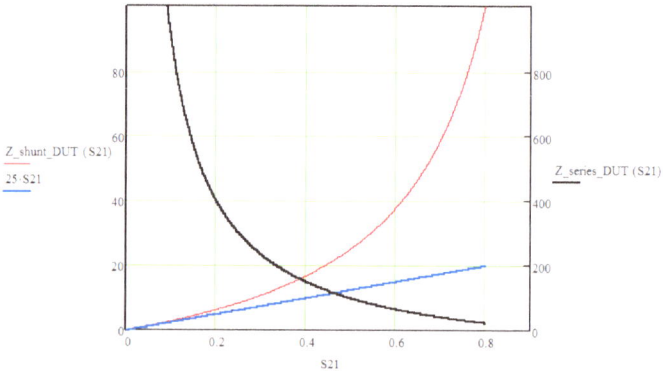

Impedance vs S21 for Shunt-Thru (Red), Series-Thru (Black) and Shunt Approximation (Blue)

Figure 7-21

Another simple approximation is to perform the THRU calibration with a 1Ω shunt calibration resistor. This allows the VNA to directly read in Ohms, and results in a displayed impedance of 25*S21. The resulting error from this simplified calibration is shown in Figure 7-22. Note that the error is zero at the calibration point of 1Ω.

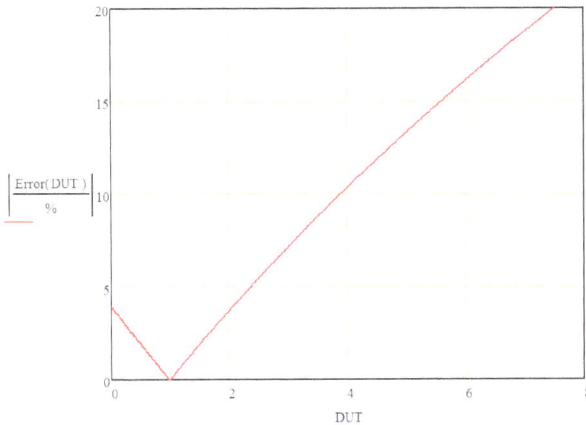

Error vs DUT Impedance for a 1Ω THRU Calibration.

Figure 7-22

While the error increases rapidly as the DUT impedance increases, the error is quite tolerable for very low impedance values, typical of PDN application.

The error is within 5% for impedance values below 2.3Ω.

Of course, applying equation 7.14 to determine the impedance from the S21 measurement eliminates the majority of this error.

There is one other issue with measuring very low impedance values using a VNA.

Looking at the schematic in Figure 7-23, we included a common-mode transformer.

This is required for most VNA's when measuring low impedance values at low frequency.

The schematic shows resistance and inductance representing each cable connecting the VNA to the DUT, while the VNA front panel is ground.

This means that the resistances from the ground braids of the two cables are in parallel with each other and in series with the DUT.

There are two ways to remedy this. One way is to include a semi-floating or differential input to the VNA, as in the case of the E5061B.

The other method is to include a coaxial 50Ω common mode transformer, such as the Picotest J2102A, on either port 1 or port 2.

The circuit is simulated both with and without the common mode transformer included.

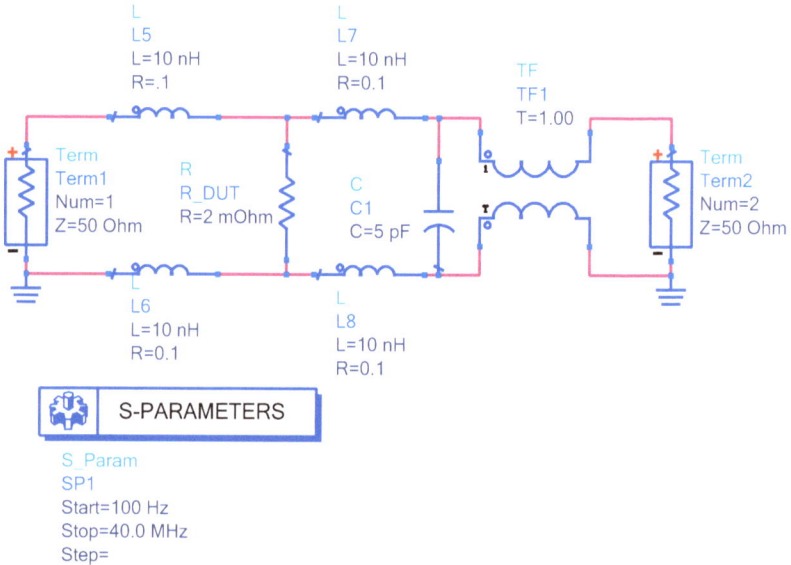

Schematic of Two-Port Impedance Measurement including Cable Resistance and 2mΩ DUT

Figure 7-23

The results of this simulation are shown in Figure 7-24. Note that without the common mode transformer the low frequency resistance is $52m\Omega$, which is the parallel resistance of the two ground braid resistances in series with the DUT. Also note that at higher frequencies, the braid inductances are in parallel and also in series with the DUT. The addition of the common mode transformer results in the correct simulation value of $2m\Omega$.

The common mode transformer is necessary for all measurements where the ground braid resistance is significant unless the VNA has a differential or semi-floating 50Ω input. In order to quantify the significance, the approximate error is the parallel ground braid resistance divided by the DUT. In the case of our simulation a 5% error would occur at a DUT value of 1Ω as the parallel ground resistance is $50m\Omega$ or 5% of 1Ω.

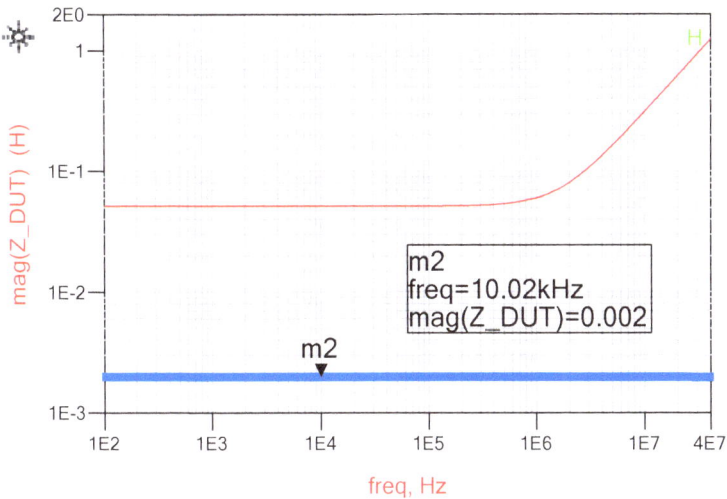

Simulation Results of the Two-Port Shunt-Thru Impedance Measurement including Cable Resistance and 2mΩ DUT with and without the Common-Mode Transformer

Figure 7-24

Pros

• The frequency range of the measurement is limited only by the accuracy of the calibration and the bandwidth of the VNA.

• In many applications a single THRU calibration is sufficient, making calibration simpler than for the single port measurement.

• The measurement can be made at zero volts and also biased, allowing measurements in the powered off state as well as the powered on state.

• The impedance range is ideal for measuring very low impedances, such as power distribution (PDN) applications including measuring motherboards, or FPGA power busses as a few examples.

• The measurement can be AC coupled using a DC blocker in order to eliminate the DC voltage limitations of the measurement.

• The most accurate low impedance measurement.

Cons

• The DC coupled measurement is generally limited to 5V, due to the power ratings of the 50Ω ports.

• Unless the equipment inputs float, a DC ground loop exists requiring a coaxial current transformer.

• Requires 2 coaxial cables or a multi-port probe.

• Not all instruments can display impedance in the two port measurement.

• The shunt thru measurement is limited to 5V max unless DC blockers are added to the setup.

Tips and Tricks

1. The physical construction of the calibrator must precisely match the construction of the DUT. For example, if the DUT is a 0805 surface mount chip device, the calibrator should also be constructed using similar 0805 devices.

2. Despite the limitations discussed above, I often use a 1Ω shunt resistor for the THRU calibration. This resistance establishes the approximation that DUT=25*S21, which is within 5% for impedance measurements between 1mΩ and 2.35Ω. This method also sets the results to display directly in Ohms for instruments that do not offer that capability.

3. It is nearly impossible to create a zero impedance "SHORT", so there is a fundamental low impedance, high frequency limitation of the measurement. This is one reason we try to measure a known value in order to quantify the measurement limitations.

4. At high frequencies even short traces and wires can contribute significant inductance, so be sure to make the measurements at the point of interest. For example, if you want to measure the ESL of a capacitor, be sure you measure <u>at the capacitor</u> as even very short traces or leads will dominate the measurement at higher frequencies.

5. <u>Always</u> measure a known device when creating a new setup to validate both the setup and the calibration.

6. Much larger oscillator signals can generally be used compared with the single port measurements. The larger signals improve the SNR when measuring very low impedance levels. You can verify the signal level by reducing it and checking whether the results change.

7. Adding a very low noise preamplifier, such as the J2180A can improve both the dynamic range and SNR

by 20dB in the 0.1Hz to 100MHz frequency range. Be sure to perform the calibration with the preamplifier connected.

8. I try to include SMA edge connectors on my designs making it simple to connect equipment for measuring the impedance.

Device Setup

The setup for the two-port measurement is shown pictorially in Figure 7-25.

The small PCB on the left side of the common mode transformer is a 1Ω shunt resistor for calibration.

Typical Setup for Two-Port Measurement with J2102A Common-Mode Transformer and 1Ω Shunt Calibrator Connected and Test Resistors in the Foreground

Figure 7-25

Calibrating the Two Port Measurement

The general method for calibrating the two port measurement is to connect the two ports together and perform a THRU calibration. The results in a very accurate measurement, though the result is in dB and requires calculation of the impedance to be performed using Equation 7.12 or Equation 7.14 for series or shunt measurement respectively.

One calibration technique, shown in Figure 7-26, is to connect the two ports through the J2102A common mode transformer (if required) to a 1Ω shunt resistor.

The shunt resistor is seen here mounted on a PCB and connected to the left side of the J2102A. The calibration resistor and DUT can be connected to either side of the transformer, but the calibrator and DUT should be connected to the same side. As discussed earlier, this technique will provide very good results for impedances from $1m\Omega$ to approximately 2.3Ω. Below $1m\Omega$ a preamplifier, such as the J2180A is required. If the preamplifier is used, then it is connected in line with the output connected to channel 2 of the VNA and the calibration is performed with the preamplifier in place.

The Agilent E5061B is capable displaying impedance in the 2-port measurement. The E5061B also supports the OPEN, SHORT and LOAD calibration in the 2-port series and shunt measurements for improved accuracy. These will also likely be possible with the Bode 100 in the future.

AC-Coupled, Two-Port Shunt-Thru Measurement

While the 2 port measurement is capable of measuring the impedance of a POL or PDN, the measurement is limited by the capability of the VNA. In general the 50 Ω ports are limited to 500mW or 1W. The OMICRON Lab Bode 100 is limited to

0.5W on the source port and 1W on each of the channel ports. This limits the absolute maximum voltage for the 2 port shunt thru measurement to 5V.

In some cases it may be desirable to measure very low impedance values at voltages above 5V. A few examples might include battery impedance and intermediate bus voltages, which are typically 10V-30V for computers and commonly 48V in telecommunications.

One method of overcoming this obstacle is to add a DC blocker, such as the J2130A to each of the two ports. The voltage limitation is then limited by the blocking capability of the J2130A or 50V. A simplified setup schematic is shown in Figure 7-26.

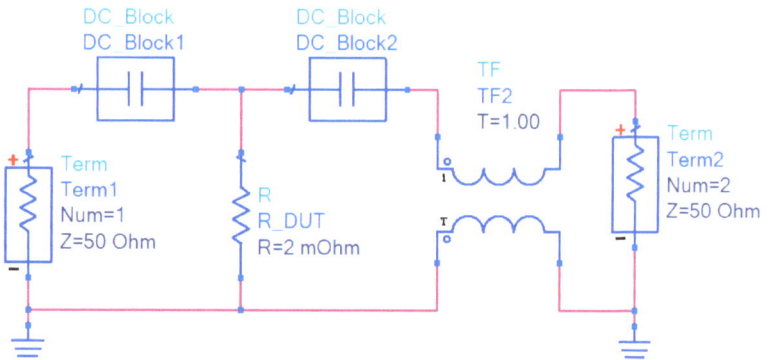

Simplified Schematic of AC-Coupled, Two-Port Shunt-Thru Measurement

Figure 7-26

Connecting to the DUT

The two port measurement is very sensitive to the connection of the VNA to the DUT. It is possible for the two cables to couple to each other as well, so the loop area must be minimized as should the lengths of all exposed wires.

The connection impedance should match the VNA all the way to the point of measurement. For this reason, the J2102A common mode transformer is also a matched 50Ω coaxial transformer, up to approximately 250MHz.

Soldering the two 50Ω coax cables directly to the DUT provides the highest fidelity connection as it eliminates contact resistance.

This is often difficult as the decoupling capacitors can be quite small making it difficult to solder two cables to the capacitor without shorting to other components on the PCB. It is also possible for the weight of the cables to destroy the PCB, by ripping the decoupling capacitor off along with the PCB pads.

A picture of two coaxial cables soldered to a CPU decoupling capacitor is shown in Figure 7-27. It is not pretty and as you can see the solder bridges several decoupling capacitors. In many applications the accessibility to these capacitors is much more limited as well.

By the time this book is published, Picotest will offer a multiport coaxial probe that will allow this measurement to be made very easily for impedance levels greater than 5mΩ and up to several hundred MHz or 1GHz.

In any case it is always a good idea to test the measurement on a known component in order to validate the measurement including the probe and interconnect methodology.

In Figure 7-27, ideally, these cables should be lying flat and at an 180 degree angle, rather than 90 degrees as shown.

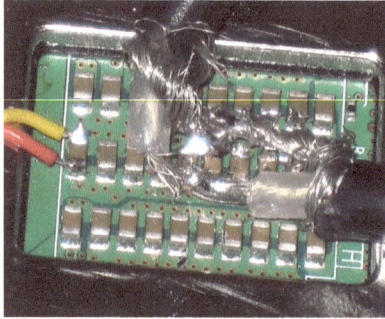

Two Coaxial Cables Soldered to a Decoupling Capacitor Bank

Figure 7-27

Examples

1. Measuring a 1mΩ Planar Current Sense Resistor

The first example shows the measurement of a planar 1mΩ resistor mounted between two SMA connectors on a PCB as shown in Figure 7-28.

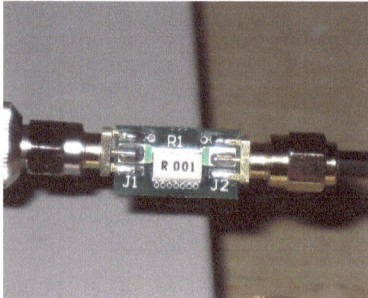

A 1mΩ Planar Resistor PCB Mounted between Two SMA Connectors for Measurement

Figure 7-28

The resulting measurement is shown in Figure 7-29 with and without the J2102A common mode transformer inserted.

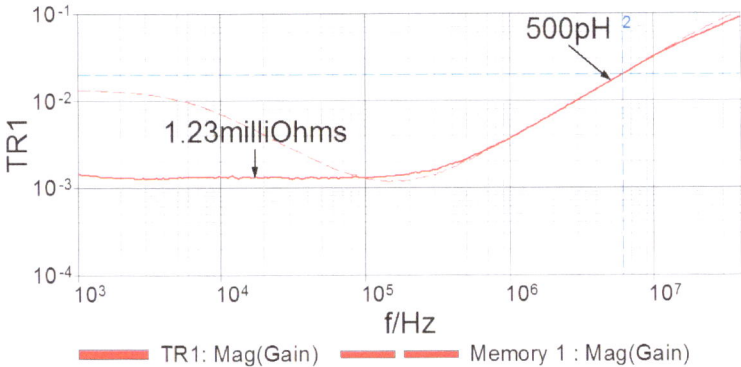

Impedance Measurement of the 1mΩ Resistor with and without the J2102A Common-Mode Coaxial Transformer using the OMICRON Lab Bode 100

Figure 7-29

As I have repeatedly stated that each setup should be verified by measuring a known value, I have also determined the correct value of this resistor.

An external power supply is connected to one of the SMA connectors via a low noise 6.5 digit DMM, thereby shorting the power supply through the 1mΩ resistor while accurately measuring the resistor current.

The other SMA is connected to a low noise 6.5 digit DMM measuring the voltage at the other side of the resistor. This replicates the VNA measuring using DC test equipment that is more accurate than the VNA.

The measurement setup is shown in Figure 7-30.

The resistor can be seen in the upper right corner of the image, while both the current and voltage displays can also be seen in the image.

The exact value of the mounted resistor is calculated as:

$$R = \frac{V}{I} = \frac{0.807mV}{0.68878A} = 1.1716m\Omega$$

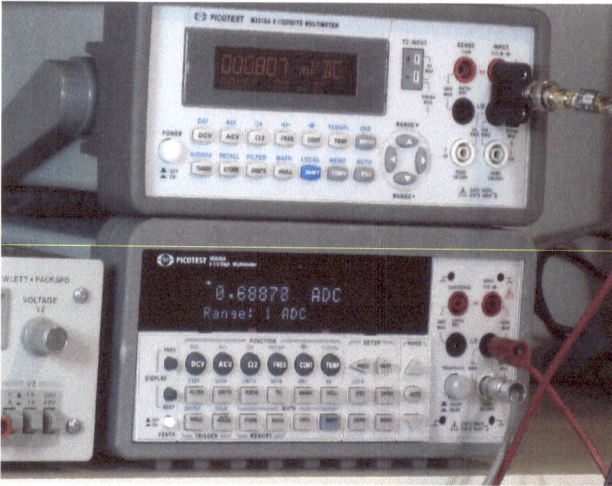

1mΩ Resistor Measurement Verification

Figure 7-30

The same resistor was also measured, using the same setup and an E5061B VNA shown in Figure 7-31.

The results show the resistor measurement as 1.142mΩ.

Each of these measurements is within 5% of the DC value, which is quite good for such a low impedance measurement.

Recalling that a +4% error is attributed to the 1Ω thru calibration of the Bode 100, the corrected measurement is within approximately 2% as is the E5061B.

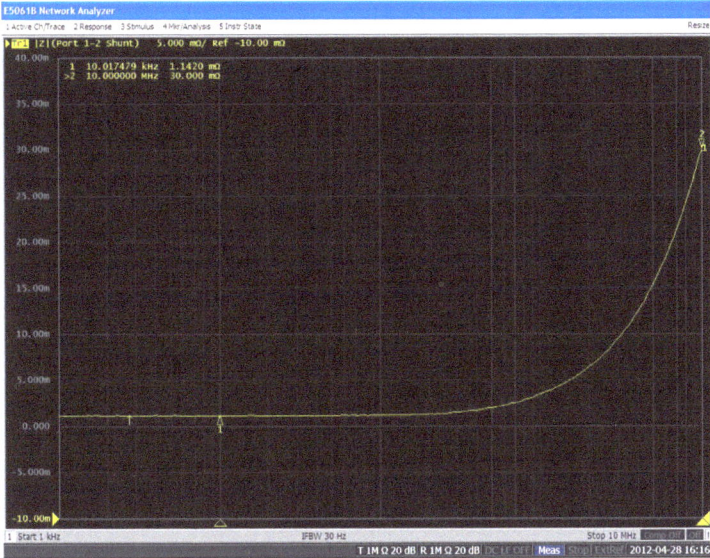

Impedance Measurement of the 1mΩ Resistor using the Agilent E5061B

Figure 7-31

2. Measuring a 250μΩ Planar Current-Sense Resistor

The range of the measurement can be improved further by adding the low noise preamplifier to the measurement to improve the signal to noise ratio and dynamic range of the measurement.

A J2180A low noise preamplifier is added to CH2 prior to the J2102A transformer and the thru calibration repeated, again using a 1 Ω shunt resistor for calibration. The resulting measurement is shown in Figure 7-32.

The resistor, mounted between two SMA connectors is shown in Figure 7-33.

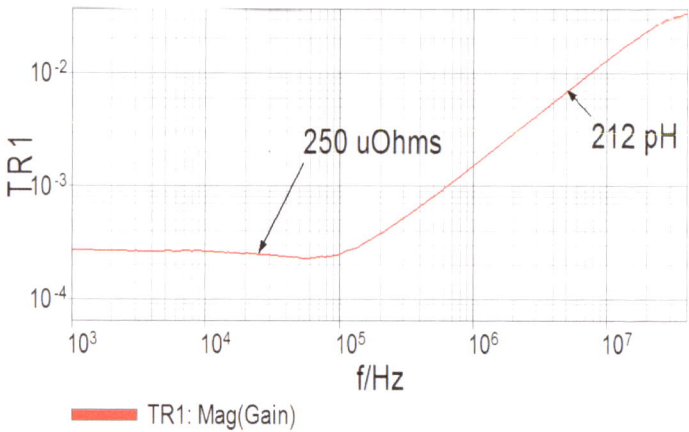

250μΩ Precision Resistor Impedance Measurement

Figure 7-32

Precision 250μΩ Resistor Mounted between Two SMA Connectors

Figure 7-33

3. Measuring a Computer Motherboard

The two port shunt thru method is applied to a motherboard. Two 50Ω coaxial cables are connected from the VNA ports to the output decoupling capacitors as shown in Figure 7-27. As in the previous measurements, the J2102A is included on CH2 of the VNA. The measurement is performed with the power applied and also with the power off. The results are shown in Figure 7-34.

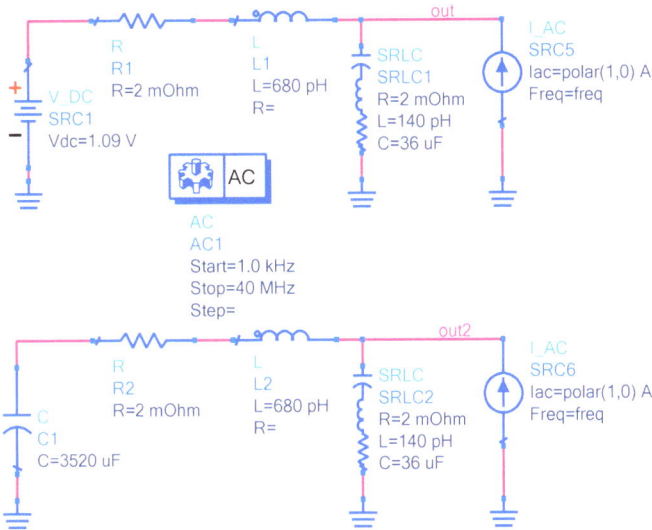

Motherboard Impedance Measurement with Power On (Solid Trace) and Off (Dashed Trace)

Figure 7-34

Based on the impedance measurements, it is simple to extract the equivalent simulation model for both the powered on and powered off states.

Each flat section or minima represents a resistance while increasing slopes are calculated as inductors and decreasing slopes are calculated as capacitors.

The source resistance can be read directly as approximately $2m\Omega$. And the decoupling cap ESR can also be read as approximately the same value of $2m\Omega$.

The remaining elements are calculated from measurement results in Equations 7.17 through 7.19.

$$Decoupling\ cap = \frac{1}{2\pi f Z} \qquad 7.17$$

$$= \frac{1}{2\pi(1.46MHz)(0.003\,\Omega)}$$

$$= 36\mu F$$

$$Decoupling\ cap\ ESL = \frac{Z}{2\pi f} = \frac{0.009\,\Omega}{2\pi(10MHz)} \qquad 7.18$$

$$= 140pH$$

$$Bulk\ cap = \frac{1}{2\pi f Z} \qquad 7.19$$

$$= \frac{1}{2\pi \cdot 0.0025 \cdot 575kHz}$$

$$= 3520\mu F$$

$$Bulk\ cap\ ESL = \frac{Z}{2\pi f} = \frac{0.003\,\Omega}{2\pi(700kHz)} = 680pH \qquad 7.20$$

The simulation model is shown in Figure 7-35 and the simulation results from this model are shown in Figure 7-36.

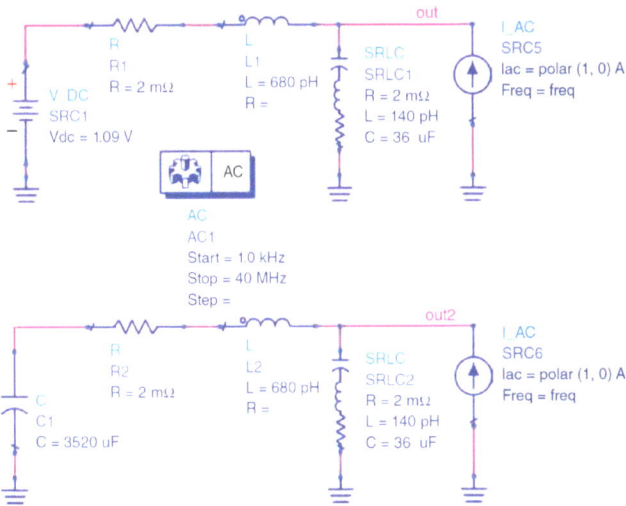

Motherboard Extracted Model Powered On (Top) and Off (Bottom)

Figure 7-35

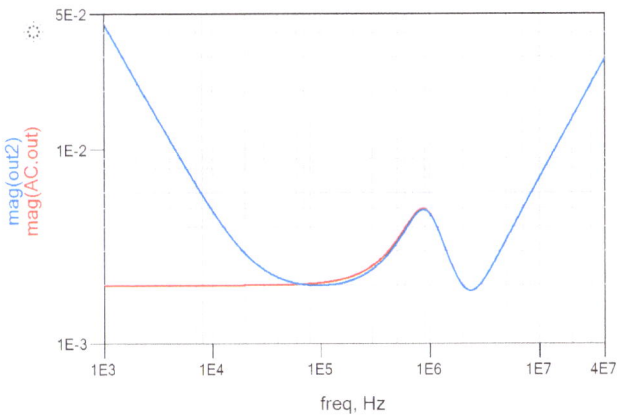

Motherboard Extracted Model Simulation Results Powered On (Red) and Off (Blue)

Figure 7-36

And other than the frequency dependent ESR, the model is in good agreement with the measured results and provides a usable model for most applications.

Of course, with additional effort this frequency dependent model can also be created if desired.

4. Measuring an Inductor using a Series-Thru Measurement Technique

A toroidal inductor is used to demonstrate the low impedance limitations of the series thru measurement.

The inductor mounted between two SMA connectors is shown in Figure 7-37.

330μH Inductor Mounted between Two SMA Connectors for Measurement

Figure 7-37

A measurement of the inductor was performed using a single port impedance measurement with a shorting plug at the end of one cable, as shown in Figure 7-37, to establish the inductance value and the low frequency resistance.

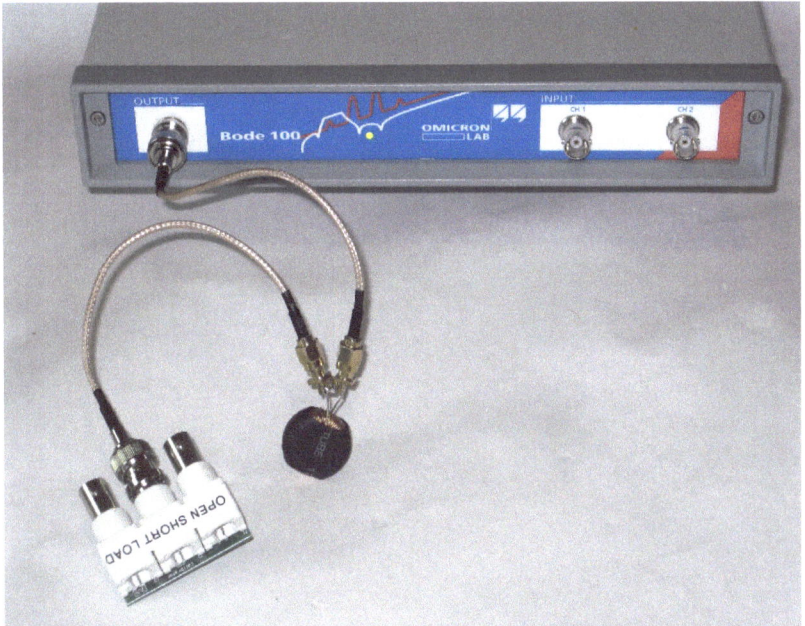

330µH Inductor Mounted for Single-Port Impedance Measurement including the Same Cables used in the Two-Port Measurement

Figure 7-38

The single port impedance measurement results are shown in Figure 7-39.

This measurement establishes the baseline inductance at 324µH and the low frequency resistance at 0.184Ω.

This resistance is less accurate than the DC measurement, primarily because it is far below the minimum recommended measurement impedance for the single port measurement.

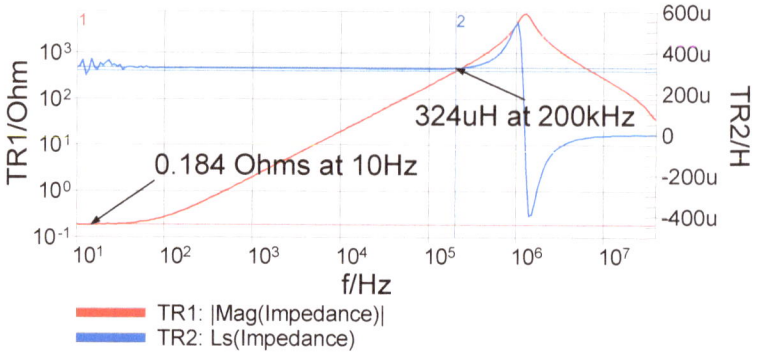

Single-Port Impedance Measurement used to Baseline the Inductance and Compare the Low-Frequency Resistance

Figure 7-39

The same setup is connected to a low noise 6.5 digit DMM, as shown in Figure 7-40 to obtain an accurate DC resistance measurement of the inductor, including the same cables used in the VNA measurements.

DC Resistance Measurement of the Inductor Including the Same Cables used in the Two-Port Measurement

Figure 7-40

The DC resistance of the inductor, including the connecting cables, is 0.156Ω as shown in Figure 7-40. This DC measurement is more accurate than the single port measurement, due to the low value of the resistance.

Finally, the inductor is measured using the two port series thru method, as shown in Figure 7-41. In Figure 7-41, note the barrel adapter used to replace the inductor for the thru calibration in the foreground.

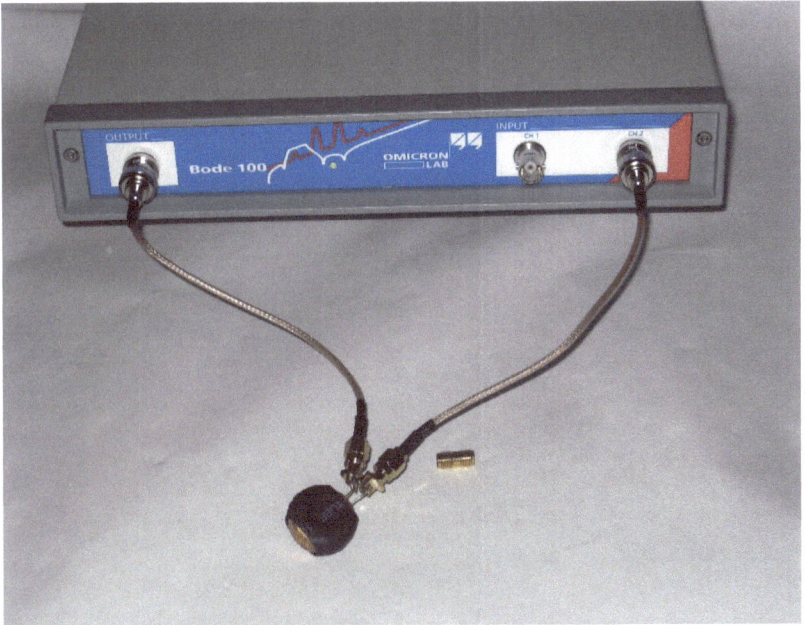

Inductor Connected for the Two-Port, Series-Through Measurement

Figure 7-41

The series thru measurement result is shown in Figure 7-42.

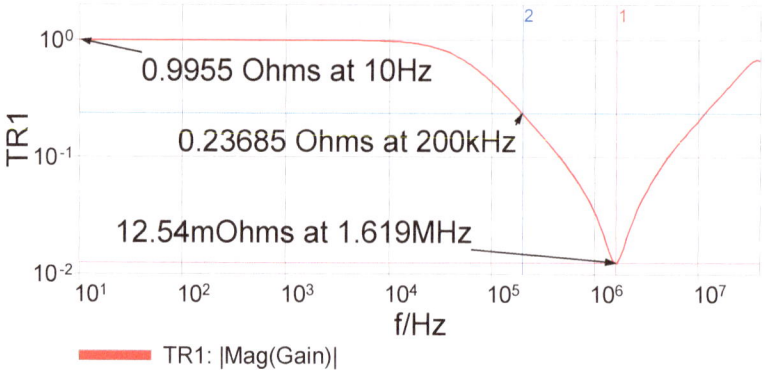

Series-Thru Measurement of an Inductor

Figure 7-42

Using Equation 7.11 to determine the inductor impedance at 200kHz results in:

$$S21 = \left| \frac{2 \cdot RL}{RL + RS + DUT \cdot i} \right| = 0.23685 \qquad 7.21$$

$$S21 = \frac{2 \cdot RL}{\sqrt{(RL + RS)^2 + DUT^2}} = 0.23685 \qquad 7.22$$

$$7.23$$

$$DUT =$$

$$\frac{\sqrt{2 \cdot RL - RL \cdot S21 - RS \cdot S21} \cdot \sqrt{2 \cdot RL + RL \cdot S21 + RS \cdot S21}}{S21}$$

$$= 410.195$$

$$L = \frac{DUT}{2\pi \cdot 200kHz} = 326\mu H \qquad 7.24$$

The resulting inductance is in agreement with the value measured using the single port method.

Applying Equation 7.11 to determine the DC resistance results in:

$$S21 = \frac{2 \cdot RL}{RL + RS + DUT} = 0.9955 \qquad 7.25$$

$$so\ DUT = 0.452$$

And as expected the low value resistance is not close to the correct result as it is far below the minimum impedance for the two port series thru measurement method.

Current Injection Measurements

One method of performing a current injector or voltage/current impedance measurement is shown in Figure 7-43.

The source signal is AC coupled by a signal transformer and AC coupled to the DUT by a coupling capacitor. A current sensing resistor in the return leg of the transformer is used to sense the current while the voltage is measured directly at the DUT either using a coaxial 50 Ω connection or a high impedance scope probe.

In some simpler implementations the 48.7 Ω resistor is eliminated and the current sense resistor is changed to 1Ω with the current signal being measured with a high impedance probe.

Yet another adaptation is to use a clamp on current probe for current sensing rather than the sense resistor.

Each of these implementations has advantages and disadvantages, though in general this method is limited at low frequency and high frequency by the parasitic effects of the transformer and capacitor.

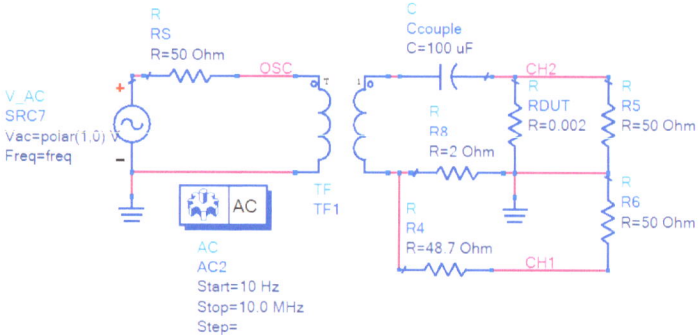

Transformer-Coupled Voltage-Current Injection

Figure 7-43

Another voltage/current injection alternative is a solid state device, such as the Picotest J2111A or the J2112A shown in Figure 7-44. The devices provide wide bandwidth closed loop transconductance, converting the modulated voltage at the input to a modulated current at the output. A precision 50Ω current monitor port is provided.

As always there are tradeoffs associated with each of these alternatives, though some are common.

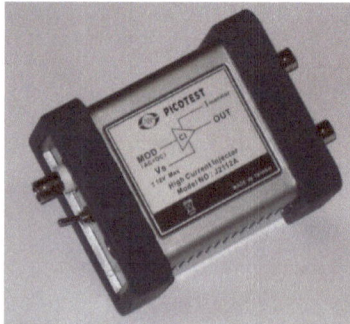

Solid State Voltage/Current Injector

Figure 7-44

Pros

• Can operate at higher voltages and over wider impedance ranges than the single or two port measurements.
• Solid state
• Can operate over a very wide frequency range of DC and up to 40MHz, depending on interconnects.
• Supports small signal step loading for time domain measurement
• Transformer coupled
• Can operate at zero volts

Cons

• Dependent on interconnect impedance
• Have a DC ground loop requiring the current port to include a J2102A or similar common mode transformer, differential probe or current probe for accurate low impedance measurements.
• Requires 2 coaxial cables or a multi-port probe

Solid State

• Cannot operate between -1V and +1V (J2111A)
• Only single polarity >1V (J2112A)

Transformer Coupled

• Limited bandwidth at both high and low frequency
• Requires a large, non-polarized, high frequency capacitor
• Only usable in the frequency domain

Tips and Tricks

1. I try to include SMA edge connectors on my designs making it simple to connect equipment for measuring the impedance
2. A multi-port probe provides a simple, compact and wideband method of connecting equipment for these measurements.

Device Setup

The basic connection of the J2111A solid state injector is shown in Figure 7-45.

The J2102A common mode transformer is not included in the setup for clarity, but can be inserted at CH1 for low impedance measurements. The semi-floating input of the E5061B eliminates the need for the common mode transformer.

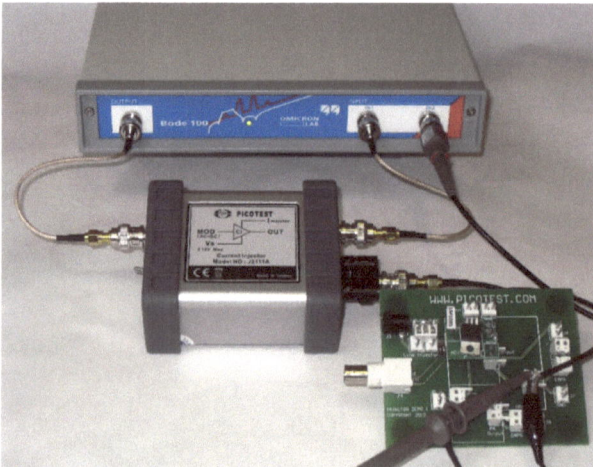

Solid State J2111A Connected to OMICRON Lab Bode 100 and a Picotest VRTS1 Demonstration Board

Figure 7-45

Examples

1. Low-Impedance Regulator Output Impedance

A comparison between the 2-port shunt thru-impedance, considered the measurement gold standard, and the J2111A current injector is shown in Figure 7-46. The measured range spans from 2mΩ to 700mΩ and from 10Hz to 10MHz with excellent agreement. The low frequency limit is DC, or further restricted by the minimum frequency of the VNA used for the measurement. The E5061 can go as low as 5Hz and the OMICRON Lab Bode 100 can go as low as 1Hz.

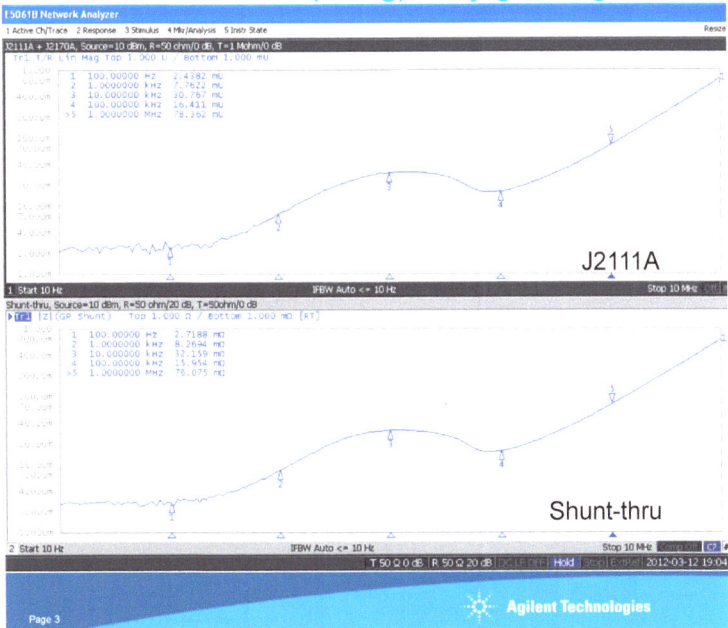

Comparison between J2111A and Shunt-Thru Impedance

Figure 7-46

2. **Voltage Reference SC4437 with Output Capacitor at 50μA and 500μA Load Current and 1μF Ceramic Output Capacitor**

The voltage reference impedance measurement in Figure 7-47 shows the sensitivity of the output impedance to the operating bias current and also shows the resolution and fidelity of the J2111A current injector.

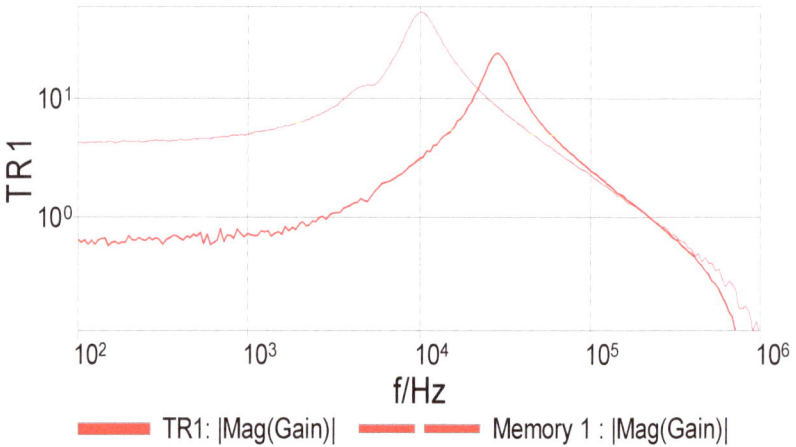

Semtech SC4437 Voltage Reference with 1μF output Capacitor at 50μA (Dashed) and 500μA (Solid) Load Currents

Figure 7-47

3. **Switching Regulator Output Impedance with 10Ω Load Resistor**

The same setup is used to measure a switching POL regulator and the impedance result is shown in Figure 7-48.

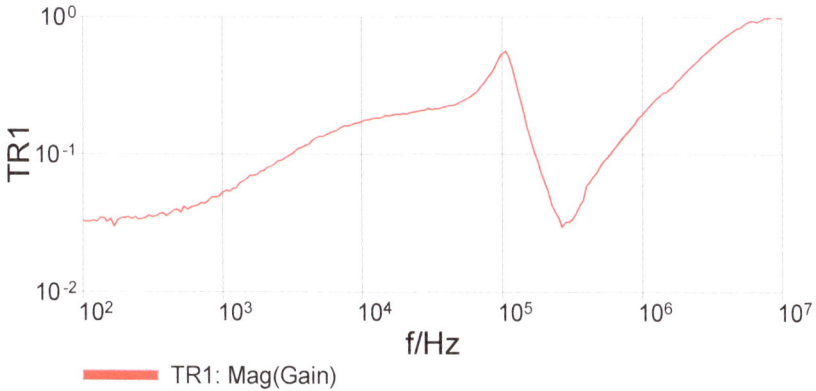

1.8V POL Output Impedance Measurement at 5 Vin and 10 Ω Load

Figure 7-48

Impedance Adapters

Impedance adapters or impedance fixtures offer a simple method of measuring passive devices, including capacitors, resistors and inductors over a wide frequency and a wide impedance range. These fixtures are also ideal for testing many devices, as devices can quickly and easily be swapped without soldering or unsoldering the DUTs. The fixtures also offer excellent repeatability and are generally available for both leaded and surface mount components.

Pro

•   Very accurate measurement over a wide impedance range and a wide frequency range.

•   Simple fixture allows quick measurements of many devices as opposed to soldering to a PCB.

Cons

- Does not allow DC Bias.
- Requires the part to be removed for measurement out of circuit.

Tips and Tricks

1. Make sure the calibration devices are the same physical size as the DUT, especially when measuring very low values.
2. I use a $20M\Omega$ resistor in the same package size as the DUT for the OPEN calibration.

Device setup

A pictorial setup of the impedance adapters connected to the OMICRON Lab Bode 100 is shown in Figure 7-49.

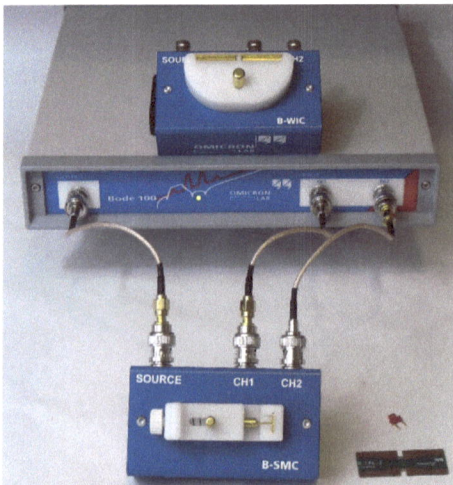

OMICRON Lab Bode 100 and Impedance Adapters

Figure 7-49

Calibrating the Impedance Adapters

As noted in the tips, it is best to calibrate the fixtures using devices that are physically similar to the DUT. In the example shown in Figure 7-50 0402 chip resistors are used in order to perform the OPEN, SHORT and LOAD calibrations. The 0402 package is quite small and so the pictures are greatly magnified so that the calibration elements can be seen in the images.

I generally use a 20MΩ resistor to represent an OPEN, as it will attempt to replicate the physical positioning of the contact pins and the device capacitance, while the 20MΩ is sufficiently large to keep it from impacting the measurement. A 0Ω resistor used for the short and a 49.9Ω resistor for the load. These are all shown in Figure 7-50. The completed calibration screen shown in Figure 7-51 indicates that the LOAD resistor value was set to 49.9Ω and the calibration has been completed for OPEN, SHORT and LOAD conditions. The short delay time was adjusted by measuring a precision inductance value and iteratively adjusting it until the known inductance measurement was correct.

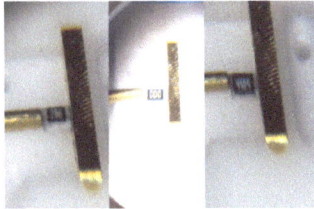

The 0402 OPEN, SHORT and LOAD Calibration Devices

Figure 7-50

User Calibration - Impedance Adapter

Gain/Phase

Replace DUT by thru cable. Afterwards press Start to perform Calibration

| Thru | Start | Not Performed |

Impedance

Connect the corresponding part and perform the calibration by pressing the start button.

Open	Start	Performed
Short	Start	Performed
Load	Start	Performed

Advanced

Load Resistor 49.90 Ω

Short Delay Time 50 00 ps

| OK | Cancel | Help |

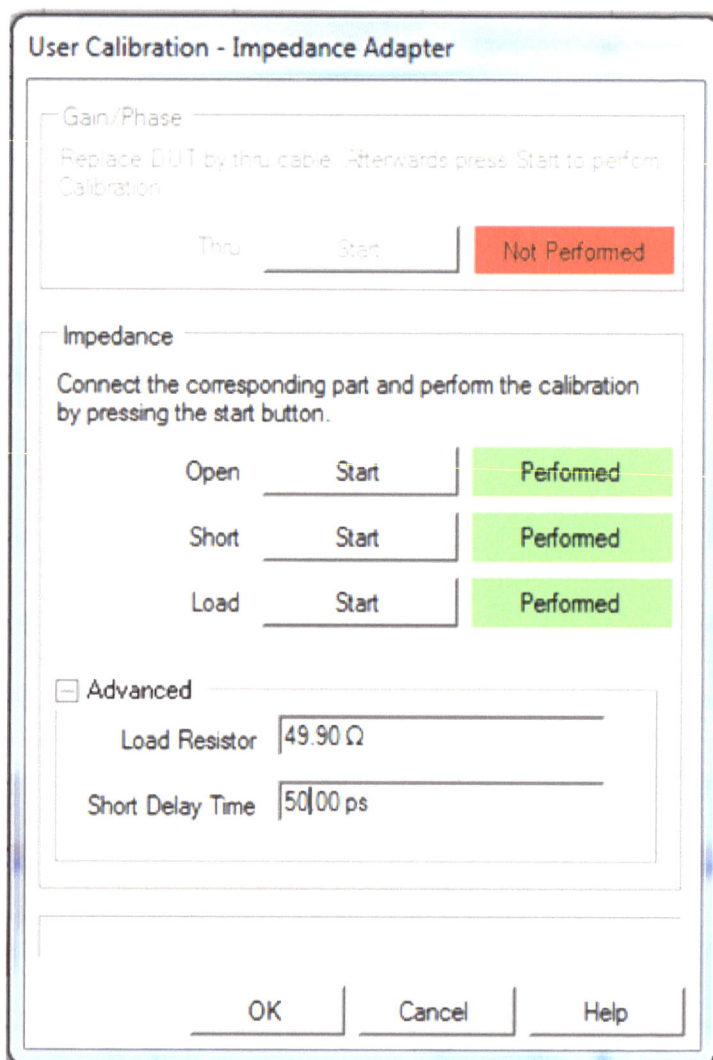

Calibration Screen Showing the 49.9Ω LOAD Setting and Showing the OPEN, SHORT and LOAD Calibrations as Completed

Figure 7-51

Examples

1. Measuring a 1.2nH Inductor

This example demonstrates the ability to measure a 1.2nH chip inductor. The inductor is a 0402 chip device, explaining the 0402 calibration elements used earlier.

With the calibration completed, the 0402 chip inductor is seen placed in the OMICRON Lab B-SMC impedance adapter in Figure 7-52.

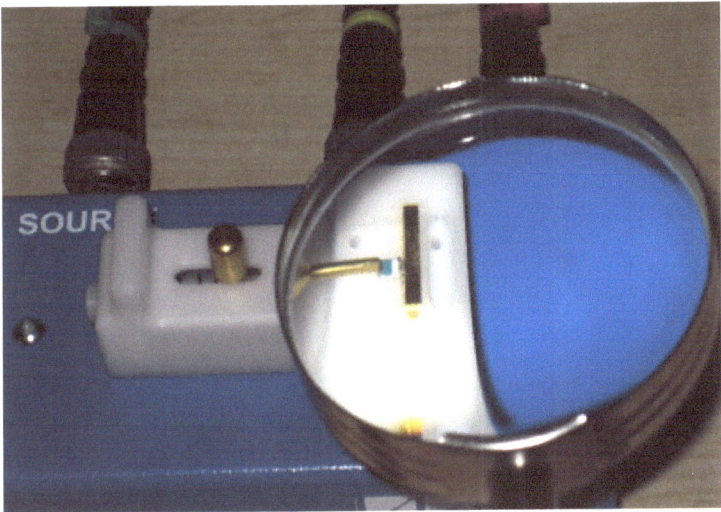

1.2nH 0402 Chip Inductor Inserted into the OMICRON Lab B-SMC Impedance Adapter

Figure 7-52

The result of the impedance measurement is shown in Figure 7-53 indicating a measured value of 1.22nH, very close to the nominal value. At lower frequencies, the measurement is noisy, though still close to 1.2nH.

The reason for the noise is that at these low frequencies,

the impedance of the inductor is very small compared with the DC resistance of the inductor (20mΩ).

The inductive impedance and the DC resistance are equal valued at a frequency of approximately 2.45MHz.

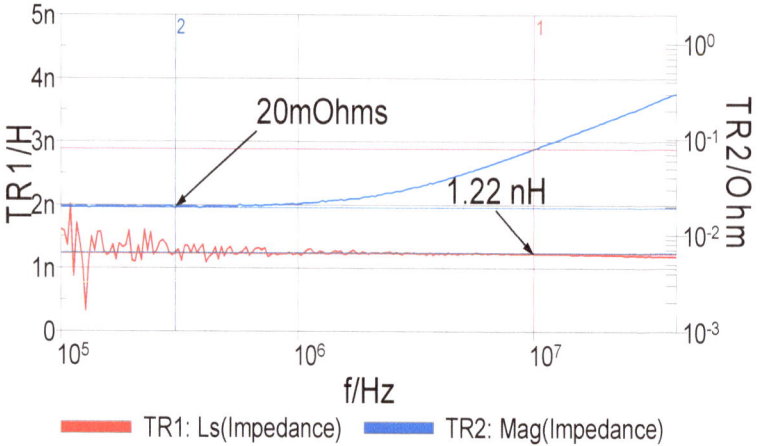

Measured Result for the 1.2nH Inductor Showing Impedance Magnitude and Inductance

Figure 7-53

2. Measuring a Ferrite Bead

Ferrite beads are common for isolating noise sources.

They are typically inductive at low frequency and resistive at high frequencies as seen in Figure 7-54.

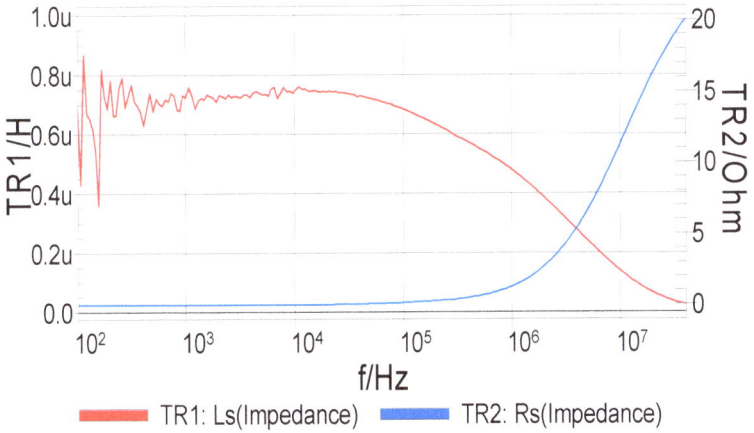

LI0805G201R-10 250Ω Chip Ferrite Bead Resistance (Blue) and Inductance (Red)

Figure 7-54

3. Measuring Impedance and ESR of a Tantalum Capacitor

The ESR of tantalum capacitors is not a constant, but generally quite frequency dependent as seen in Figure 7-55.

Impedance adapters are ideal for measuring these parameters as they offer excellent accuracy over a very wide range of impedance magnitudes and also over a very wide frequency range.

Some impedance adapters do allow DC Bias to be included, while others do not

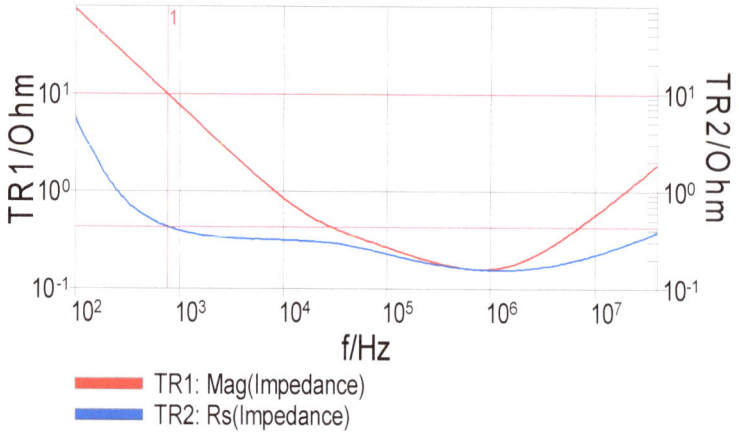

Impedance and ESR of a 22µF Tantalum Capacitor Impedance and ESR

Figure 7-55

Chapter References

1. Agilent 5990-5902, *Evaluating DC-DC Converters and PDN with the E5061B LF-RF Network Analyzer*
   cp.literature.agilent.com/litweb/pdf/5990-5902EN.pdf
2. Agilent 5968-4506E *New Technologies for Accurate Impedance Measurement*
   literature.agilent.com/litweb/pdf/5968-4506E.pdf
3. Agilent 5989-5935 *Ultra-Low Impedance Measurements Using 2-Port Measurements*
   cp.literature.agilent.com/litweb/pdf/5989-5935EN.pdf
4. *DC Biased Impedance Measurements*, OMICRON-Lab
   http://www.omicron-lab.com/fileadmin/assets/application_notes/App_Note_DC_Bias_Impedance_V1_0.pdf
5. *Using the Bode 100 and the Picotest J2130A DC Bias Injector*, PICOTEST
6. *Capacitor ESR Measurement* OMICRON-Lab
   http://www.omicron-lab.com/fileadmin/assets/application_notes/App_Note_ESR_Measurement_V1_1.pdf
7. *Using the Bode 100 and the B-WIC Impedance Adapter*, OMICRON Lab
   http://www.omicron-lab.com/fileadmin/assets/application_notes/App_Note_ESR_Measurement_V1_0.pdf
8. Steven M. Sandler, *An Accurate Method For Measuring Capacitor ESL*, How2Power.com April 2011
9. Steven M. Sandler, *How to Measure Ultra-Low Impedances, Electronic Design, June 15, 2012*
   http://electronicdesign.com/boards/how-measure-ultra-low-impedances
10. Steven M. Sandler, *Simple Method to Determine ESR Requirements for Stable Regulators*, Power Electronics Technology Aug 2011 pages 42-44
11. Steven M. Sandler, Tom Boehler, Charles E. Hymowitz, *Network Analyzer Signal Levels Affect Measurement Results,*

Steven M. Sandler

Power Electronics Technology January 2011 pages 26-28

12. *Measuring Optocouplers with the J2130 DC Bias Injector and the OMICRON Lab Bode 100 VNA*

13. *Measuring MOSFET Gate Resistance*, PICOTEST www.picotest.com/blog

14. Agilent Technologies Impedance Measurement Handbook, literature.agilent.com/litweb/pdf/5950-3000.pdf

15. OMICRON Lab Bode 100 User Manual *http://www.omicron-lab.com/fileadmin/assets/manuals/Bode_100_Manual_AE4_HR.pdf*

16. Steven M. Sandler, *How to Make Higher Voltage Power-Distribution-Network Measurements*, EDN.com, January 23, 2013

    http://www.edn.com/design/test-and-measurement/4405531/How-to-make-higher-voltage-power-distribution-network-measurements

17. *Simple and Systematic Modeling Procedure of Capacitors Based on Frequency Response*,

    http://www.deepdyve.com/lp/institute-of-electrical-and-electronics-engineers/simple-and-systematic-modeling-procedure-of-capacitors-based-on-L1FS4ul6sc

18. Valdivia, V., et al, *Simple and Systematic Modeling Procedure of Capacitors Based on Frequency Response*, *Applied Power Electronics Conference and Exposition (APEC), 2011 Twenty-Sixth Annual IEEE*. IEEE, 2011.

19. Panov, Yuri, and Milan Jovanovic, *Practical Issues of Input/Output Impedance Measurements in Switching Power Supplies and Application of Measured Data to Stability Analysis*, Applied Power Electronics Conference and Exposition, 2005. APEC 2005. Twentieth Annual IEEE. Vol. 2. IEEE, 2005.

Chapter 8
Measuring Stability
Stability and Why it Matters

NEGATIVE FEEDBACK IS used in voltage regulators, POL regulators, voltage references, op-amps and other electronic circuits in order to produce an accurate output in response to an input.

In the case of a voltage regulator, the input is an accurate voltage reference and negative feedback is used to compare a sample of the output to the reference.

The difference, or error signal, is amplified and used to correct the output.

If the feedback loop is stable, then the output converges to the correct voltage. If the feedback loop is not stable then the output diverges or results in a sustained oscillation.

The performance within this range of convergence and divergence is a function of the stability margins.

The relative stability is traditionally determined from an

open-loop Bode (gain-phase) plot in the frequency domain.

The Phase Margin (PM) is equal to 180° minus the phase shift measured at the circuit's open loop cross-over frequency (0 dB point), while the Gain Margin (GM) is the gain when the phase crosses 0 degrees.

In space applications, where fidelity is critical, an end-of-life phase/gain margin of 30 degrees/10dB is usually the absolute minimum requirement.

The stability of the control loop matters because poor stability degrades output noise, PSRR, reverse transfer, output impedance and step load response.

These localized degradations result in further system performance degradation including clock jitter, reduced SNR and Bit Error Rate (BER) to name a few. It is for these reasons that we are so concerned with the stability of all control loops.

The data sheets for linear regulators, POL regulators and voltage references rarely provide the necessary information to produce a stable design and therefore measurement is a necessity.

Control Loop Basics

The goal of most circuits is to produce a clean and well-defined output for a given input.

The output is generally controlled by an amplifier gain loop that compares the output with an input signal. The difference or error voltage is amplified.

This amplified signal is then used to control the output. The accuracy of the closed loop output can be calculated from the performance without the amplifier loop.

A simple amplifier feedback circuit is shown in Figure 8-1.

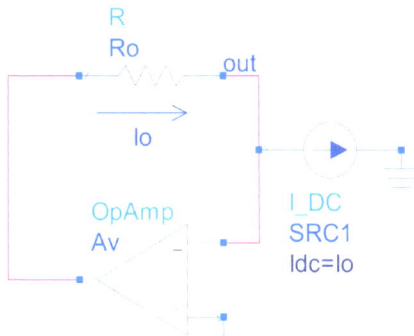

Amplifier Loop to Keep the Output Voltage at Zero

Figure 8-1

If the feedback loop is removed and a fixed zero volts is connected in place of the amplifier output, as shown in Figure 8-2, then the open loop output voltage changes dependent on the output current, I_o, as in Equation 8.1.

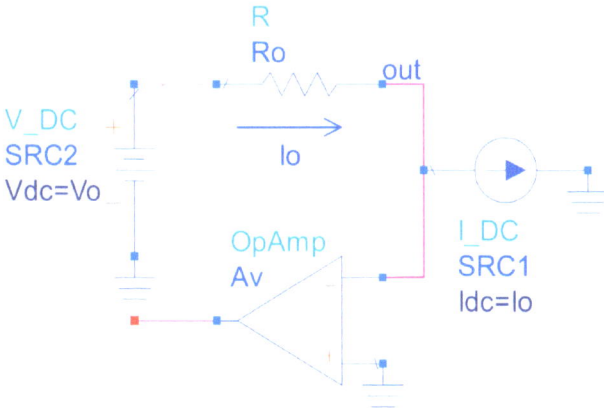

Amplifier with the Loop Opened and a Fixed Voltage Source Representing the Open Loop Amplifier Output

Figure 8-2

$$V_{open\_loop} = -I_o \cdot Ro \qquad 8.1$$

If the feedback amplifier is inserted in the loop then the amplifier will increase the voltage at the output by the amplifier gain, A_v.

Since no current flows into or out of the amplifier inputs, the current through Ro must be equal to the output current, I_o.

$$\frac{(V_{inv} \cdot -A_v) - V_{inv}}{R_o} = I_o \qquad 8.2$$

Solving for the closed loop output voltage, which is equal to the inverting input voltage results in:

$$V_{closed\_loop} = \frac{-I_o \cdot Ro}{A_v + 1} = \frac{V_{open\_loop}}{A_v + 1} \qquad 8.3$$

If the loop gain incurs a 180 degree phase shift at unity gain, the A_v is equal to negative one and the denominator becomes zero resulting in an infinite output.

This is the definition of oscillation.

It can be seen in Equation 8.3 that for values of A_v close to, but not exactly, negative one, the closed loop response is greater than the open loop response.

This also holds true for all noise paths discussed in the distributed systems chapter.

As A_v approaches negative one, all noise paths are degraded. If it is equal to negative one, the circuit will oscillate.

The relative stability is, therefore, a measure of the proximity of the total loop gain to the single point, which has unity gain and 180 degrees of additional phase shift within the loop.

Gain Margin, Phase Margin, Delay margin and Stability Margin

The most common stability margins are gain margin, phase margin and delay margin.

Margin	Definition
Gain margin ('GM')	The smallest change in gain that results in instability
Phase margin ('PM')	The smallest change in phase that results in instability
Delay margin ('DM')	The smallest change in time delay that results in instability
Stability margin ('SM')	The closest distance between the loop gain function and the singular unstable point (-1,0)

Summary of Several of the Most Common Measures of Stability Margin

Table 8.1

In the assessment of the open loop gain, the singular unstable point is located at (-1,0). If we include the negative feedback of the amplifier, this point is moved to the opposite side at (1,0). In this book, we include the amplifier feedback and so the convention used here is that the singular unstable point for the complete loop is at (1,0).

It is possible for higher order control loops to have multiple stability margins. In this case, each of the margins must be assessed. In addition, many regulators have multiple loops working together. The overall stability of the device must at least include an assessment of the "combined" performance (all loops superimposed).

If the gain function is second order, there may not be a GM, only a PM. This is because a 2nd order system incurs a maximum 180 degree phase shift, 90 degrees for each pole. The sum of the 180 degrees and the 180 degrees incurred due to negative feedback is exactly 0 degrees, thus there is no point at which the phase crosses through 0 degrees. This is shown using a simulation in Figure 8-3.

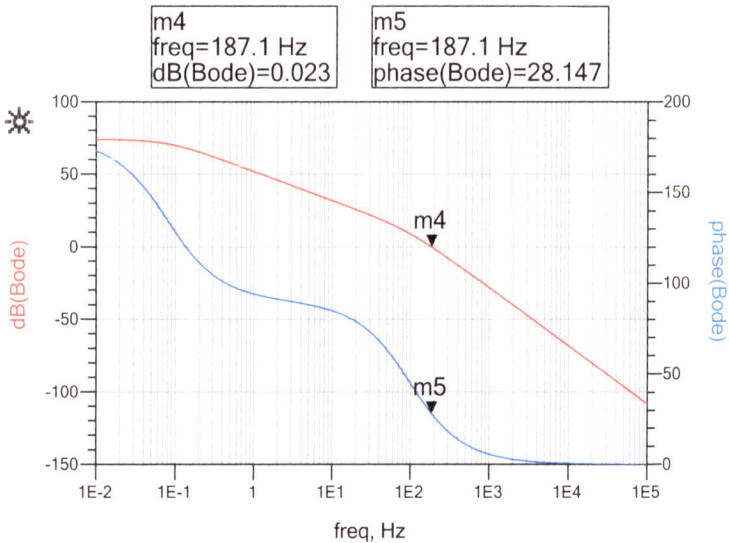

Simulation of a 2nd Order System showing the Phase Asymptotically Approaches 0 Degrees and so Never Crosses Zero Degrees

Figure 8-3

Bode Plots and Nyquist Charts

There are many methods of assessing stability, including Routh-Hurwitz, Root Locus, Nyquist, Nichols and Bode. The most popular is the Bode plot, most likely because it is simple to interpret, is usually an easy measurement to make, and the asymptotic diagrams can easily be hand drawn, allowing the use of the Bode plot to assist in designing a stable loop. A good example of the Bode plot is the operational amplifier open loop gain and phase measurement shown in Figure 8-4. Most op-amp datasheets include a figure similar to this.

In Figure 8-4, the phase margin is marked with the red cursor and the gain margin is marked with the blue cursor

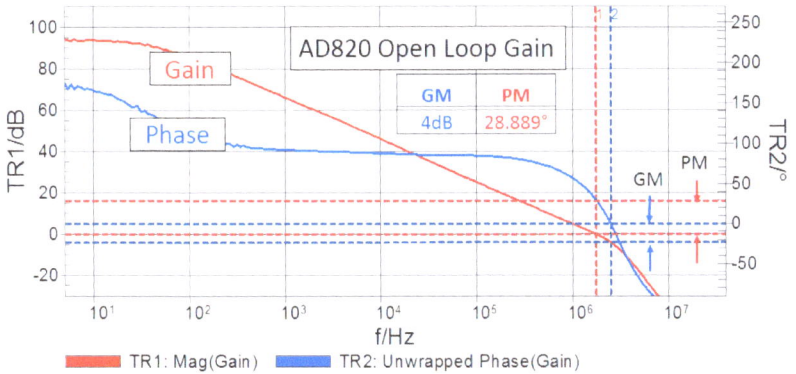

Open-Loop Bode Plot of an AD820 Operational Amplifier

Figure 8-4

The Bode plot assessment addresses stability using two margins, the gain margin and the phase margin. If both the gain margin and phase margin are greater than zero, then the circuit does not oscillate. The larger the margins are the more stable the circuit is.

The PM in Figure 8-4 is measured as 28.89 degrees using the red cursor, which is set to 0dB gain. The gain margin is measured as 4dB using the blue cursor, which is set to 0 degrees.

In actuality, almost all circuits are higher than 2nd order due to high frequency limitations in the transistors used to create the devices. The phase shift in Figure 8-4 clearly falls far below 0 degrees indicating that there is at least one additional pole. Higher order systems may have several GM and/or PM solutions in which case each GM and PM is independently assessed to assure stability.

The Nyquist plot is a better method of stability assessment for higher order systems. It is measured using the same method and instrumentation as the Bode plot. The operational amplifier Bode plot from Figure 8-4 is shown as a Nyquist chart in Figure 8-5.

The Nyquist chart shows the open loop gain as real (horizontal) and imaginary (vertical) terms. The singular unstable point in this case is at the point (1,0). The unstable point is shown on the right side of the graph. We often see it on the left hand side at the point (-1,0) as stated above. The difference between these two conventions is whether the 180 degrees from negative feedback is included in the measurement. Since we measure the complete loop, which includes the 180 degrees for the negative feedback and the additional phase shift incurred within the loop, the unstable point is at the point (1,0).

In the Nyquist chart, the stability is assessed by the proximity of the loop gain function to the singular unstable point at (1,0). The gain margin is assessed as the length along the horizontal axis between the unstable point at (1,0) and the curve as shown in the green arrow, indicating a gain margin of 0.371. The gain at this point is (1-0.371), or 0.629, which is -4dB.

The phase margin is assessed as the angle formed by the horizontal axis and the unity gain length connection from the point (0,0) to the loop gain curve. The stability margin, SM, is defined by the closest distance between the unstable point at (1,0) and the loop gain curve.

In Figure 8-5, the GM (green line), PM (blue line) and SM (back line) are annotated.

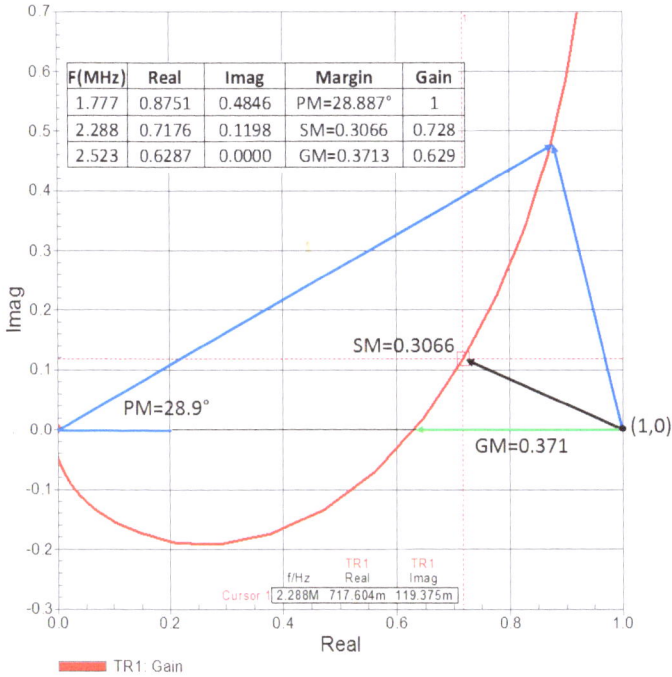

F(MHz)	Real	Imag	Margin	Gain
1.777	0.8751	0.4846	PM=28.887°	1
2.288	0.7176	0.1198	SM=0.3066	0.728
2.523	0.6287	0.0000	GM=0.3713	0.629

Nyquist Representation of the Bode Plot of Figure 8-4

Figure 8-5

The distance from any point along the loop gain curve to the singular unstable point (1,0) can be calculated as:

$$Distance = \sqrt{(1 - Real)^2 + Imag^2} \qquad 8.4$$

Applying this to the phase margin measurement the distance from the point (1,0) is:

$$PM \ Distance = \sqrt{(1 - 0.8751)^2 + 0.4846^2} \qquad 8.5$$
$$= 0.500$$

The gain at any point along the loop gain curve is calculated as:

$$Gain = \sqrt{Real^2 + Imag^2} \qquad 8.6$$

Applying this to the point at which the PM is assessed we get:

$$Gain = \sqrt{0.8751^2 + 0.4846^2} = 1 \qquad 8.7$$

This confirms that this is the unity gain length connection between (0,0) and the loop gain curve. The phase angle formed by the unity gain length connection and the horizontal axis is calculated from:

$$PM = atan \left(\frac{Imag}{Real}\right) \qquad 8.8$$

Applying this to the point at which the PM is assessed we get:

$$PM = atan \left(\frac{0.4846}{0.8751}\right) = 28.9 \; deg \qquad 8.9$$

The gain margin is calculated directly from the length along the horizontal axis as:

$$GM = (1 - 0.371) = 0.629 = 4dB \qquad 8.10$$

The results of Equations 8.9 and 8.10 confirm that the Nyquist chart assessment provides the same GM and PM as the Bode plot. Figure 8-5 also shows us that SM result is less stable than either the GM or the PM.

In some cases, the circuit adds a delay, either due to sample and hold functions, such as in a digitally controlled power supply, or due to the discrete-time nature of switching power supplies. While different topologies have different performance, the result is that as the frequency approaches the sampling rate, or switching frequency, additional phase shift is incurred while the gain remains unchanged.

This effect is clearly illustrated in the Bode plot measurements shown in Figure 8-6. The DC/DC converter being tested is measured with the switching frequency set for 250 kHz in the red traces and 500 kHz in the blue traces. The transfer function at low frequencies is relatively unaffected. At higher frequency, the slope of the gain does not significantly increase, though the phase curve rolls off quickly incurring an additional 180 deg. phase shift at the switching frequency.

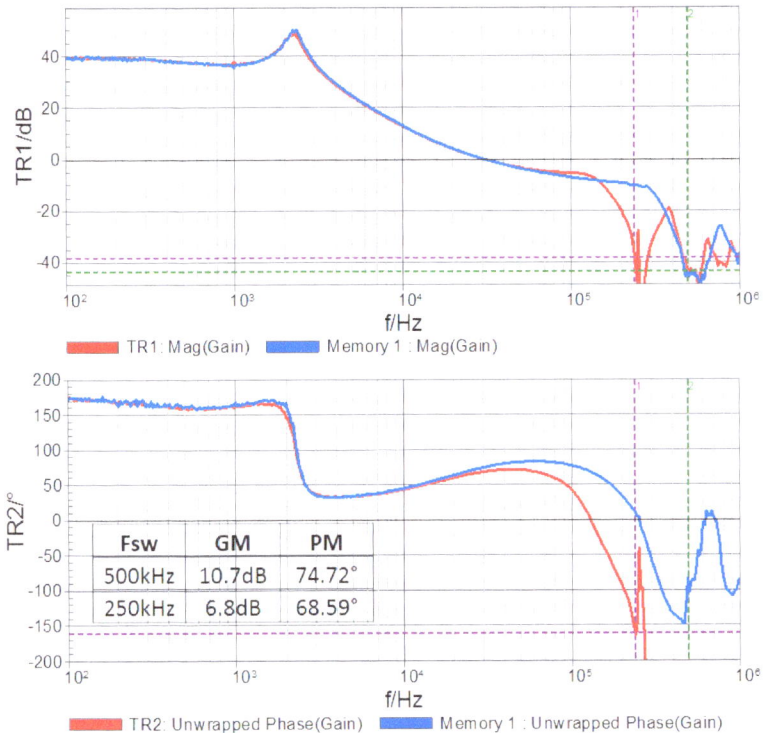

Fsw	GM	PM
500kHz	10.7dB	74.72°
250kHz	6.8dB	68.59°

Bode Plots for a DC-DC Converter Switching at 250kHz (Red) and 500kHz (Blue)

Figure 8-6

The same measurement data is displayed as a Nyquist chart in Figure 8-7. In the Nyquist chart, it is obvious that the stability is reduced at the 250kHz switching frequency since the entire curve shifts towards the singular unstable point (1,0). At the control loop bandwidth of 30.72 kHz, the time delay has an effect on the phase margin as evidenced by the reduction from 74.7 degrees at 500 kHz to 68.6 deg. at 250 kHz. The closest distance between the unstable point at (1,0) and the loop gain curve is shortest at the SM in both cases, and not the GM or PM, again showing that the SM is a better assessment of relative stability.

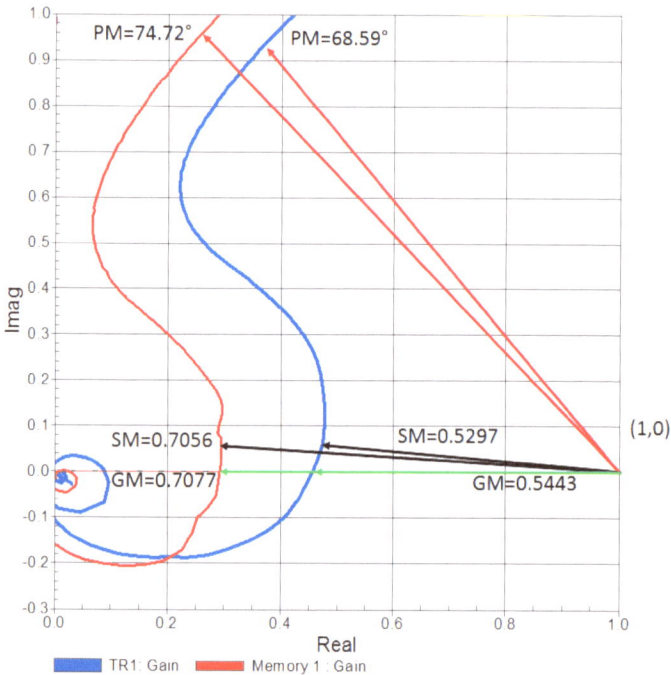

The Same Data Displayed as a Nyquist Chart at 500kHz (Red) and 250kHz (Blue)

Figure 8-7

The real and imaginary gain terms are extracted from the VNA measurement. These terms are used to calculate the PM, GM and SM, as well as the gain magnitude. The results are shown in Table 8.2. Note that the gain at the PM is unity.

250kHz Switching Frequency				
F(kHz)	Real	Imag	Margin	Gain
30.72	0.3651	0.9311	PM=68.59°	1
125.9	0.4740	0.0630	SM=0.5927	0.5298
130.44	0.4557	0.0000	GM=0.5443	0.4557

500kHz Switching Frequency				
F(kHz)	Real	Imag	Margin	Gain
30.72	0.2637	0.9648	PM=74.72°	1
245	0.2949	0.0304	SM=.7056	0.2965
252.8	0.2923	0.0000	GM=0.7077	0.2923

Real and Imaginary Gain and Margins for 250 kHz and 500 kHz

Table 8.2

Open Loop Measurement

The DUT control loop must be operating in its normal linear region in order to measure the loop gain. This may be difficult due to small offset voltages and the high DC gain of the feedback amplifier.

The operational amplifier shown in Figure 8-8 illustrates

the issue. Using the open loop gain of 95dB from the measurement in Figure 8-4, and allowing a 1mV offset in the operational amplifier, results in a 56V output.

Since the operational amplifier is not likely connected to a supply voltage that can allow a 56V output, the amplifier is not in its linear region (i.e. saturated or "railed").

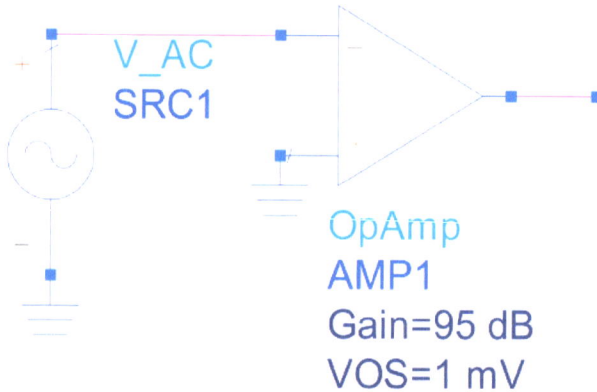

V_AC
SRC1

OpAmp
AMP1
Gain=95 dB
VOS=1 mV

In an Open-Loop Measurement , the High Open-Loop Gain makes it Difficult to Maintain the Amplifier in the Linear Region

Figure 8-8

In order to avoid this situation we can measure the open loop gain within a closed loop circuit. Maintaining the DUT in its closed loop state assures the device is in its linear operating region, assuming the input voltage and load current are within the normal operating range.

Modifying the amplifier circuit to allow the DC path to remain intact, while injecting an AC signal, allows the device to operate in its closed loop state.

Measuring the operational amplifier output divided by the inverting input results in the open loop gain.

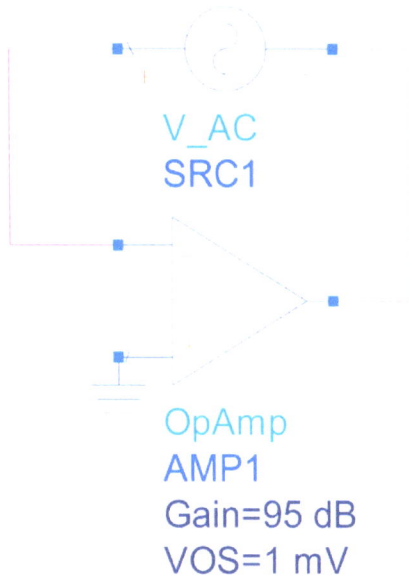

V_AC
SRC1

OpAmp
AMP1
Gain=95 dB
VOS=1 mV

In a Closed-Loop Measurement the Amplifier Needs to Operate in the Linear Region—Accomplished by Providing a Unity-Gain DC Path in the Feedback Loop along with the Injection Signal

Figure 8-9

Selecting an Injection Point

The signal can be injected anywhere within the loop, but should be a place where the impedance is a much lower impedance on one side of the injection point than the other.

An ideal place to break the loop is between the regulator or amplifier output and the resistive feedback divider as shown in Figure 8-10.

The regulator output generally has a much lower impedance than the top divider resistor, so this is often a good injection point. An alternate injection point would be between the amplifier and the MOS gate. The impedance at the MOS gate is likely much higher than the output impedance of the feedback amplifier.

The injection resistor is often included in the design and is, therefore, always available as an injection point. Terminal posts can also be included on either side of the injection resistor in order to facilitate the connection of the measurement probes.

In Figure 8-10, the most common injection location is between the feedback divider and the output.

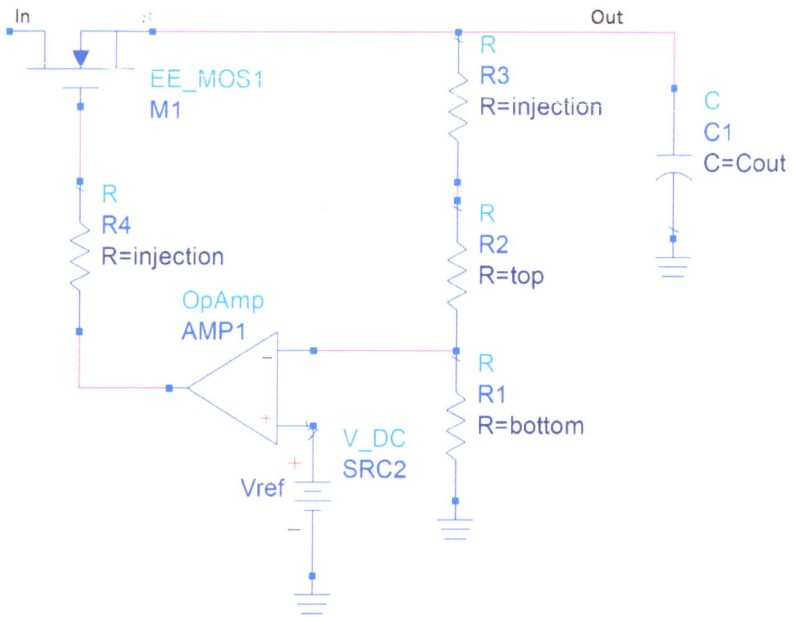

Two Possible Injection Locations

Figure 8-10

Measuring High Voltage Loops

Many offline power supplies include a Power Factor Correction (PFC) circuit that typically regulates a voltage of approximately 400V.

It is possible to inject at the top of the voltage divider using a 600V certified injection transformer. The high voltage requires high voltage probes and the large attenuation of the probes significantly degrades the SNR of the measurement.

A much better solution is to add an operational amplifier to buffer the divider impedance as shown in Figure 8-11.

The gain setting resistor, Rin, should be selected to be much greater than the buffer output impedance. The injection signal is then placed in series with Rin.

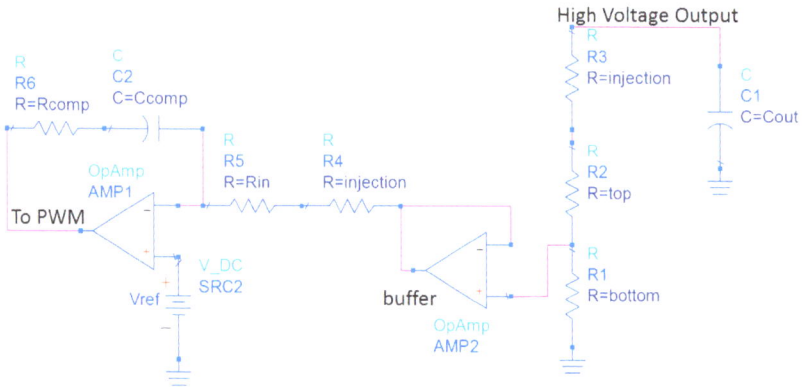

The Best Method of Measuring High Voltage Power Supply Control Loops is to Buffer the Voltage Divider with an Op-Amp and Inject between the Buffer and the Feedback Amplifier

Figure 8-11

Injection Devices

Transformers

An injection transformer is a precision transformer that connects the VNA to the injection resistor in the DUT. The VNA is then swept over a wide frequency range and the ratio of the voltages on each side of the injection resistor provides the magnitude and phase information. The result can be plotted either as a Bode plot or as a Nyquist plot.

In order to achieve a very wide measurement bandwidth and flat low frequency performance, special materials and annealing processes are used to manufacture the transformer core. The winding technique must also be carefully controlled. The Picotest J2100A, J2101A and the OMICRON Lab B-WIT 100 are all CE certified to CAT II 600V and all three offer outstanding performance with slightly different frequency ranges.

While the very high permeability of the core provides excellent performance it is also very sensitive to DC bias and so these transformers should not be subjected to currents above the few milliamps generally associated with the output voltage divider.

Injection Resistor

The lower -3dB point occurs at the frequency at which the reactive impedance of the injection transformer is equal to the real resistance of the injection resistor. It is for this reason that we need very high inductance to operate at very low frequencies.

Reducing the injection resistance for a given transformer inductance, therefore, results in a lower -3dB frequency. The measurements of an injection transformer with 50Ω and 5Ω injector resistor values are shown in Figure 8-12.

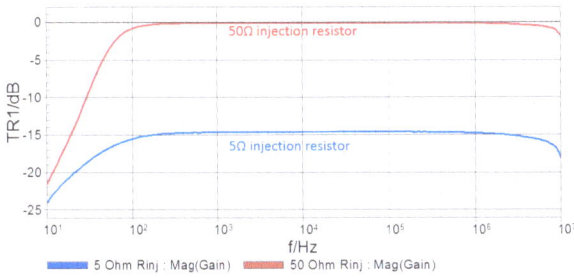

*The Same Injection Transformer with 5Ω and 50Ω
Injector Resistors shows the Improved Low-Frequency
Performance with the Smaller Injection Resistor*

Figure 8-12

Some additional benefits of using the lower value injection resistor are that the resistor value will be less likely to influence the measurement and will reduce the possibility of saturating the injection transformer. If the signal gets large enough at low frequency, the BH Loop will enter its non-linear saturation region.

The result is degraded low frequency response as seen in Figure 8-13.

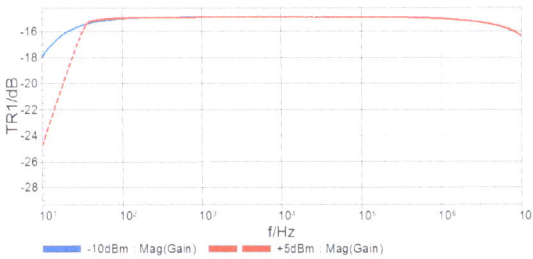

*Injection Transformer with -10dBm (Dashed) and
+5dBm (Solid) Inputs showing the Impact of the
Amplitude Reaching the Curve in the BH Loop*

Figure 8-13

Solid State Injectors

While it is possible to obtain high quality injection transformers with bandwidths as wide as 1Hz to 5MHz or more, in some cases this is still insufficient. For example, typical temperature control loops have bandwidths in the mHz range while some linear regulators and op-amp circuits can have bandwidths of up to 100MHz. For these applications, a solid-state injector can provide the necessary bandwidth.

A solid-state injector can perform over a frequency range of DC up to the upper frequency limit as dictated by the components selected, the printed circuit board material and the circuit board layout. The Picotest J2110A solid-state injector has a typical bandwidth of DC to greater than 100MHz. A frequency response measurement of the J2110A solid-state injector is shown in Figure 8-14.

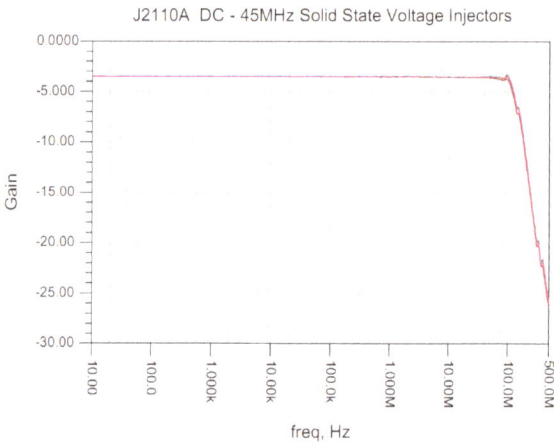

Measurement of the Picotest J2110A Solid-State Injector shows the Flat Response and a -3dB Frequency of 109MHz

Figure 8-14

It is essential that ripple from the solid-state injector's power supply does not dramatically degrade the dynamic range or the signal to noise ratio of the measurement. For this reason, a low noise power supply is included with the J2110A injector.

The selection of a valid injection point in the circuit is more critical when using a solid-**state** injector than with the transformer injector. The solid-**state** injector presents a high impedance input and a 25 Ohm impedance output with an infinite impedance between the points of injection. In most cases, the solid-state injector can be used at the same injection point as the injection transformer. The solid-state injector does not use an injection resistor, so if one is installed it should be removed.

The Picotest J2110A solid-state injector was used to measure the operational amplifier open loop gain plot shown in Figure 8-4.

Probes

In general, it is best to use oscilloscope probes for the gain and phase measurements. If the DUT is sensitive to capacitance, it is best to use a 10X probe, in order to minimize the capacitance. Typical oscilloscope probes present a capacitive load of 80-100pF in the 1X setting and 10-15pF in the 10X setting. The tradeoff is the lost SNR due to the probe attenuation vs. increased capacitive loading of the probe. Most 40MHz or 60MHz 1X/10X oscilloscope probes made for BNC inputs work with typical high input impedance FRAs and VNAs, including the OMICRON Lab Bode 100.

Calibrating the Setup

Prior to making the measurement the probes and interconnects should be calibrated to correct for any frequency response differences between the two probes in either magnitude or phase. This calibration is completed by connecting both probes to the VNA oscillator and performing a THRU calibration as shown in Figure 8-15. Details for the calibration of the VNA can be found in the user manual for your particular VNA.

In Figure 8-15, the probes are shown connected to a demo board for the thru calibration.

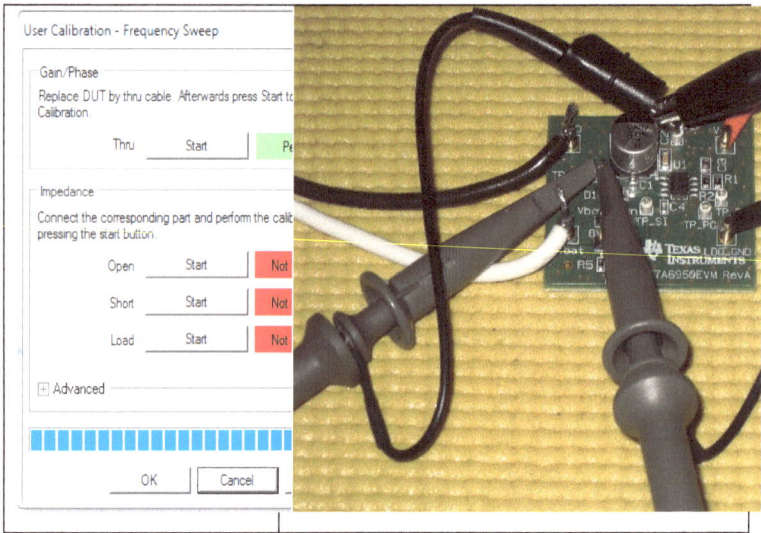

The THRU Calibration is Located on the Calibration Menu for the Bode 100

Figure 8-15

Small Signal vs. Large Signal

The stability assessment is a small signal measurement. It is essential to make sure the injection signal level does not cause any silicon junctions in the DUT to enter their non-linear regions or to cause the DUT output to become saturated. If this occurs the Bode plot will become distorted due to large signal effects, which are often noted by sharp discontinuities in the resulting plots.

We can monitor the DUT output signals to be sure it does

not saturate, but we cannot monitor the signals internal to the DUT. The injection signal level is deemed sufficiently low if increasing the level by 3dB does not change the resulting measurement.

If the results change due to increasing the signal amplitude by 3dB, the signal level should be continually reduced further until a 3dB increase does not affect the result. This maximum signal level is usually impacted by the operating load current and so this process is repeated for each operating condition.

Operating at a lower output current requires a reduced maximum injection signal level. This signal level effect will be demonstrated in the first example.

Example

1. TPS40200 DC/DC Converter Board

A commercial demonstration board is used to demonstrate the measurement of a Bode plot. This particular demo board is a 3.3V/2.5A output non-synchronous buck regulator. The EVM includes a 50Ω injection resistor. We did not change this resistor to the recommended 4.99 Ohms, however, we also did not make the measurement down to the 1Hz transformer minimum.

The test setup is shown in Figure 8-16. The external 20V input is provided via the red and black grabber hook on the left side of the image. The Picotest J2100A injection transformer connects to the printed circuit board using banana leads with grabber hooks on the bottom of the board. Two scope probes are also visible connected to either side of the injection transformer.

In Figure 8-16, the J2112A current injector is used to provide a constant current load. The current injector is similar to an electronic load but is non-capacitive and, hence, does not impact the measurement (non-invasive)

TPS40200 EVM Board Connected for Bode Plot

Figure 8-16

With the load current set to 500mA, the Bode plot is measured for various injection signal levels as shown in Figure 8-17.

Only the region near zero crossing is shown for clarity. The -8dBm signal level, and the -11dBm signal level, severely distort the measurement. -14dBm has a slight impact and the -17dBm and -20dBm levels are identical, so the largest injection signal for a 500mA output current is -17dBm.

The complete Bode plot is then measured using a -17dBm injection signal level, as well as a -20dBm injection signal level in Figure 8-18.

In Figure 8-17, note that -8dBm and -11dBm are both significantly different than the -14dBm, -17dBm and -20dBm signals.

The largest signal should be less than -17dBm at 500mA.

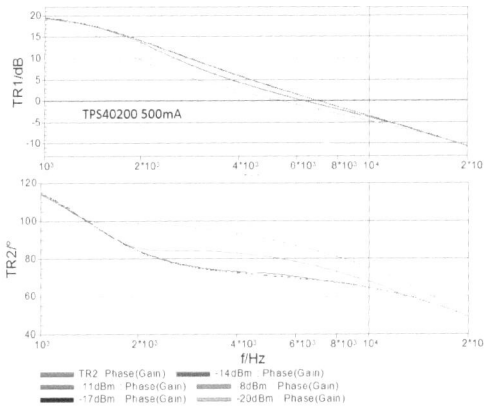

Zoom of Bode Plot at 500mA Load Current showing the Effects of Signal Amplitude

Figure 8-17

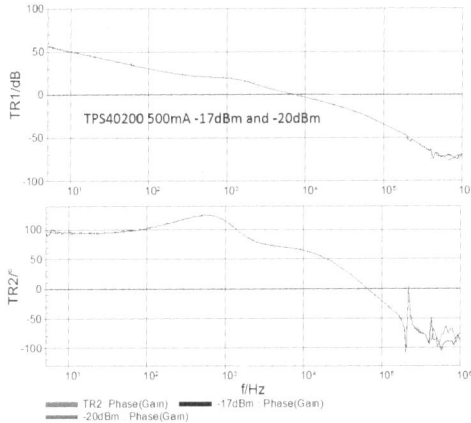

A Complete Frequency Sweep at -17dBm and -20dBm shows no Large Signal Effects at Either Level—the Inflection at 200kHz is due to the Switching Frequency of this PCB

Figure 8-18

The load current is reduced to 250mA and the narrow frequency range sweep is repeated. The resulting plot is shown in Figure 8-19. In this case, the large signal effects are clearly seen with the injection signal levels of -21dBm and -24dBm and can be seen subtly at -27dBm.

The results at -30dBm and -33dBm are identical with no signs of any large signal effects. The largest injection signal at a 250mA load current is -30dBm, which is significantly lower than the injection level at the 500mA load current. This process should be completed for each new test condition in order to validate the small signal measurements.

In Figure 8-19, the large signal effects are clearly evident at -21dBm and perceivable at -27dBm. The signals at -30dBm and -33dBm are identical. The largest signal should be less than -30dBm at 250mA.

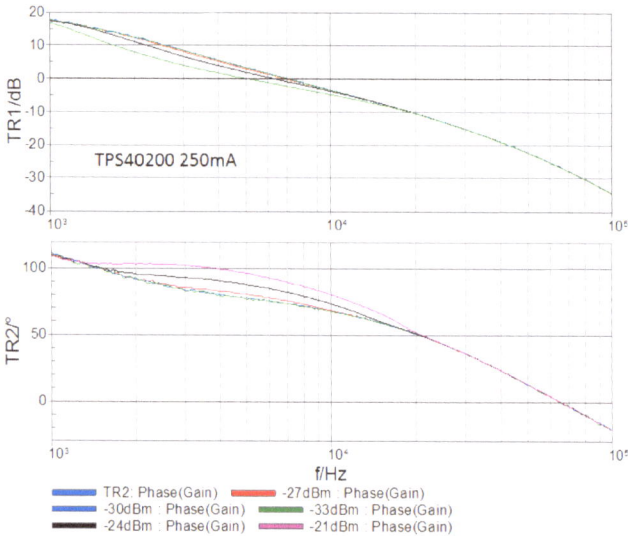

Zoom of Bode Plot at 250mA Load Current showing the Effects of Signal Amplitude

Figure 8-19

We can better evaluate the overall stability of the switching regulator by evaluating the Nyquist plot as shown in Figure 8-20. As shown previously, the PM indicates phase angle between the real axis and the unit length magnitude. The GM is the distance along the horizontal axis connecting the loop gain curve to the singular unstable point (1,0). The SM is the closest distance between the loop gain curve and the singular unstable point (1,0). As in the previous cases, the SM is worse than either the PM or GM and so is a better measure of the relative stability.

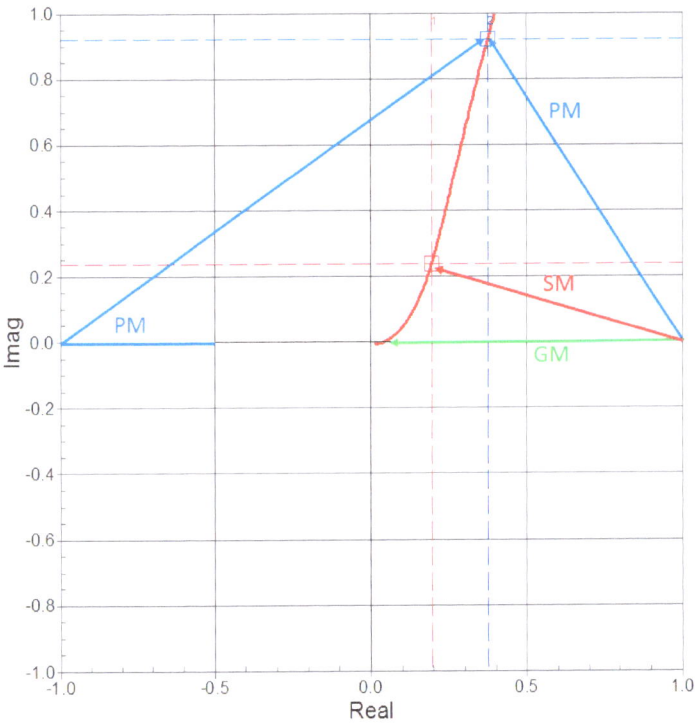

Nyquist Plot for the Bode Plot of the TPS40200 EVM Operating at 500mA as in Figure 8-18

Figure 8-20

The real and imaginary gain measurements as well as the distance between the singular unstable point (1,0) and the loop gain curve at the PM, SM and GM are summarized in Table 8.3. Note that the closest point to (1,0) occurs at the SM and not either the PM or the GM.

F(kHz)	Real	Imag	Distance to (1.0)
7.10	0.3762	0.9258	PM=1.19
19.50	0.1950	0.2396	SM=0.8399
69.09	0.0466	0.0000	GM=0.9933

Real and Imaginary Gain Terms for the Nyquist Plot in Figure 8-20

Table 8.3

Tips and Tricks

1. Unless you are in need of the low frequency gain information, don't waste too much time on reducing the low frequency noise. Focus on accurate measurements within a decade of the crossover frequency.

2. Always perform the measurement at different signal levels to make sure the crossover is not affected by the injection signal level.

3. Make sure that the load is not impacting the result. This will often be the case with electronic loads and low power devices. Resistive vs. constant current loading often produces dramatically different results.

4. The scope probes only need to be good for the frequency range of the measurement and even then, most errors can be calibrated out using the VNA THRU calibration function. Therefore, 40MHz scope probes are generally acceptable and available at very low cost.

5. Place the ground wires of both scope probes at the same point, either close to the bottom of the voltage divider, or close to the output capacitor in order to minimize noise.
6. When measuring very sensitive circuits, such as the op-amp in Figure 8-4, the scope probe tips and grounds all need to be very close to the op-amp pins. Tack soldering a short lead wire to the pin will give you something to clip the probes on.

Closed Loop Measurement

In many cases, it is not possible to measure the open loop gain response using a Bode or Nyquist plot. This may be because access to the feedback loop is not available at all for direct measurement or only one of several feedback loops is available. This is true of many device types including fixed voltage regulators and voltage references. In other cases, it may not be feasible to open the loop as it would require cutting traces or lifting wires. In many high reliability applications this would not be acceptable. Yet in other cases, the bandwidth of the device may be too high to allow signal injection without impacting the circuit's performance. This is often the case with operational amplifier circuits which can have bandwidths up to several GHz.

It is still possible to assess stability from a closed loop measurement of either output impedance or PSRR. The output impedance measurement is generally the most convenient as the output is usually accessible. The PSRR is based on the input and output, which are both frequently available.

On and Off Measurements

One method of determining the stability is to measure the output impedance with the device or circuit both powered and unpowered. This is not always the most accurate method. For example, in a voltage regulator the OFF measurement is a measurement of the output capacitance, but with the capacitor(s) at 0VDC. The capacitance at the nominal operating voltage may be significantly different than when measured in the OFF state as illustrated in Figure 8-21.

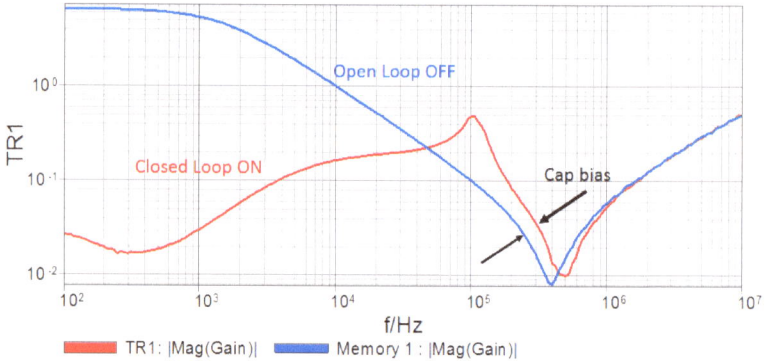

ON and OFF Output Impedance Measurement of a POL Regulator—note the Capacitive Impedance is approximately 50% Lower with Power Applied

Figure 8-21

This method of measuring with the power ON and OFF can also aid in determining which responses are passive resonances and which are active resonances (i.e. ones you can adjust through the feedback compensation and ones you may not be able to).

An output impedance measurement of the DC/DC converter from Figure 8-6 is a good example of both a passive resonance and an active resonance. The measurement, shown in Figure 8-22, indicates a time delay stability issue, since the closed loop ON measurement has a peak that is above the open loop OFF state measurement.

The output capacitor resonance just below 1MHz appears in both the ON and OFF state, so it is a passive resonance and not related to the loop gain.

This resonance is due to the ESL in series with the ceramic

decoupling capacitors.

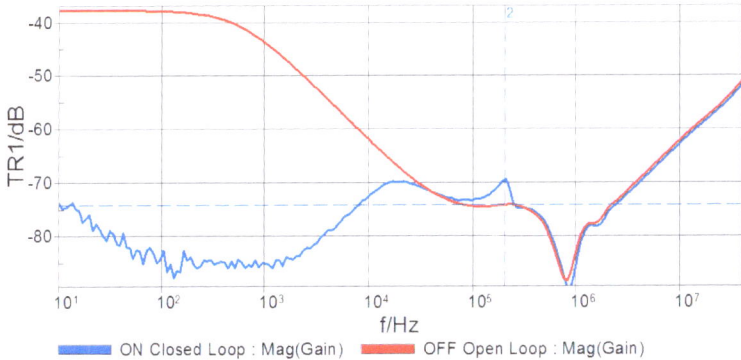

ON (Blue) and OFF (Red) S-Parameter Measurement
of the Output Impedance using the DC/DC Converter
of Figure 8-6

Figure 8-22

Forward Measurements

The relationship between the open loop Q and the PM is another method that can be used to determine the phase margin. This method was first published in *Fundamentals of Power Electronics*. It is listed in the references at the end of this chapter.

Two limitations prevent this method from being applied as a general solution.

The first limitation is that the measurement requires the ability to measure and modulate the internal voltage reference within the regulator. This is because the derivation is from the Q of the open loop response and not the Q of the closed loop response.

In most cases, this voltage reference is not available, and if it were, then the loop could be directly measured using a Bode or Nyquist plot.

The second limitation is that the mathematical derivation does not include the equivalent series resistance (ESR) in the output capacitor. It is generally the case that the ESR dominates the control loop stability of the device.

The result is that this measurement is not very useful in practice. An improved method is demonstrated later in this chapter.

Minor Loop Gain

The minor loop gain assessment is very popular for assessing the stability of an input filter combined with the negative resistance of a switching converter. This is also the method generally used to verify system level stability involving independent circuits.

A practical example of such a black box stability assessment is the interaction between a DC/DC converter and an electronic load. Each can be considered an independent circuit and each can be stable by itself, yet the combination of the two may not be as stable as either one.

Output impedance measurements of a DC/DC converter connected to a B&K model 8540 electronic load, and the same converter connected to a Picotest J2112A current injector (the load current), are shown in Figure 8-23.

This measurement is an S-parameter measurement and must be transformed to Ohms. This transformation is detailed in the Measuring Impedance chapter, though the result is included here for convenience.

For a 50Ω VNA, the transformation is defined by Equation 8.11.

$$Ohms = 25 \cdot \frac{S21}{1 - S21}$$

8.11

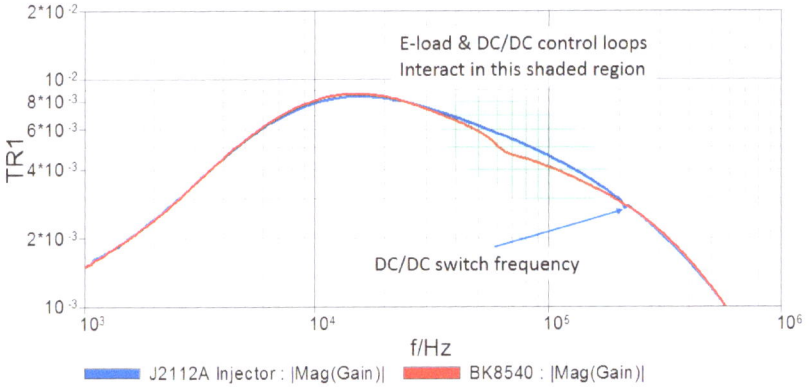

Picotest J2112A Control Loop does not interact with the DC/DC Converter Control Loop like the BK8540 Electronic Load Control Loop—note: the Y Scale S-Parameter must be transformed to Ohms

Figure 8-23

The minor loop gain is evaluated by separating the connection between the two systems and defining one as the source and the other as the load. Each system is defined as a frequency dependent vector, consisting of real and imaginary terms as shown in Figure 8-24.

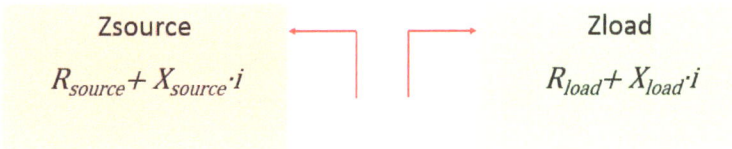

The Minor Loop Assessment is made by Separating the Source and Load then Measuring the Impedance looking into the Source Output and the Load Input

Figure 8-24

The minor loop gain is assessed as the source and load impedance ratios.

$$T_m = \frac{Z_{source}}{Z_{load}} = \frac{R_{source} + X_{source} \cdot i}{R_{load} + X_{load} \cdot i} \qquad 8.12$$

The unity gain crossover frequency is the frequency at which the magnitude of the source and load terms are equal.

It is possible to have several crossover frequencies if the source and load magnitudes are equal at more than one frequency.

$$|R_{source} + X_{source} \cdot i| = |R_{load} + X_{load} \cdot i| \qquad 8.13$$

The minor loop gain T_m is solved at the crossover frequency as a real part (Re) and an imaginary part (Im) and the PM is evaluated as:

$$PM = \mathrm{atan}\left(\frac{Im}{Re}\right) \qquad 8.14$$

The minor loop gain method is well established, though it requires separating the two systems, measuring the impedance of each system, and then computing the crossover frequency and the PM.

The impedance magnitudes for both the electronic load and the DC/DC converter are extracted from the VNA measurement and imported into EXCEL. Within EXCEL, a graph is produced including two impedance measurements and the minor loop gain, as shown in Figure 8-25.

Zsource and Zload are abbreviated in the figure as ZS and ZL respectively for simplicity.

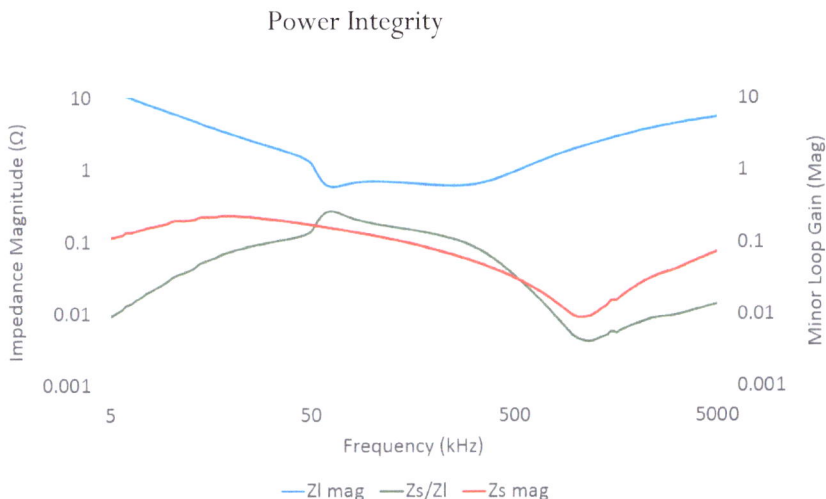

TPS40200 Output Impedance and BK8540 Eload Impedance along with the Minor Loop Gain, Zsource/Zload

Figure 8-25

In this case, the gain does not reach one and, therefore, does not have a PM since the PM is defined as the phase at a gain of one.

The highest gain occurs at approximately 62 kHz, which coincides with the maximum impact seen in Figure 8-23.

The magnitudes of the source impedance, load impedance and the source and load combined are tabulated for three frequencies in Table 8.4.

F(kHz)	Zs	Zl	|Zs|	|Zs/Zl|	Zs | |Zl
30.1	0.2046- 0.0548i	0.4733- 2.154i	0.212	0.096	0.202
62.2	0.1317- 0.0873i	0.3322- 0.4900i	0.158	0.267	0.126
201.4	0.0472- 0.0669i	0.6314- 0.0065i	0.082	0.13	0.076

Source and Load Impedance, Minor Loop Gain and the Source and Load Combined

Table 8.4

The peak gain is 0.267, occurring at 62.2 kHz, while at 30.1 kHz and 201 kHz the gain is much lower. While we are trying to measure the impedance of the DC/DC converter, we are actually measuring the impedance of the DC/DC converter in parallel with the impedance of the electronic load.

At 30.1kHz and 201.4kHz, the measured impedance is reasonably close to the impedance of the DC/DC converter, while at the peak gain point the measured impedance of $126m\Omega$ is 25% lower due to the interaction of the DC/DC converter with the impedance of the electronic load.

While we are generally concerned with the stability of a filter interacting with the negative resistance of the switching converter, this instability is also the basis of many quartz crystal and resonator oscillators.

Consider the circuit in Figure 8-26, which includes a resonator (source) and a negative resistance (load). The series resistance of the resonator and the negative resistance are set to the same value. At the resonator series resonant frequency the source impedance consists of only the real resistance, Rs and no imaginary terms.

The negative resistance load impedance is -Rs.

Resonator

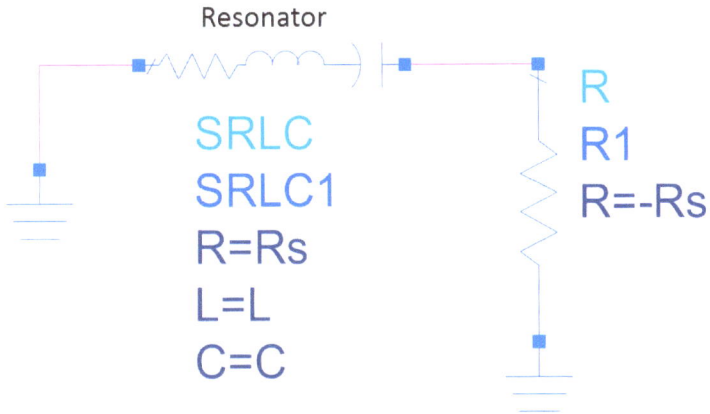

Applying the Minor Loop Gain to a Resonator Combined with a Negative Resistance equal to the Series Resistance of the Resonator

Figure 8-26

From Equation 8.13 we can see that the magnitudes of the source and load impedance are equal at the resonant frequency of the resonator, defining the zero crossing frequency. The phase margin at this frequency using Equation 8.14 is zero degrees, defining an oscillator.

Non-Invasive Closed Loop Measurement

The non-invasive closed loop measurement is simple to perform from a single impedance sweep and provides an exact PM and crossover frequency consistent with a Bode plot or Nyquist measurement for 1st or 2nd order systems. The mathematical derivation includes the ESR of the output capacitors, though this information is derived from a single impedance sweep and does not require a separate measurement. For higher order systems, the method determines the stability margin which is not necessarily the same as the PM.

This mathematical solution to the impedance to phase margin conversion is embedded in the OMICRON Lab Bode 100 operating software, Bode Analyzer Suite and demonstrated in example 1 below. The Bode Analyzer Suite allows the phase margin to be directly obtained as a cursor measurement on the output impedance curve for PMs <= 70 degrees.

The mathematical solution applies to output impedance and PSRR.

In addition, the mathematical solution also applies to minor loop gain assessment, such as the stability of an input filter combined with the negative resistance of a switching converter.

The details regarding the measurement techniques for impedance and PSRR can be found in their respective chapters of this book.

Examples

1. A Demo Board Based on the TPS40222

A Texas Instruments demo board is used as an example of a POL switching regulator assessment using a non-invasive impedance measurement. The two-port shunt thru measurement (S21) is discussed in detail in the Measuring Impedance chapter and is shown in Figure 8-27.

After placing the cursors on the impedance peak and the Q peak, the phase margin is read directly as 32.585 deg.

This measurement is made using the two-port method in conjunction with a multiport probe that is placed on the output capacitor. One reason this particular demo board is selected is that it also provides feedback loop access. This allows the non-invasive result to be compared directly to the results from the

Bode plot.

In Figure 8-27, note that the frequency associated with the phase margin measurement is not the crossover frequency, though it is possible to calculate the crossover frequency.

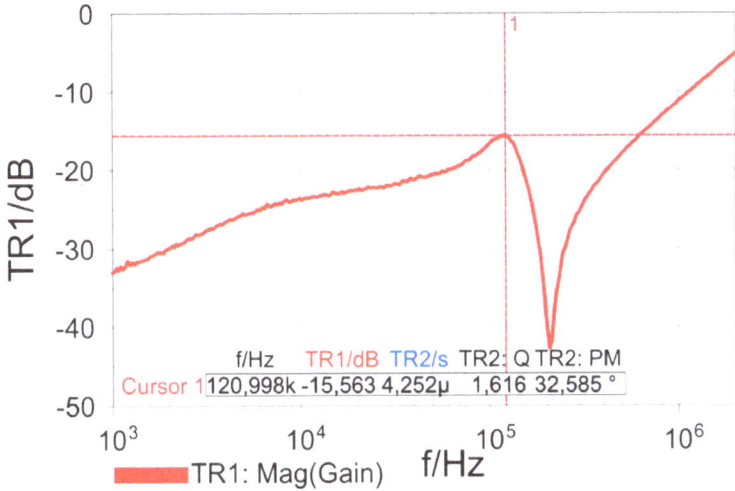

Two-Port Shunt-Thru Measurement (S21) of the Output Impedance of the TPS40222 with the Cursors Placed and the Non-Invasive Phase Margin Result Displayed

Figure 8-27

Using the same conditions of input voltage and load current, the traditional Bode plot measurement is performed and the result is shown in Figure 8-28.

The non-invasive measurement reports a PM of 32.585 degrees, while the Bode plot result reports a PM of 32.581 degrees, nearly equivalent as expected.

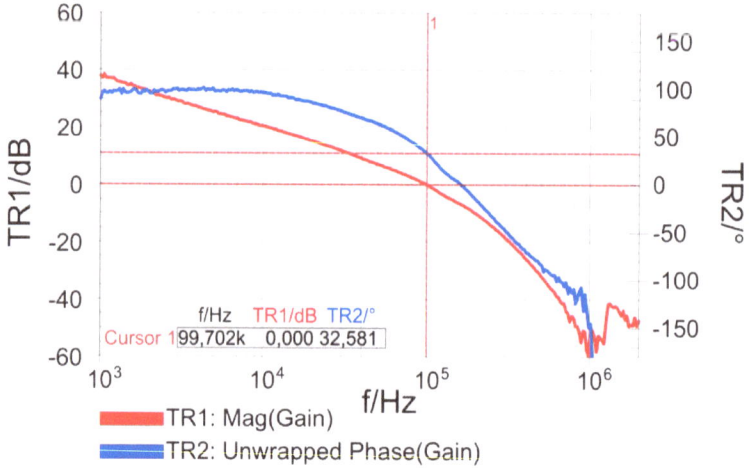

Bode Plot for the TPS40222 Demo Board under the Same Conditions

Figure 8-28

2. A Ref-03 Voltage Reference with the Manufacturer-Recommended 0.1 µF Output Capacitor

An Analog Device REF-03 voltage reference is used to demonstrate the phase margin resulting from the addition of the manufacturer recommended 0.1 µF ceramic output capacitor.

The non-invasive method is performed from both PSRR and from output impedance data.

Each measurement is shown without the output capacitor in blue and with the output capacitor in red.

In this case, the device does not have feedback loop access and so a Bode plot cannot be obtained for comparison.

In Figure 8-29, note the phase margin of this vendor-recommended configuration is only 8.83 degrees. In Figure 8-30, note the phase margin of this vendor-recommended configuration is only 8.95 degrees.

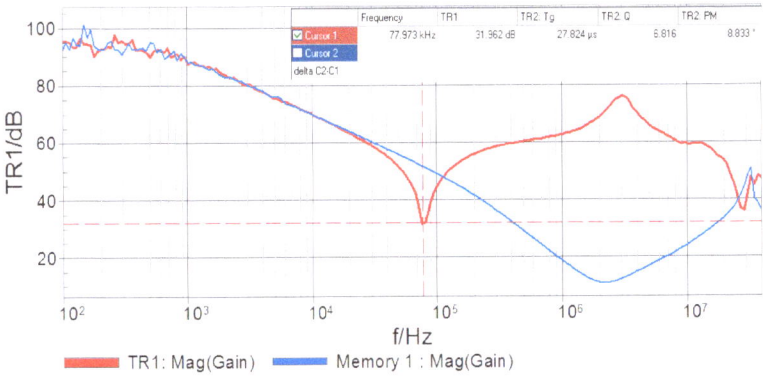

REF-03 Voltage Reference PSRR with (Red) and without (Blue) the Manufacturer-Recommended 0.1μF Output Capacitor

Figure 8-29

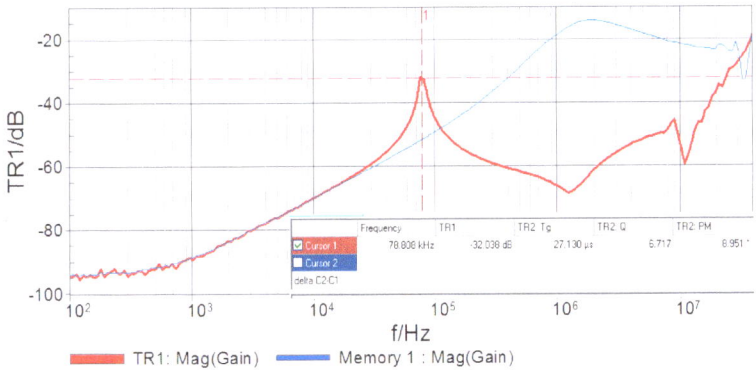

REF-03 Voltage Reference Output Impedance with (Red) and without (Blue) the Manufacturer-Recommended 0.1μF Output Capacitor

Figure 8-30

A 500µA load step is applied to the voltage reference using a Picotest J2111A current injector.

The output impedance of the current injector is much higher than the 0.1µF capacitor, and the bandwidth of the injector is much higher than the 78 kHz response, so the current injector does not alter the measurement result.

The response to the step load is shown in Figure 8-31. While the step load response does not provide a quantitative result, it does corroborate the poor PM indicated by the non-invasive measurement.

In Figure 8-31, the ringing corroborates the very poor phase margin.

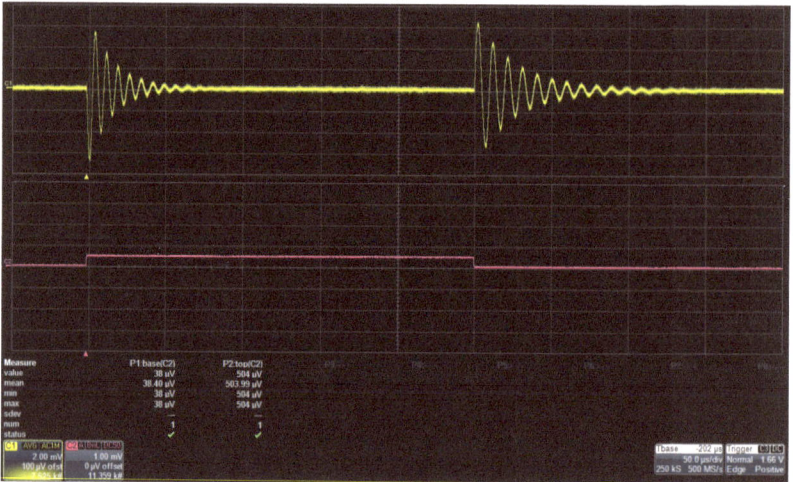

A Picotest J2111A Current Injector is used to apply a 500µA Step Load to the Voltage Reference with the 0.1µF Output Capacitor Installed

Figure 8-31

3. LM317 Voltage Regulator with a 22µF Moderate-ESR Output Capacitor

The Picotest VRTS01 VNA Test Standard kit includes a 3.3V LM317 voltage regulator and a selection of output capacitors. A 22uF moderate ESR capacitor labeled Cap 5 is installed at the output of the regulator. A J2111A current injector is used to provide a high impedance 25mA constant current load, while the Bode plot is measured.

In Figure 8-32, note The PM is displayed as 51.097 degrees.

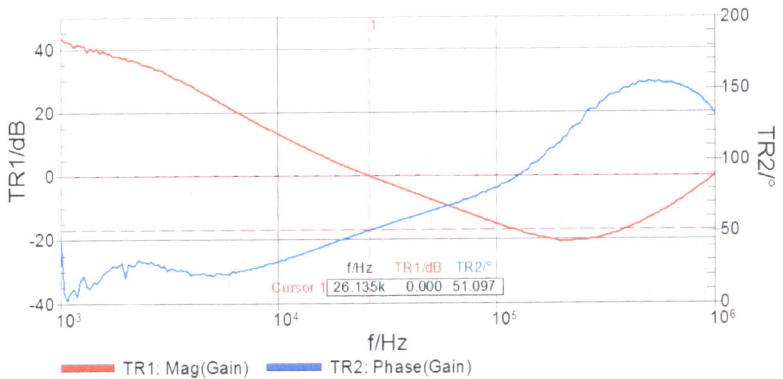

Bode Plot of the 3.3V LM317 Voltage Regulator with a 22µF Moderate ESR Tantalum Output Capacitor operating at a 25mA Constant Current Load

Figure 8-32

The closed loop output impedance measurement is obtained using the Picotest J2111A current injector at the load, while monitoring the output voltage and the injector current. The impedance result is shown in Figure 8-33.

The details of this impedance measurement method can be found in the *Measuring Impedance* chapter.

The OMICRON Lab Bode 100 software calculates the PM

directly using the output impedance measurement along with the group delay function. The PM is reported as 51.307 deg. and the two methods are virtually identical. At this time, the Bode 100 is the only instrument that has the impedance to PM software embedded in it and can make the automatic conversion.

In Figure 8-33, note the PM is displayed as 51.307 deg.

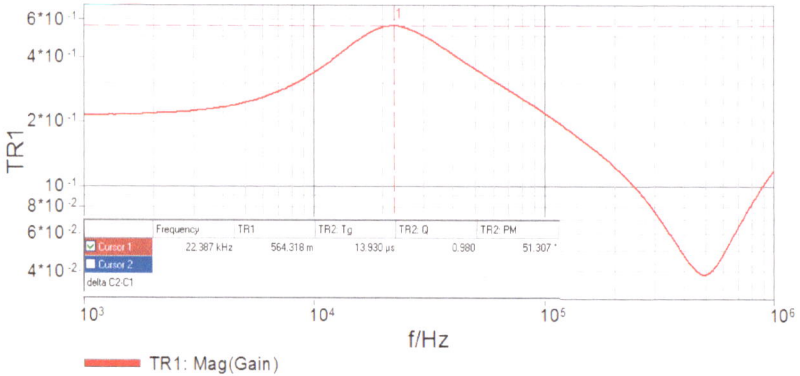

Output Impedance Plot of the 3.3V LM317 Voltage Regulator with a 22µF Moderate ESR Tantalum Output Capacitor operating at a 25mA Constant Current Load

Figure 8-33

Tips and Tricks

1. The impedance can be measured using any of the methods in the Measuring Impedance chapter, though for low power devices you will want to AC couple the probes so as not to load the DUT.

2. The use of a multiport probe expedites the process of measuring the stability of many devices. Details for this probe are included in the *Probe* chapter.

3. Measure all voltage references, linear regulators and

unity gain op-amps. They are the most common circuits which have poor stability. Most voltage references that use a ceramic output capacitor will also have poor stability.

4. Voltage references are generally extremely sensitive to operating output current. LDOs are less sensitive, but still sensitive. Op-amps are the least sensitive to output current.

Chapter References

1. Karl Johan Åström, Richard M. Murray. *Feedback Systems: An Introduction for Scientists and Engineers*, Princeton University Press, 2008.
2. Agilent 5990-5902 *Evaluating DC-DC Converters and PDN with the E5061B LF-RF Network Analyzer,* cp.literature.agilent.com/litweb/pdf/5990-5902EN.pdf
3. R.W Erickson, D. Maksimovic *Fundamentals of Power Electronics, 2nd ed.* Springer, 2001
4. *DC/DC Converter Stability Measurement OMICRON Lab,* http://www.omicron-lab.com/bode-100/application-notes-know-how/application-notes/dcdc-converter-stability-measurement.html
5. S.M. Sandler, C.E. Hymowitz, P. Ho *When Bode Plots Fail us*, Power Electronics Technology, May 2012 http://powerelectronics.com/power-electronics-systems/when-bode-plots-fail-us
6. S.M. Sandler, T. Boehler, C.E. Hymowitz, *Why Network Analyzer Signal Levels Affect Measurement Results* Power Electronics Technology, Jan. 2011 http://powerelectronics.com/power-electronics-systems/why-network-analyzer-signal-levels-affect-measurement-results
7. Picotest, VRTS01 Kit https://www.picotest.com/products_VRTS01.html

Chapter 9

Measuring PSRR

SIMPLY STATED, POWER Supply Rejection Ratio (PSRR) is the magnitude of the input signal divided by the output signal.

This can be a source of confusion as some manufacturers show PSRR as a positive log magnitude (input divided by output), while others show it as a negative log magnitude (output divided by input).

In this book, we use the positive rejection ratio convention result, determined by the input divided by output. The input is an AC modulation signal superimposed on the power supply voltage.

The output is the AC output signal of the device at the same modulation frequency.

$$PSRR = \frac{\Delta Vin}{\Delta Vout} \qquad 9.1$$

While most power conversion favors switching regulators for their high efficiency, the performance of today's electronics often require very low noise. Combining a switching regulator with a low dropout (LDO) linear regulator is a compromise offering good efficiency and low noise.

The PSRR of the LDO is used to eliminate the switching ripple and noise, providing clean power to the circuit. This method is popular enough that some manufacturers produce integrated circuits that include the switching regulator and the LDO in the same package.

PSRR is an applicable benchmark to many types of devices, including op-amps, linear regulators and voltage references to name a few.

PSRR performance is significantly dependent on control loop stability. Poor control loop phase margin will be seen as poor PSRR. Conversely, we can also assess stability from the PSRR performance. Since the stability of a regulator is often dependent on the operating current, the PSRR can also be dependent on the operating current.

The PSRR can also be dependent on the voltage between the regulator input and output.

Measurement Methods

In selecting the methodology for the measurement, one decision is whether to make the measurement in-circuit or out-of-circuit. Another choice is whether to measure directly or indirectly. Yet another is which measurement domain (time, frequency, or spectrum) to use.

Each of these choices is essentially independent of the others.

For example, a measurement can be made either in or out-of-circuit and independently be either a direct or an indirect measurement, made in either the time domain or the spectrum domain.

These choices are explained in the following sections.

In-Circuit or Out-of-Circuit

The in-circuit vs. out-of-circuit selection is based, in part, on the goal of the measurement.

In general, it is best to measure in-circuit, as this is the closest you can get to typical circuit performance.

The PSRR performance of a circuit, especially in the case of very high rejection ratios, can be easily impacted by printed circuit board effects including cabling, interconnect impedances, crosstalk, decoupling parasitics and control loop stability.

In order to measure the PSRR performance of a device, such as a voltage regulator, it is better to make the measurement out-of-circuit. This allows better control over interconnects and isolates the performance from the rest of the circuit.

This might be the case if the goal is to compare the device's measured performance with the datasheet specified performance or to compare several devices.

If the goal is to troubleshoot or optimize the completed circuit it is often better to perform the measurement in-circuit. One common issue with measuring in-circuit is that the regulator often has large capacitors connected at the input, making it difficult to inject a high frequency signal.

In those cases, the input capacitors may need to be removed for the measurement.

Direct or Indirect Measurement

A direct measurement is a measure of the ratio of the power supply modulation signal and the resulting output signal magnitudes either at a particular frequency or over a wide range of frequencies.

This is an absolute measurement of the ratio.

Oscillators and ADC clocks also have significant sensitivity to power supply noise.

They too can be assessed for PSRR.

In these cases, the assessment considers oscillator jitter or phase noise in response to a power supply modulation signal. The jitter, or phase noise, is typically measured using either a high performance spectrum analyzer or a signal source analyzer. The phase noise spur generally includes both AM and FM modulation.

These can also be demodulated in signal source and spectrum analyzers.

The direct vs. indirect measurement is dependent on the goals of the measurement. If the goal is to troubleshoot the clock jitter, or to determine the sensitivity of the clock to a power supply signal, then it is best to measure indirectly.

This may also be the case if the goal is to determine the maximum signal level at the clock in order to meet a particular jitter requirement or for a jitter budget.

Modulating the Input

Independent of the method used, the PSRR measurement requires a modulation signal to be superimposed on the power supply voltage. The injected signal level is based on the performance of the equipment used to make the measurement and the PSRR of the measurement.

The minimum injected signal level can be calculated based on the minimum measurable signal, V_{floor} and the PSRR in dB as:

$$V_{inj} \geq V_{floor} \cdot 10^{\frac{PSRR}{20}} \qquad 9.2$$

The minimum injected signal vs. PSRR is shown for several values of V_{floor} in Figure 9-1.

The minimum measurable signal is dependent on the sensitivity of the equipment used and the noise management of the interconnecting signals.

The measurement of this output signal is very similar to measuring ripple and noise, so many of the techniques discussed in that chapter are applicable here as well.

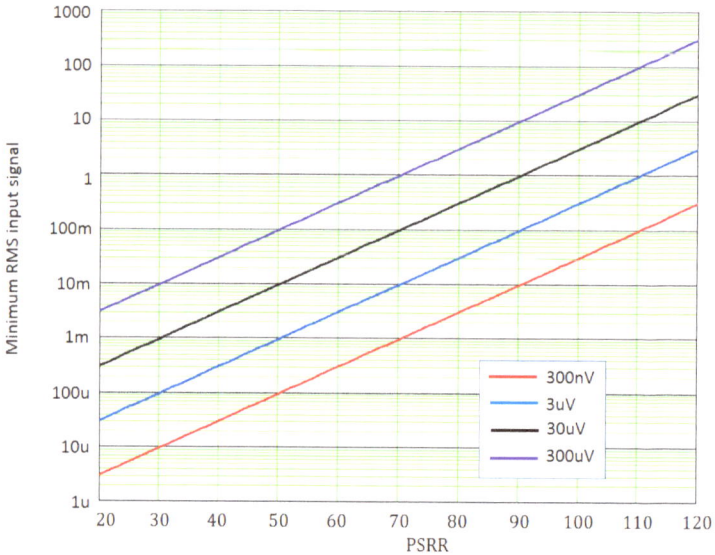

Minimum Injected Signal Level vs. PSRR for Several Minimum RMS Measurement Limits

Figure 9-1

The low frequency PSRR of modern high performance voltage references, op-amps or LDOs can be greater than 110dB. Assuming a minimum measurable signal level of $300nV_{rms}$, a minimum injection signal level of approximately $100mV_{rms}$ is required. At higher frequencies, the PSRR is reduced and, therefore, the minimum injection signal level is reduced.

For a given capacitance the signal source current required to obtain the minimum injected signal level is:

$$I_{rms} = 2\pi \cdot V_{rms} \cdot frequency \cdot capcitance \qquad 9.3$$

The signal source current required for a $100mV_{rms}$ injection signal vs. frequency is shown in Figure 9-2 for several values of capacitance.

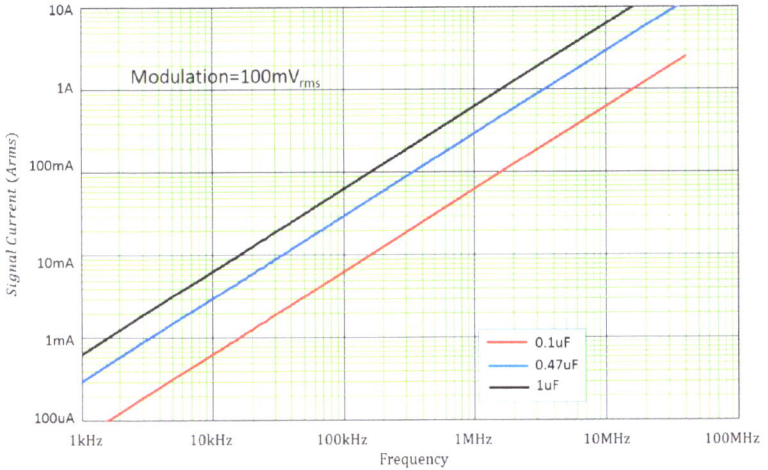

Signal Source Current vs. Frequency Required to Present 100mV_rms across Several Capacitor Values

Figure 9-2

If the input capacitance is 1μF and the PSRR is measured up to 10MHz, the signal source current is required to be $6.3A_{rms}$, while a 0.1μF capacitance measured to 10MHz requires $630mA_{rms}$.

It is clearly difficult to inject a signal if the input capacitors are installed. Even if the necessary current is provided, the resulting capacitor current could permanently damage the capacitors.

In general, it is best to remove the capacitors if at all possible. If it is not possible to remove the capacitors, make sure the capacitors can tolerate the resulting current.

Line Injector

The Line Injector is a test adapter that is connected in series between the input power supply and the input to the DUT. It allows a small signal to be superimposed on the DC power supply. The Line Injector is a passive device, resulting in a reduced voltage at the DUT. The voltage drop is a function of the operating current as is the output impedance of the injector. The typical output impedance of the Picotest J2120A line injector is approximated by:

$$R = 0.124 \cdot I_{DC}^{-0.853} + .077 \qquad 9.4$$

The output resistance vs. DC operating current is shown graphically in Figure 9-3.

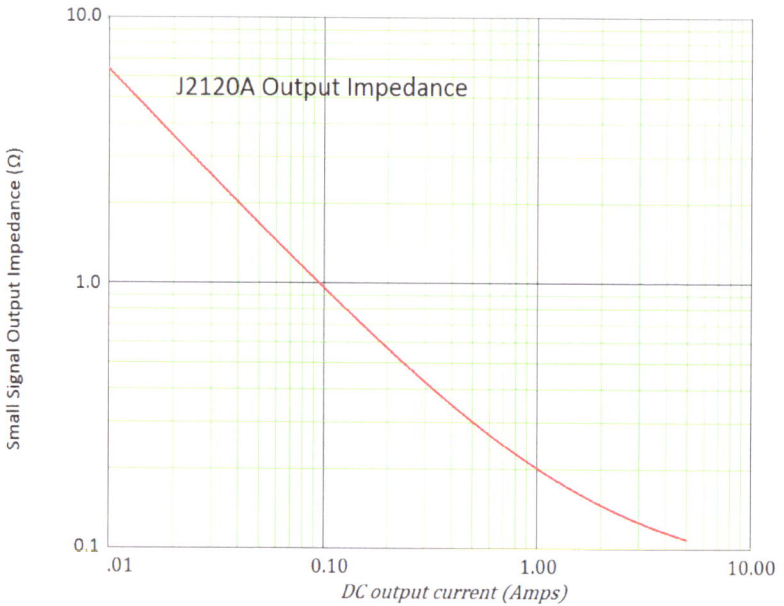

Typical Output Resistance of the Picotest J2120A Line Injector

Figure 9-3

The typical output inductance of the Picotest J2120A line injector is approximately 100nH. The interconnecting cables between the injector and the DUT can be significantly greater than the injector inductance. The output resistance and inductance of the injector and the cables combined with the DUT input impedance limit the usable bandwidth of the injector.

The Line Injector impedance at higher frequencies is dominated by the injector and interconnect inductance. The transfer function of the Picotest J2120A connected through a 50Ω coaxial cable, terminated into a 50Ω VNA input port is shown in Figure 9-4.

The 3dB bandwidth into a 50 Ohm load is 10Hz to 10MHz. Figure 9-4 confirms that the high frequency effects are not nearly as great as the mid frequency effects, since the inductance is dominant.

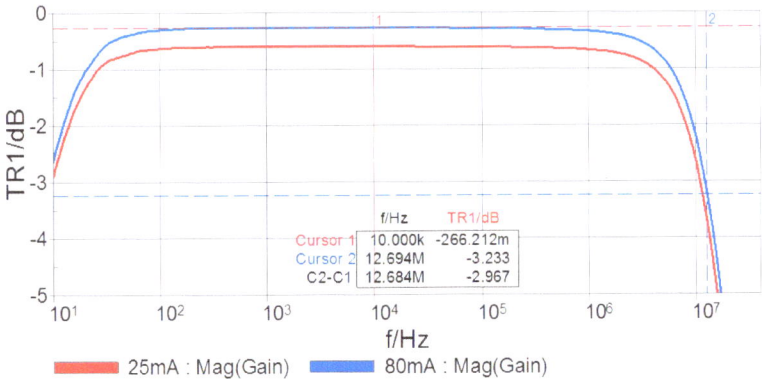

Transfer Function of the Picotest J2120A Line Injector into a 50Ω Output Port at 25mA (Red) and 80mA (Blue)

Figure 9-4

PSRR is measured as the ratio of the DUT input injected signal and the resulting output signal of the DUT.

The Line Injector is outside of the measurement and so the bandwidth of the Line Injector does not directly determine the PSRR measurement bandwidth.

We can confirm this by inserting the Line Injector and performing a through calibration with both CH1 and CH2 of the VNA connected to the output of the injector.

The setup is then used to measure the PSRR of a 20dB attenuator.

The result is shown in Figure 9-5.

The results confirm that the Line Injector does not directly limit the usable measurement bandwidth.

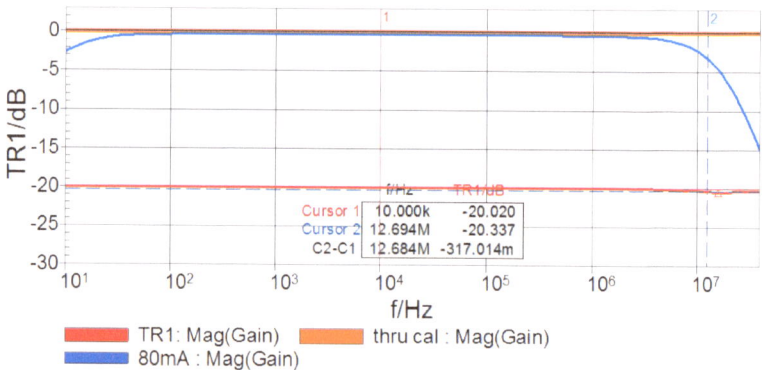

Cursor 1	10.000k	-20.020	
Cursor 2	12.694M	-20.337	
C2-C1	12.684M	-317.014m	

TR1: Mag(Gain) thru cal : Mag(Gain)
80mA : Mag(Gain)

The Picotest J2120A Transfer Function at 80mA (Blue), Thru Cal Measurement (Orange) and the Attenuator (Red)

Figure 9-5

The usable bandwidth limit is primarily based on the ability to maintain the minimum injected signal level, which is calculated in Equation 9.2.

The Line Injector is a low distortion device at signal levels up to approximately $1V_{rms}$ (13dBm).

The Line Injector output signal voltage, and the associated spectrum content for a -7dBm, 1MHz modulation input signal are shown in Figure 9-6. The center frequency display measurement shows -8.23dBm and the second harmonic is approximately 44dB lower than the fundamental, or -44dBc.

The signal generator output measured alone with the same settings is shown in Figure 9-7. The fundamental output of the signal generator is -7.92dBm and the second harmonic is approximately -50dBc confirming the 0.3dB loss shown in Figure 9-5.

This also indicates that the distortion is due primarily to the harmonics of the signal generator. In any case, the distortion is low enough to use in a time domain measurement, while the VNA and spectrum analyzer are narrowband devices and, therefore, not influenced by the signal distortion.

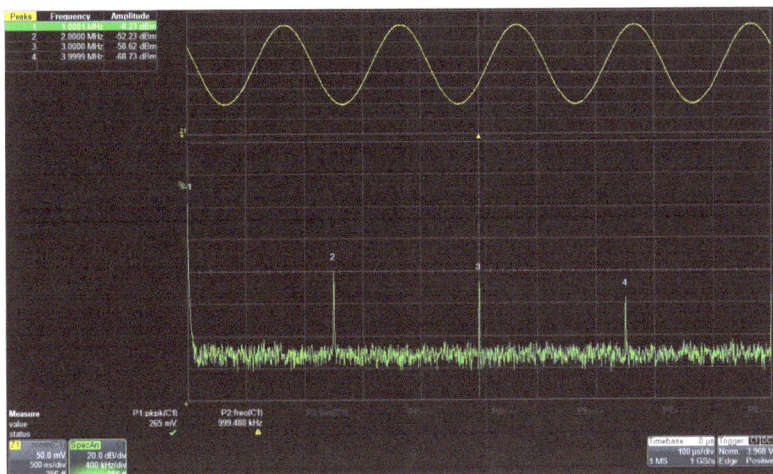

Line Injector Output Signal at 80mA for a -7dBm 1MHz Modulation Input Signal

Figure 9-6

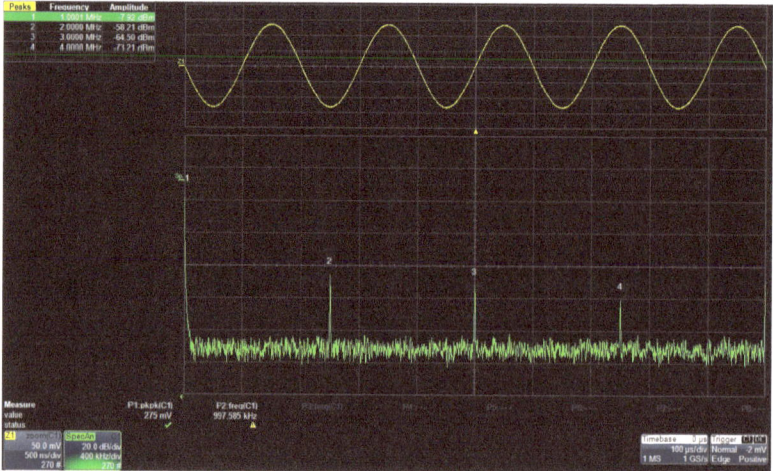

The Signal Generator Measured with the Line Injector Bypassed

Figure 9-7

Current Injector

Another method of signal injection uses a current injector to provide an input signal for the PSRR measurement. The current injector allows the measurement to be made as a non-invasive measurement. This is useful in cases where the input power signal is not accessible for insertion of the line injector.

The Current Injector uses the impedance at the DUT input to generate a modulation voltage from the modulated current injector. Every circuit presents a finite, non-zero impedance, meaning that a signal can be generated. The modulated voltage is determined by the amplitude of the injected current and the circuit impedance at the DUT input.

$$V_{inj} = I_{inj} \cdot Z_{DUT\_in} \qquad 9.5$$

Substituting Equation 9.2 into Equation 9.5 results in the

minimum injector current for a given noise floor, PSRR and DUT input impedance.

$$I_{inj} \geq \frac{V_{floor} \cdot 10^{\frac{PSRR}{20}}}{Z_{DUT}} \qquad 9.6$$

The minimum injected current vs. PSRR is shown for several values of V_{floor} assuming a 100mΩ DUT input impedance, in Figure 9-8.

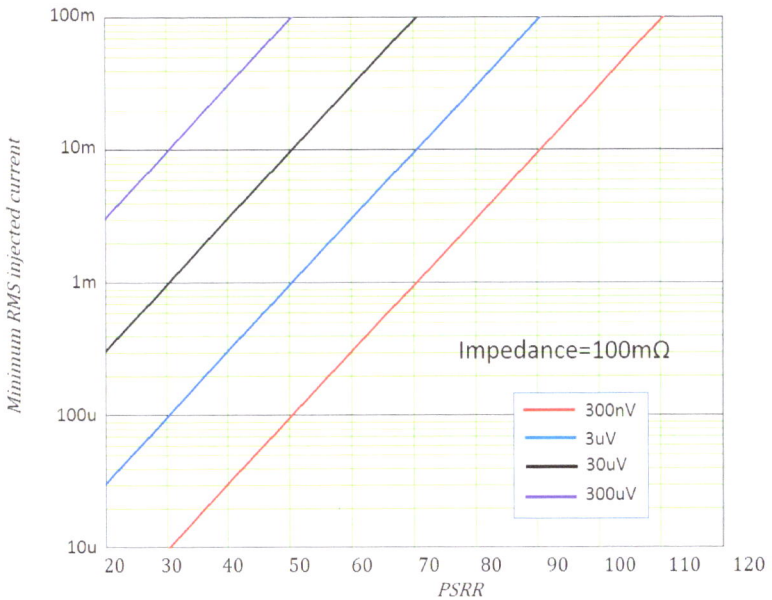

Minimum Injected Current Level vs. PSRR for Several Minimum RMS Measurement Limits and a 100mΩ Impedance

Figure 9-8

The Picotest J2111A Current Injector can modulate up to 15mA_{rms} at a maximum voltage of 40V.

The Picotest J2112A is capable of modulating up 150mA_{rms} at a maximum voltage of 5V.

They both offer flat response from DC to more than 10MHz.

DC Amplifier

It is also possible to provide the input modulation and power from a DC coupled power amplifier, though since many are designed for speaker loads, they may be unstable when connected to a power supply input, which can be either capacitive or inductive depending on whether the input includes an input filter.

One source for high powered DC amplifiers and four quadrant power supplies is NF Electronics in Japan.

Choosing the Measurement Domain

In the case of direct measurement, the domain selection may be determined by the magnitude of the measurement.

Oscilloscopes generally use 8-12 bit ADCs, resulting in much lower dynamic range than a VNA or spectrum analyzer, which often have 24 bits or more of dynamic range.

The VNA and spectrum analyzer also have much greater sensitivity and selectivity, which are often needed for measuring higher performance devices.

Another benefit of the VNA is that it can easily correct for errors due to the probe's frequency response.

VNA

The VNA is the instrument of choice for direct PSRR measurement. The VNA has excellent dynamic range, high sensitivity, and selectable measurement bandwidth.

In addition, the VNA can directly measure gain over a wide frequency range in a single measurement sweep.

The OMICRON Lab Bode 100 and Agilent E5061B can typically measure signals as small as $1\mu V_{rms}$ and can measure both the input modulation signal and the resulting output signal simultaneously in one sweep, making the measurement very simple and fast.

Spectrum Analyzer

The spectrum analyzer can typically measure signals at least as small if not smaller than the VNA.

The E5052B, N9020A and RSA5106A can all measure a $100nV_{rms}$ signal. The spectrum analyzer generally starts at a higher frequency than the OMICRON Lab Bode 100 (1Hz) or Agilent E5061B (5Hz) and typically measures a single frequency at a time. Since the spectrum analyzer does not measure gain, the input modulation signal and resulting output signal have to be measured separately at each frequency.

Some spectrum analyzers can be used as a Scalar Network Analyzer, with the addition of an optional tracking generator. A Scalar Network Analyzer provides magnitude information, but not phase information. This option allows the spectrum analyzer to measure the amplitude vs. frequency of the input modulation signal and resulting output signal.

The PSRR can then be computed externally from the two sweeps.

Oscilloscope

A typical oscilloscope can measure a signal amplitude of 100s of µVs. Most oscilloscopes today offer spectrum analyzer functionality. The spectrum analyzer is a narrow band instrument resulting in a significantly improved signal-to-noise ratio than the oscilloscope's wide-band time domain measurement.

The spectrum analyzer can also generally display a log y-axis greatly improving the dynamic range, but not a log frequency axis, making wideband measurements difficult to interpret.

An oscilloscope in spectrum analyzer mode allows measurements of signals as low as 30-50uVpp.

Therefore, if we were to inject a $100mV_{rms}$ input signal, which is often specified, regulators with PSRR up to 66dB can be measured using this method.

A typical LM317 voltage regulator has a low frequency PSRR of approximately this level and could possibly be measured using this method, depending on the sensitivity of the oscilloscope.

The PSRR of current state of the art regulators is approximately 115dB, requiring much more sensitive equipment than an oscilloscope.

As with the spectrum analyzer, the input modulation signal and resulting output signal have to be measured separately at each frequency.

Probes and Sensitivity

With the injection method chosen, the next choice is how to probe the modulated input signal and resulting output signal.

The choices are in the probe selection and coupling of the probe to the equipment.

The probe and interconnect chapter offers a detailed

284

assessment of the probe selection and interaction with the circuit. A summarized version is included here, but it is highly recommended to consult that chapter as well.

The probe selection for PSRR is very similar, if not identical, to the measurement of ripple and noise. The details of the selection can be found in the *Measuring Ripple and Noise* chapter.

It is best to measure the output signal using a 50Ω AC coupled coaxial cable if the circuit can tolerate this low impedance.

If not, then a 1X passive probe is often a reasonable solution if the circuit can tolerate the 100pF loading (approximately) that the probe will present. If the circuit cannot tolerate a 100pF probe, then a 10X probe can be used. This generally presents a capacitance of approximately 10pF.

The downside to using a 10X probe is that it increases the minimum measurable signal by 20dB. A low noise amplifier, such as the Picotest J2180A, can be added to recover this 20dB loss or to reduce the minimum measurable signal when using a coaxial cable or 1X probe.

In all cases, it is recommended that you verify the measurement setup by measuring a known signal first.

The validation using a spectrum analyzer or oscilloscope is also covered in the ripple and noise chapter, so it is not repeated here.

In order to validate the VNA measurement, the modulation signal going to the DUT is bypassed and the modulation signal is measured directly by the VNA.

Perform a THRU calibration to remove any irregularities and to correct for any gain or attenuation of the probes and/or preamplifiers. A setup image of the THRU calibration, with the preamplifier installed, is shown in Figure 9-9.

Note that we use a scope probe to measure the power supply modulation signal, which is much larger than the output signal.

This measurement should be made close to the DUT.

Note that the majority of cables are shielded coaxial cables and the scope probe is using a BNC tip to eliminate the ground lead.

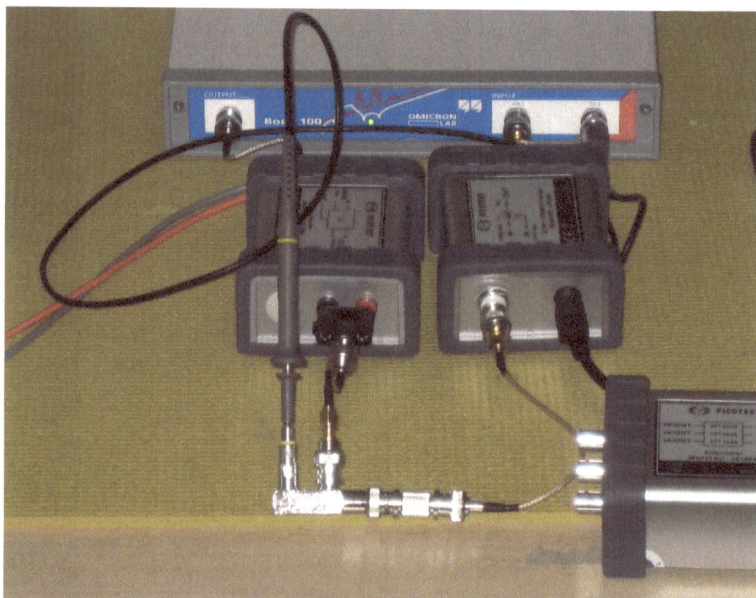

Setup of the THRU Calibration using the Picotest J2180A Preamplifier (Right) and J2120A Line Injector (Left)

Figure 9-9

The minimum measurable signal is determined by replacing the THRU connector with one or more cascaded attenuators, such as the Picotest J2140A, until the signal begins to get noisy and inaccurate.

The plot in Figure 9-10 shows the THRU calibration result at 0dB along with the measurements of 20dB, 40dB, 70dB and 80dB attenuators.

The modulation signal amplitude into the line injector is -

27dBm (10mV$_{rms}$) and the resolution bandwidth is 30Hz.

All VNA attenuators are off. Note that 70dB is the highest tolerable attenuation, since the 80dB attenuator measurement shows more than 5dB variation in the measurement.

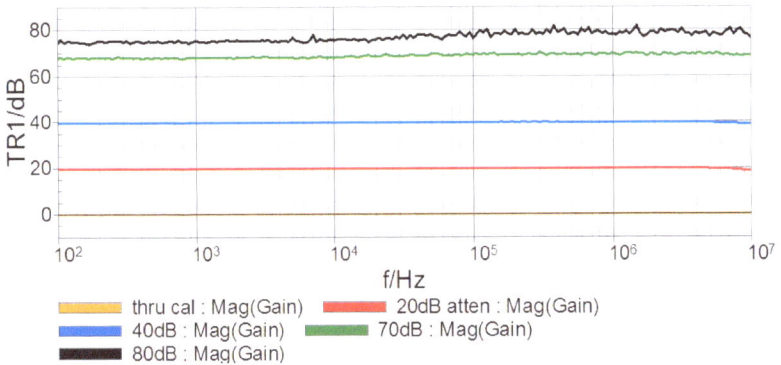

The Power Supply Modulation Signal Measured through Various Attenuators along with the 0dB THRU Calibration Result

Figure 9-10

This results in a minimum measurable signal of 3uV$_{rms.}$

The addition of the Picotest J2180A preamplifier improves this by 20dB allowing the measurement of 316nV$_{rms}$ for this setup. Once the preamplifier is installed, the THRU calibration is performed to account for the amplifier gain.

The measurement of a 90dB attenuator with the preamplifier installed is shown in Figure 9-11. The smallest measurable signal is then calculated for this setup as:

$$Vmin_{measurable} = -27dBm - 70dB \qquad 9.7$$

$$= -97dBm$$

The -97dBm can be converted to V_{rms}.

$$V_{rms} = \sqrt{e^{2.3026*dBm-2.9957}} \qquad 9.8$$

$$= \text{-97dBm} = 316nV_{rms}$$

The measurements in Figure 9-11 show the 0dB THRU calibration result and the attenuator measurement in the red trace.

Note that there is a 10dB low frequency error noted by the variation in the measured level.

Since we are measuring a fixed attenuator the signal response should be flat. This low frequency error is generally associated with a ground loop. These ground loops are discussed in detail in Chapter 7.

One simple solution is to insert a coaxial common mode transformer, such as the Picotest J2102A, in series with one of the VNA ports. This is the same method used for the impedance measurements.

Once the coaxial common mode transformer is inserted, the THRU calibration must be repeated. The result with the transformer inserted is shown in the blue trace with just a bit of noise evident at low frequency. The noise floor can be seen by removing VNA oscillator connection.

The noise floor, shown in the green trace, just touches the measurement at approximately 120Hz. The noise floor should be plotted along with the PSRR measurement to assure that the measurement is valid.

A minimum of 6dB to 10dB margin between the measurement and the noise floor is recommended.

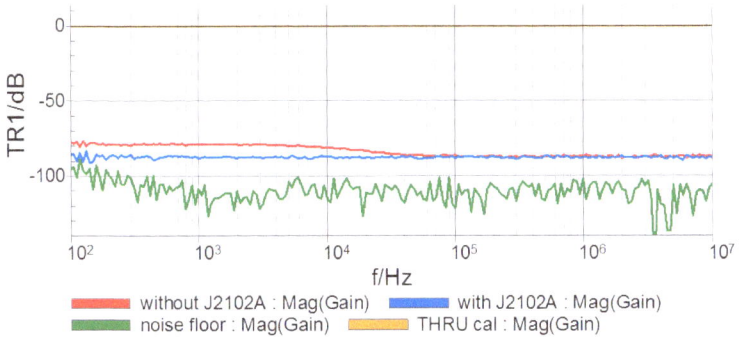

Measurement of a 90dB Attenuator with the Picotest J2180A Preamplifer Inserted

Figure 9-11

The total variation in the attenuator is just marginally acceptable with the 90dB attenuator confirming the 20dB improvement.

Tips and Tricks

1. Always perform a THRU calibration with the DUT bypassed. This removes any frequency errors associated with the interconnecting cables, probes, amplifier, and common mode transformer.

2. Validate the measurement setup. Measure the modulation signal through one or more cascaded attenuators, such as the Picotest J2140A, in order to determine the minimum measurable signal and to verify the correct result is obtained.

3. Set the equipment attenuators to the minimum level that does not overload the ADC. Measuring closer to full scale achieves the best signal-to-noise ratio.

4. Reducing the resolution bandwidth can help improve noise.

5. The J2180A preamplifier can be used to connect a high impedance scope probe to a 50Ω instrument, such as a

spectrum analyzer. It can also be used to improve the SNR and sensitivity.

6. It is possible to have low frequency ground loops in the test setup. Inserting a common mode coaxial transformer, such as the Picotest J2102A, can greatly diminish these effects. The floating inputs of the Agilent E5061B are immune from these ground loop effects. However, other instruments may not be immune to these effects.

7. When using a scope probe, keep the minimum length ground clip, or better yet, use BNC adapter as shown in Figure 9-9.

Examples

1. PSRR Measurement of a LM317 Voltage Regulator

An LM317 demo board, set for a 3.3V output is connected as shown in Figure 9-12. An external lab supply is connected to the Picotest J2120A line injector and the line injector is connected to the input pins of the voltage regulator with banana wires and grabber clips. Two oscilloscope probes are connected to the VNA to measure the modulated input signal and the resulting output signal. The oscillator output of the VNA is used to modulate the line injector.

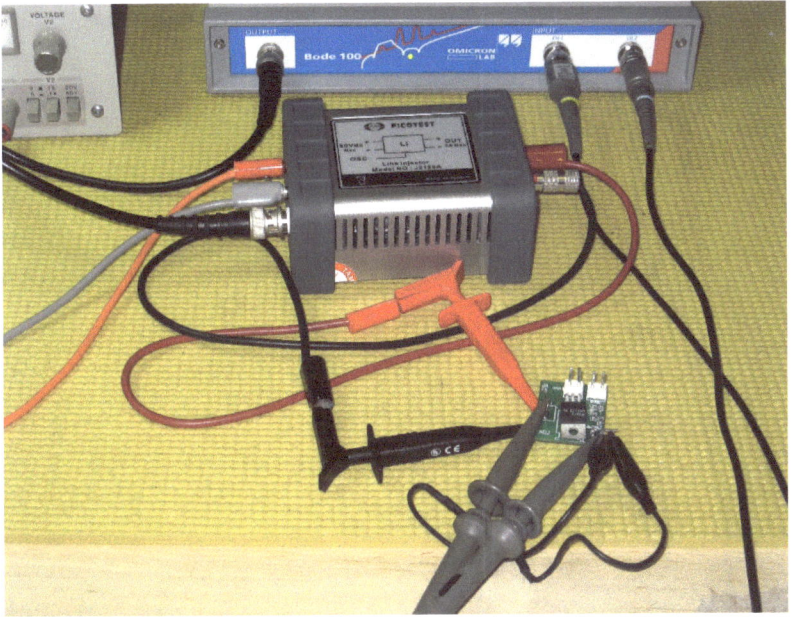

A Linear Regulator Connected to a Line Injector and a VNA for PSRR Measurement

Figure 9-12

The measurement is calibrated by connecting both probes to the highest signal level, which in this case is at the regulator input.

A THRU calibration is performed.

The grounds are connected at the ground of the smallest signal, which is the output. The output side probe is shorted to the output ground in order to measure the noise floor.

Both of these connections are both shown in Figure 9-13.

Connected for THRU calibration   Connected for noise floor

Closeup of the Voltage Regulator connected for THRU Calibration (Left) and for Noise Floor Measurement (Right)

Figure 9-13

The PSRR measurement is then performed for several different input-output differential voltages, as shown in Figure 9-14.

The lower set of traces shows the PSRR for differential input output voltages of 2V, 3V, 5V and 7.5V.

The orange traces shows the noise floor with the scope probes connected as shown in Figure 9-12.

To get the plots of Figure 9-14, use two scope probes (orange), replace the output probe with 50Ω coax cable (purple) and insert the Picotest J2102A common-mode transformer between the 50Ω coax cable and the VNA.

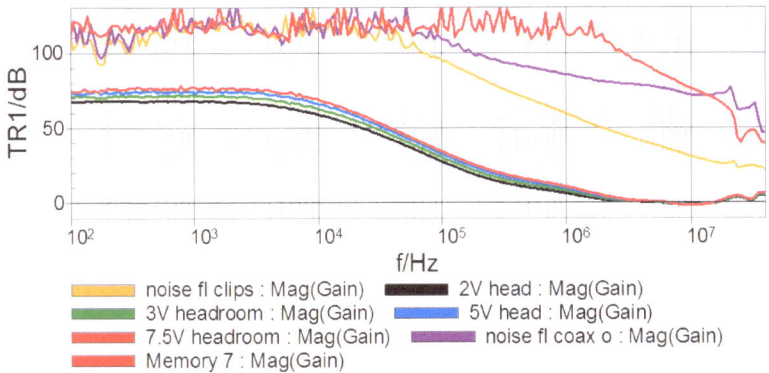

LM317 PSRR for Several Headroom voltages in Lower traces 2.5V (Black), 3V (Green), 5V (Blue) and 7.5V (Red). Noise floor in upper traces.

Figure 9-14

The noise floor can easily be improved. The purple trace in Figure 9-14 is the result of replacing the output scope probe with a 50Ω coaxial cable. The top red trace adds a Picotest J2102A coaxial common mode transformer between the 50Ω coaxial cable and the VNA.

Due to the relatively low PSRR of the LM317, even the scope probe measurement has adequate margin between the PSRR measurement and the noise floor. The injection signal level is approximately $100mV_{rms.}$

At this level it was determined that the minimum measurable signal is $3\mu V_{rms}$ corresponding with 90dB PSRR as shown in Figure 9-1.

The same test setup, using the coaxial cable and the Picotest J2102A is used to measure a much higher PSRR regulator as shown in Figure 9-15. The blue trace is the noise floor measurement and the red trace is the PSRR measurement.

The low frequency noise level is improved by increasing the amplitude of the injection signal to $1V_{rms}$. A similar noise floor improvement could be obtained by adding a Picotest J2180A, or similar preamplifier, between the 50Ω coaxial cable along with the Picotest J2102A coaxial common mode transformer.

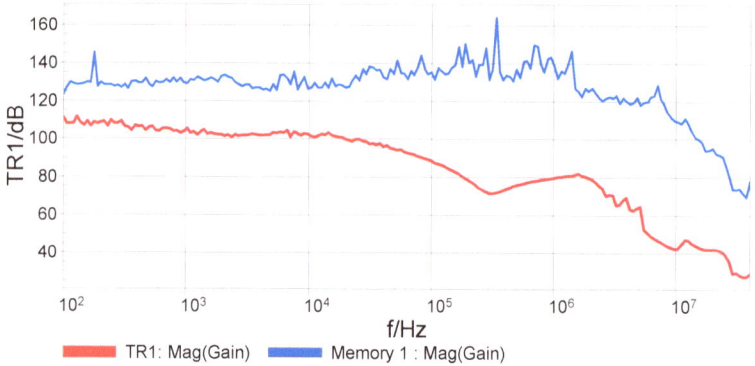

Measurement of a Very High PSRR Voltage Regulator

Figure 9-15

2. Indirect Measurement using a Line Injector for Modulation

The PSRR modulation signal can be applied to the input of a clock's power supply using either a line injector or a current injector. The setup in Figure 9-16 illustrates a Picotest J2120A line injector connected to a small surface mounted clock. The line injector modulates the power supply input while the clock output is monitored via the 50Ω coaxial cable seen in the upper right corner.

In this example, the clock is being monitored using a Tektronix RSA5106A real time spectrum analyzer.

In Figure 9-16, the clock output is connected to a Tektronix RSA5106A real time spectrum analyzer via the 50Ω coaxial cable seen exiting on the right.

A Line Injector is Connected to a 13.4MHz Clock and Modulated by a 1kHz Sine Signal from a Picotest G5100A Arbitrary Waveform Generator

Figure 9-16

The resulting modulation can be seen in two ways.

The image in Figure 9-17 shows the Tektronix RSA5106A measuring the clock spectrum, as well as the AM and FM modulation components.

The 1kHz current modulation signal is clearly seen in the spectrum sidebands, as well as the FM and AM modulation.

In Figure 9-17, note that there is both AM and FM modulation and they both reflect the 1kHz sine modulation signal.

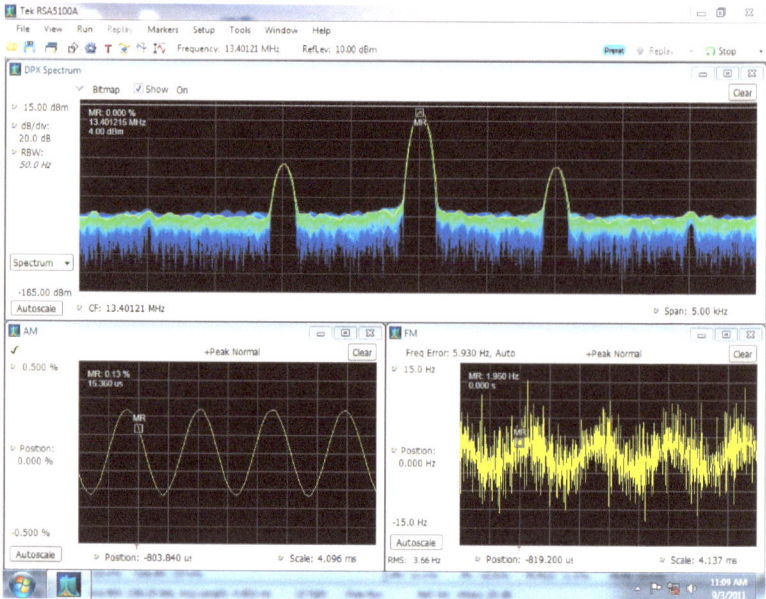

*Clock Output Spectrum Showing the Modulated Signal
as Sidebands (Top), the AM Demodulated Clock
Output (Lower Left) and the FM Demodulated Clock
Output (Lower Right)*

Figure 9-17

The measurement in Figure 9-18 shows the same measurement being displayed as phase noise.

The phase noise and associated jitter can both be easily measured as seen in the lower left corner of the display. This plot shows a spurious response at the 1kHz modulation frequency and the display reports the 1kHz spurious response to be -70.88dBc/Hz.

This measurement can be repeated at many frequencies in order to evaluate the sensitivity of the clock or to compare the performance of various voltage regulators and clocks.

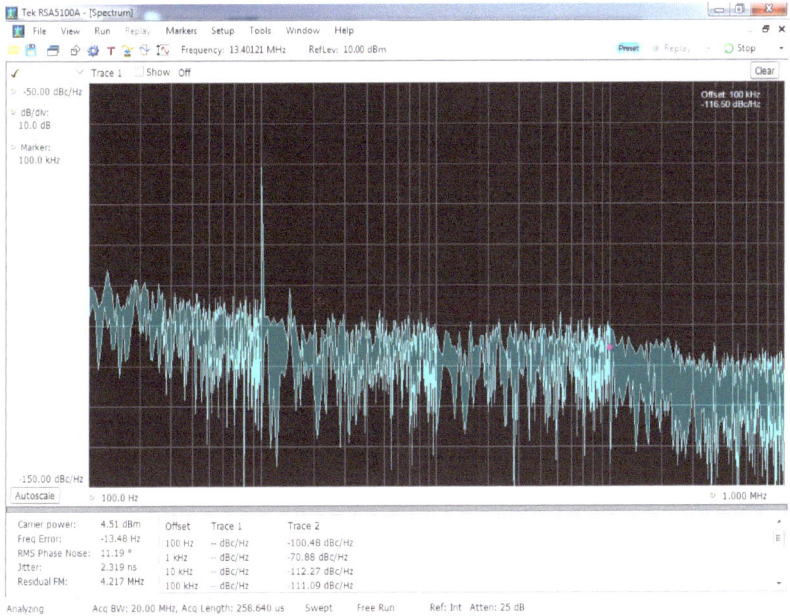

The Tektronix RSA5106A Displaying the Clock Output Phase Noise showing the 1kHz Spurious Signal resulting from the Power Supply Modulation

Figure 9-18

3. Measuring Op-Amp PSRR

The op-amp is connected in a unity gain configuration with the input grounded.

The op-amp output is connected to the VNA using a 50Ω coaxial cable terminated into a high impedance input. The op-amp's positive supply voltage is modulated using a Picotest J2120A line injector connected to the VNA oscillator output.

The setup is shown in Figure 9-19.

Op-Amp Connected for PSRR Measurement using Picotest J2120A Line Injector

Figure 9-19

The same device is measured using a Picotest J2111A current injector to modulate the op-amp's positive supply voltage as shown in Figure 9-20. The positive supply is connected to the op-amp through a 100Ω resistor.

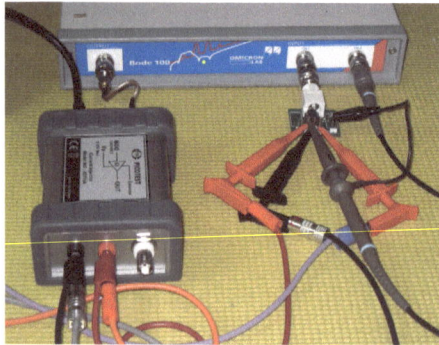

PSRR Measurement using a Picotest J2111A Current Injector to Modulate the Op-Amp Positive Supply Voltage

Figure 9-20

The measurement noise floor and the PSRR measurements for both the line injector and the current injector are shown in Figure 9-21.

In Figure 9-21, note that other than some small differences at higher frequencies, the measurements are nearly identical.

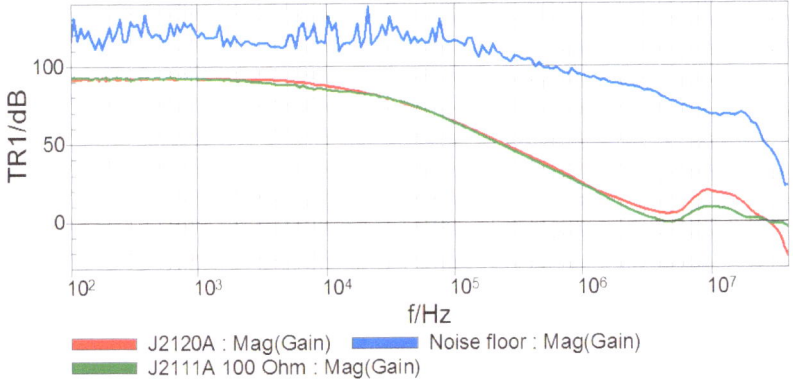

The Measurement Noise Floor (Blue) along with the PSRR using the Line Injector (Red) and using the Current Injector (Green)

Figure 9-21

Chapter References

1. S.M. Sandler and C.E. Hymowitz, *Use a Signal Analyzer to Measure Power Supply Noise,* Electronic Design, Sept 6, 2012
   http://electronicdesign.com/power/use-signal-analyzer-measure-power-supply-regulator-and-reference-noise

2. F. Hämmerle and S.M. Sandler, *Power Supply Rejection Ratio Measurement Using the Bode 100 and the Picotest J2120A Line Injector,* Omicron-Lab,
   http://www.omicron-lab.com/bode-100/application-notes-know-how/application-notes/psrr-measurement.html

3. *Measuring Opamp PSRR,* Picotest, Jan 31, 2012
   https://www.picotest.com/blog/?p=839

4. *Power Supply Measurement Techniques Instructional Videos: PSRR Measurement using the J2120A Line Injector* Picotest
   https://www.picotest.com/blog/?p=432

5. *Voltage Regulator Test Standard Manual 1.0d,* Picotest
   https://www.picotest.com/downloads/VRTS01/Voltage%20Regulator%20Test%20Standard%20Manual%201.0d.pdf

Chapter 10

Reverse Transfer and Crosstalk

REVERSE TRANSFER IS a measure of the change in input current signal resulting from a varying output current signal as noted in Equation 10.1. This measurement cannot always be made within in a system as it requires access to both the output current and the input current of the power supply. The output current can easily be modulated using a current injector, or electronic load, however, the input connection is not always accessible for a current probe.

An alternative is to measure the input voltage resulting from the modulation of the output current, which is the product of the reverse transfer and the impedance at the regulator input. The input voltage that results from the output current modulation then feeds through other regulators and circuits that are connected to the same input.

These paths were discussed in Chapter 6, but are also included in Figure 10-1 for convenience. In Figure 10-1, the regulator input current is transformed to a voltage from the finite input impedance. This voltage than passes through the second regulator as PSRR, shown in red.

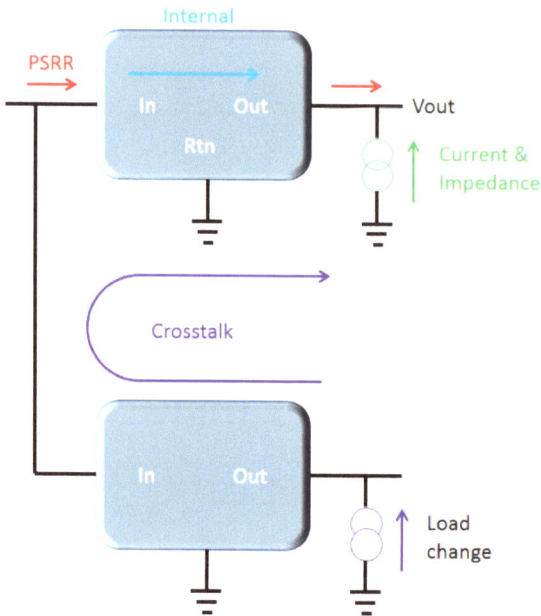

The Modulated Output Current, Shown in Purple, appears at the Regulator Input due to Reverse Transfer

Figure 10-1

$$Reverse = \frac{\Delta Iin}{\Delta Iout} \qquad 10.1$$

Reverse Transfer of Various Topologies

Series Linear Regulators

A simplified series linear regulator schematic is shown in Figure 10-2. The output capacitor is not included for clarity. In this figure, we can see that the load current flows through the emitter of the NPN pass transistor.

Due to the current gain of the transistor, the collector current, which is the regulator's input current, is approximately the same as the output current.

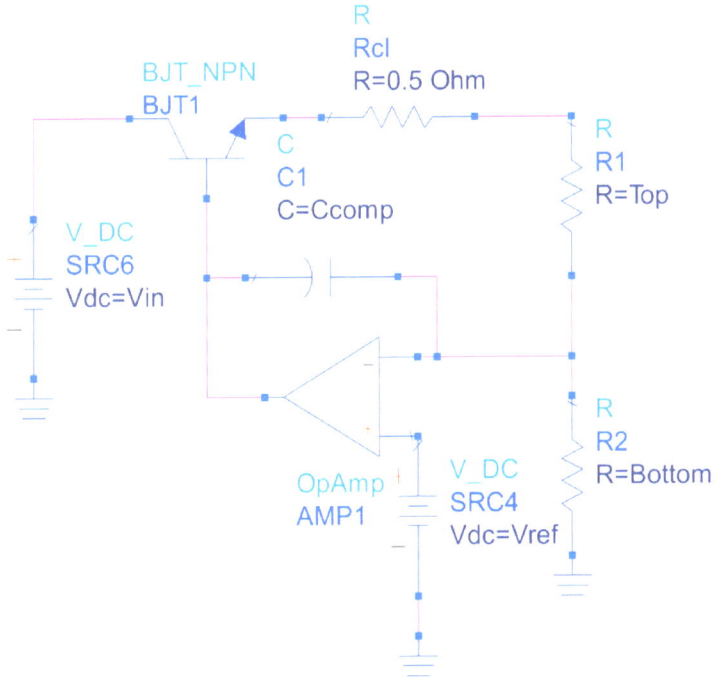

Simplified Schematic of a Series Linear Regulator

Figure 10-2

Shunt Regulators

The shunt regulator has very little reverse transfer, since the output impedance of the regulator is very low compared to the bias or source impedance which is generally very high. In many cases, the bias current is provided by a JFET current regulator diode or a constant current circuit, significantly increasing the impedance of the bias current source.

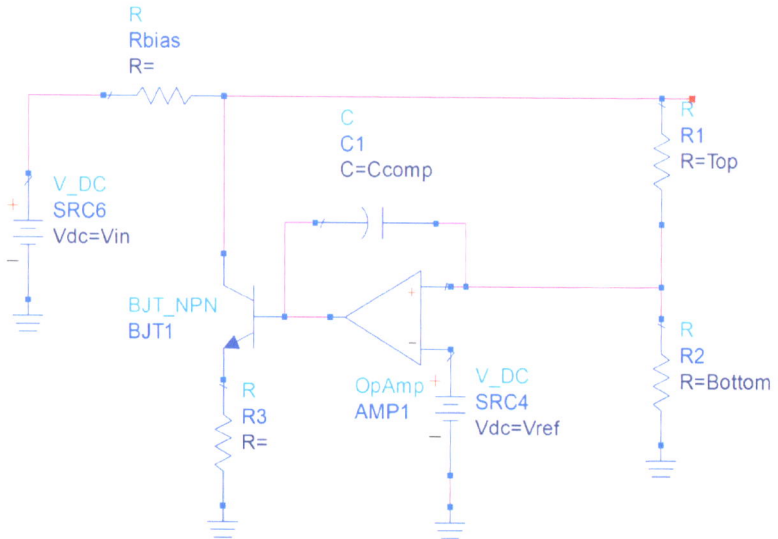

Simplified Schematic of a Shunt Linear Regulator

Figure 10-3

The low frequency shunt regulator's reverse transfer is determined by the resistance of the shunt element and the resistance of the bias current source. Assuming the shunt resistance is much lower than the bias current resistance, the approximate low frequency reverse transfer is:

$$Reverse = \frac{R_{shunt}}{R_{bias}}$$
10.2

The value of R_{bias} is either the value of the resistor used or the resistance of the JFET current regulator diode.

POL Regulators

In the case of a switching type regulator, the reverse transfer is not constant, nor is it necessarily 0dB at low frequency. The input power can be determined from the output power and efficiency, η.

$$P_{in} = \frac{P_{out}}{\eta} \qquad 10.3$$

The input current is then determined by replacing the input and output power with their respective voltage and current products.

$$I_{in} = \frac{V_{out}}{V_{in} \cdot \eta} \cdot I_{out} \qquad 10.4$$

Applying the relationship in 10.1 to the low frequency reverse transfer of a switching type regulator results in:

$$Reverse = \frac{V_{out}}{V_{in} \cdot \eta} \qquad 10.5$$

In the case of a switching regulator, the reverse transfer is related to the ratio of the output and input voltages.

Operational Amplifiers

A typical op-amp is constructed as a class AB amplifier as shown in Figure 10-4.

Positive or sourced output current variations are passed through the NPN transistor from Vcc in much the same way as in the series linear voltage regulator, while negative or sinked current is conducted through the PNP transistor to Vee.

In Figure 10-4, output current flows through the NPN transistor from Vcc and through the PNP transistor to Vcc.

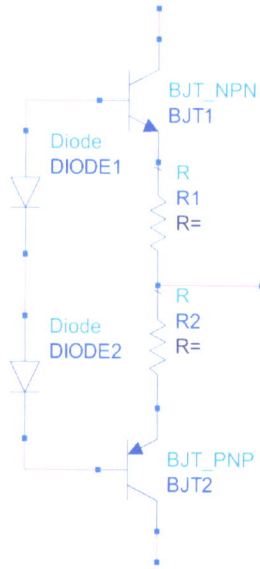

Typical Class AB Output Stage of an Opamp

Figure 10-4

Modulating the Output Current

The two most convenient methods of modulating the DUT load current are via a current injector or a direct connection of the VNA oscillator to the DUT output through a DC blocker.

Some electronic loads can also accept a modulation input, but be careful since many loads present a low impedance output as discussed in Chapter 7.

Current Injector

The Picotest J2111A current injector can be easily controlled by time (AWG) or frequency (VNA) domain sources. This allows the output current to be accurately modulated for the reverse transfer measurement.

The Picotest J2111A current injector needs to be biased using either the bias option switch or by externally biasing the current injector.

This bias puts the current injector into a class A operating mode allowing clean modulation of the regulator's output current by the VNA.

A benefit of using the current injector is that the injector includes a precision wideband current monitor that can be connected directly to one 50Ω VNA input port.

DC Bias Injector

Another method of modulating the load current in the frequency domain is to connect the VNA oscillator to the DUT output using a DC blocker, such as the Picotest J2130A Bias injector.

This DC blocker protects the input of the VNA from excessive DC input voltages and also keeps the 50 Ohm VNA port from sinking additional DC load current from the DUT. This method requires the use of a second current probe to measure the output current.

Measuring the Input Current

The input current signal is best measured using a clamp on current probe. Since reverse transfer is an AC measurement, a passive current probe can be used to measure the input current signal.

Most passive current probes are intended to connect to either a 50Ω or 1MΩ input. The correct terminating impedance for your probe can be found in the user manual. The input port impedance of the VNA should be set accordingly.

In some cases, including the Picotest VRTS01 demo kit, an input current sense resistor in the input return connection allows the current to be directly measured without the use of a current probe. A simplified schematic of this connection is shown in Figure 10-5.

In this circuit, the 2Ω sense resistor, R1 is used to measure the input current. The 50Ω matching resistor, R4 properly terminates the signal to a 50Ω cable and VNA port.

The attenuation resulting from R4, and the 50Ω port, results in a wideband signal level of $1V/A$.

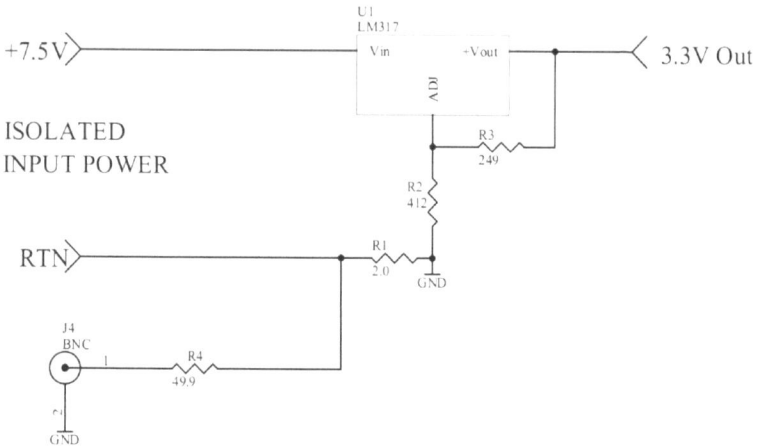

The VRTS01 Demo Board 50Ω Input Current Port

Figure 10-5

Calibrating the Measurement

VRTS01 Input Current Port

The voltage regulator is replaced by a jumper between the regulator input and output pins, thereby connecting the input voltage directly to the output. The input current and the modulated injector current are then equal and performing a thru calibration corrects the scaling, as well as correcting most frequency related measurement artifacts.

The voltage measurement should either be AC coupled to a 50Ω port or the port should be set to a high impedance so that the voltage does not exceed the ratings of the port. It is most important, whatever method is used, that the voltage remains constant for both the calibration and the subsequent measurement. The photo in Figure 10-6 shows the voltage regulator replaced with a jumper circuit board for calibration. After the thru calibration is performed, the jumper board is replaced with the voltage regulator seen in the foreground.

In Figure 10-6, after calibration, the linear regulator seen in the foreground replaces the jumper circuit board.

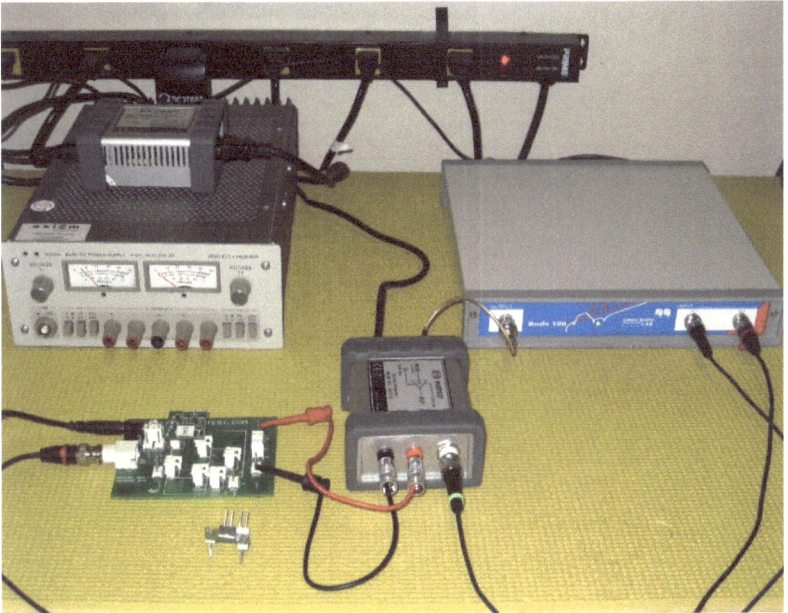

The VRTS01 Demo Kit with the Calibration Jumper Board installed for Thru Calibration

Figure 10-6

Current Probe

If a current probe is used, the current probe should be clamped onto the current injector lead used to modulate the output current as shown in Figure 10-7.

In Figure 10-7, the current probe is clamped around the current injector lead (bottom center) so that the monitor port and current probe are measuring the same signal.

A thru calibration is then performed.

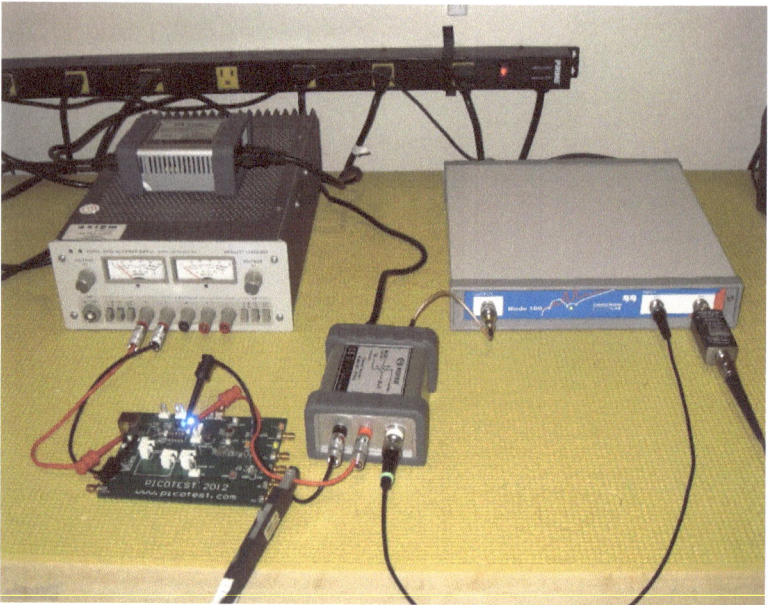

Reverse Measurement Setup for Calibration

Figure 10-7

This setup results in the current injector monitor port and the current probe measuring the same signal.

Performing a THRU calibration of the VNA in this state corrects for the low frequency performance limitations of the current probe and also assures that the measurement scaling is independent of the current probe scaling factor.

With the thru calibration completed, the current probe is moved to the input wire and the reverse measurement is performed as shown in Figure 10-8.

The pre-calibration, post-calibration and reverse measurement responses are shown in Figure 10-9.

With the Calibration Completed, the Current Probe is Moved to the Input Current to Perform the Measurement

Figure 10-8

In Figure 10-9, note that despite the 20kHz low frequency bandwidth of the current probe, the calibration has corrected the frequency response.

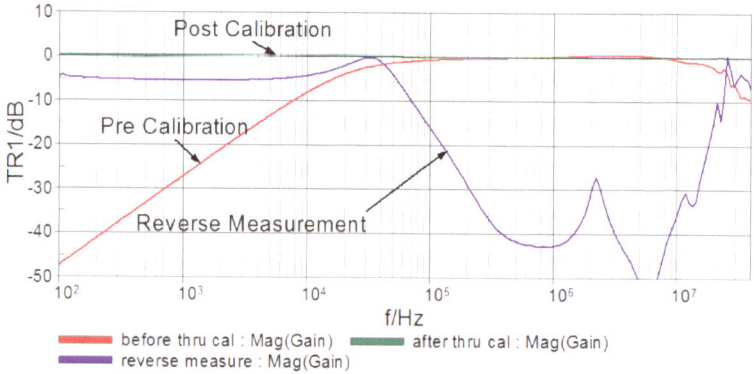

Pre- and Post-Calibration Plots along with the Reverse Transfer Measurement

Figure 10-9

Measuring the Input Voltage

Since the magnitude of the impedance at the regulator input is finite and not zero, the reverse transfer current transforms to a voltage at the regulator's input.

It is this transformed voltage that permeates the system through the PSRR paths connected to this input as crosstalk.

With this in mind, another option is to measure the input voltage resulting from the load current modulation, or transimpedance, rather than measuring the ratio of the input current to output current.

$$Transimpedance = \frac{V_{in}}{Iout} = \frac{I_{in}}{Iout} \cdot Z_{in} \qquad 10.6$$

The transimpedance reflects the product of the reverse transfer and the finite input source impedance, Z_{in}.

Calibrating the Measurement

A Tee adapter is inserted into the current monitor port of the VNA. A scope probe is connected to the other VNA input port and the tip is inserted into the Tee so that both channels are measuring the same signal and a thru calibration is performed. This calibration setup is shown in Figure 10-10.

In Figure 10-10, the scope probe and the current injector monitor port are connected via a Tee adapter. A thru calibration is performed.

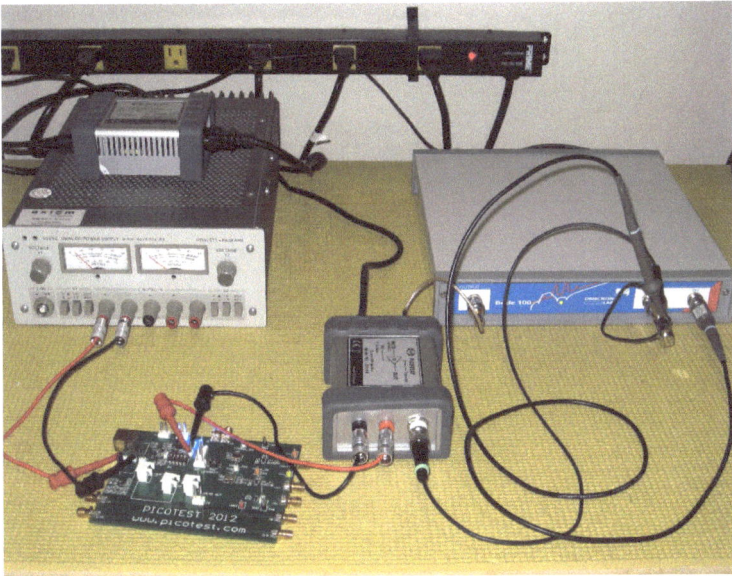

Setup for Calibrating the Voltage Measurement

Figure 10-10

With the calibration completed the frequency effects of the voltage probe and the scaling are corrected. The voltage probe is then moved to the input voltage for the transimpedance measurement as shown in Figure 10-11.

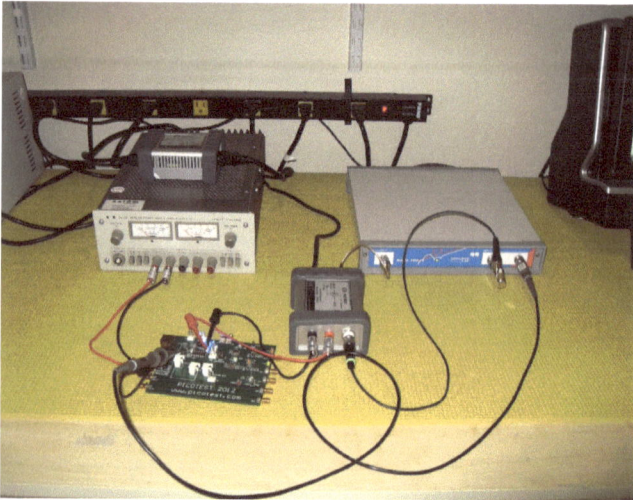

After the Calibration, the Scope Probe is Connected to the Input Voltage to Measure the Transimpedance

Figure 10-11

The resulting transimpedance measurement, and the thru calibration plots, are shown in Figure 10-12.

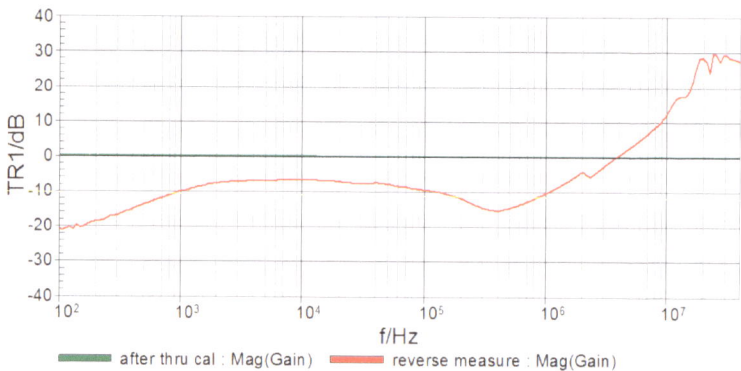

Transimpedance Measurement of Figure 10-11

Figure 10-12

Indirect Measurement

It is also possible to measure the effects of reverse transfer using indirect methods, similar to the methods used to measure ripple and noise.

In this scenario, the load current of the regulator is modulated while the phase noise of a clock connected at the regulator's input is monitored.

Tips and Tricks

1. At small signal levels, it is possible to have DC ground loops in the current injector and VNA connections. As with other measurements, the addition of a coaxial common mode transformer, such as the Picotest J2101A, can greatly improve the measurement. The Agilent E5061B uses semi-floating input ports eliminating the possibility of such ground loops for the low frequency measurement set.

2. The thru calibration corrects for the attenuation of the current probe signal below its -3dB frequency. The SNR is degraded as a result of the attenuated signal at very low frequencies, but is generally sufficient.

Examples

1. LM317 Linear Regulator

An example of the series linear regulator confirming the 0dB reverse transfer is shown in Figure 10-13.

This measurement is performed using an LM317 regulator, operating at a 25mA load with and without an output capacitor installed.

In Figure 10-13, the regulator is loaded with 25mA from a Picotest J2111A current injector.

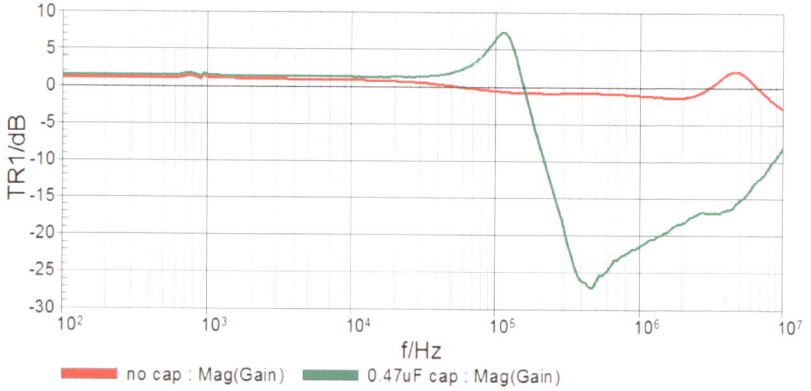

LM317 Reverse Transfer with and without a 0.47µF Ceramic Output Capacitor

Figure 10-13

The addition of the output capacitor reduces the bandwidth, and in this case, degrades the stability of the voltage regulator resulting in a significant gain peak at 100kHz.

Above the bandwidth, the capacitor improves the reverse transfer.

2. Shunt Linear Voltage Regulator

The reverse transfer measurement of a linear shunt regulator is shown in Figure 10-14—confirming the significantly improved (lower gain) performance for this characteristic.

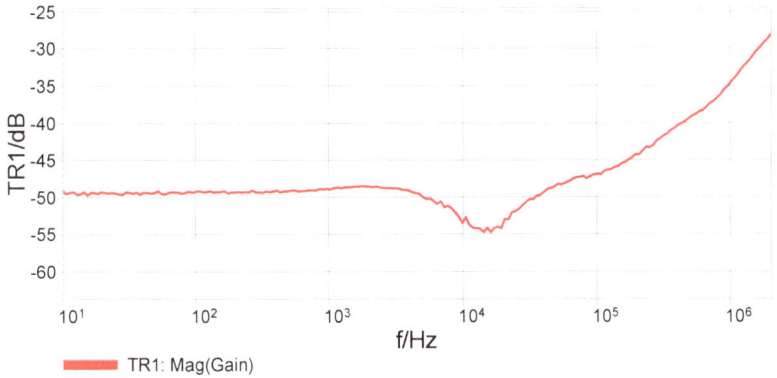

*Reverse Transfer of a Linear Shunt Regulator showing
the Significantly Improved Reverse Transfer
Performance*

Figure 10-14

A step load is applied to both an LM317 linear regulator and a linear shunt regulator, while the step load current and input current are monitored on an oscilloscope.

The LM317 measurement is shown in Figure 10-15, while the shunt regulator measurement is shown in Figure 10-16.

This once again confirms the superior reverse transfer performance afforded by the shunt regulator topology.

The input current ringing in Figure 10-15 is indicative of poor phase margin.

In Figure 10-15, the step load current (blue) and resulting input current (purple) are both shown.

In Figure 10-16, the step load current (blue) and resulting input current (purple) are both shown.

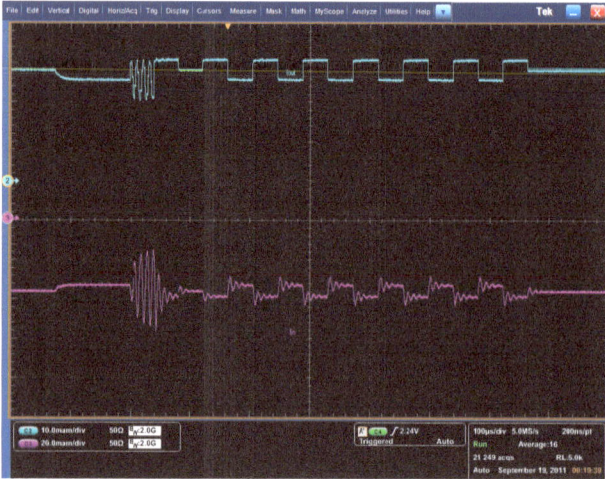

Step Load Response of LM317 Linear Regulator with a Ceramic Output Capacitor

Figure 10-15

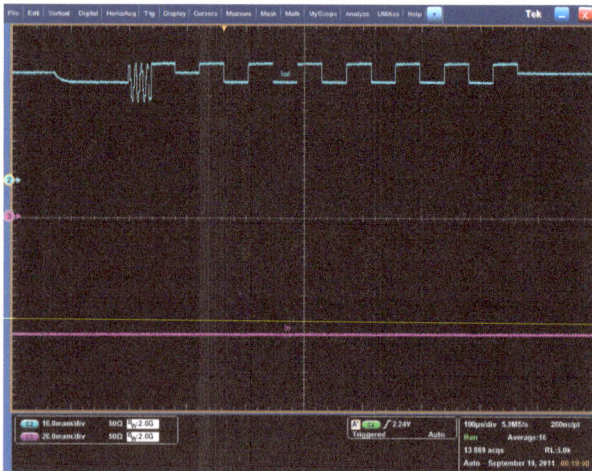

Step Load Response of a Linear Shunt Regulator with the same Ceramic Output Capacitor

Figure 10-16

3. POL Switching Regulator

The input voltage dependency, derived in 10.5, is illustrated in Figure 10-17. The reverse transfer measurement of a 5V input 3.3V output POL included on the Picotest VRTS2 demo board is measured at three different input voltages. For this reason, it is best to measure over a range of operating conditions, or with the circuit's actual loading.

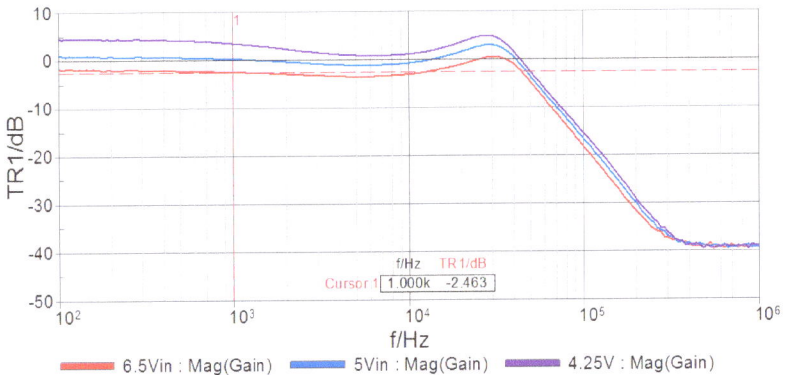

PSRR Measurements of a POL Regulator at Three Input Voltages confirming the relationship in Equation 10.5

Figure 10-17

4. Indirect Measurement

The POL used in example three, and shown in Figure 10-10, is used to demonstrate the indirect reverse measurement.

The 5V input to the POL is connected to a 10MHz clock, as well as to the 3.3V output POL regulator, as shown in Figure 10-18. The current is modulated at the 3.3V output, while the clock phase noise is measured as shown in Figure 10-22.

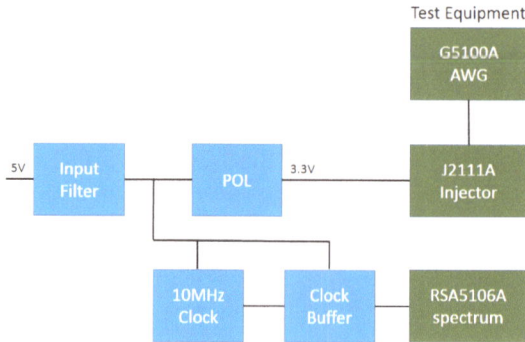

Block Diagram showing the 5V Input, Filter and 3.3V POL Regulator being modulated, the Clock and the Test Equipment Connections

Figure 10-19

The setup is shown pictorially in Figure 10-20. The input filter, POL, clock and clock buffer are all on the demonstration PCB.

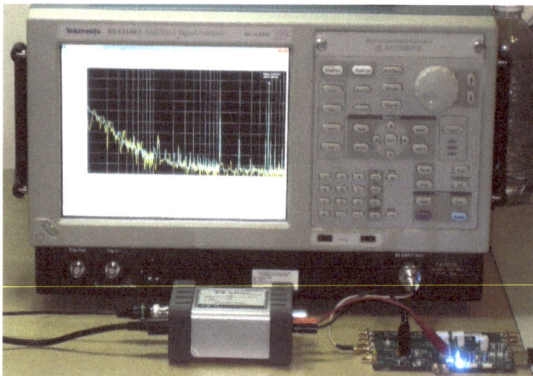

The Demo Board shown Connected to the Picotest J2111A Current Injector and Tektronix RSA5106A Analyzer with Phase Noise Capability

Figure 10-20

The baseline phase noise is shown in Figure 10-20 without the modulated J2111A . The baseline measurement allows us to determine the phase noise that is from sources other than the reverse transfer modulation we have inserted. In this case, the baseline jitter is 2.07ps and the signal level at 5.5kHz is -129.4dBc.

In Figure 10-21, note the spurious signals that appear in both the modulated and unmodulated measurements, such as at 120Hz and 1kHz.

These noise signals are from the power supply, clock and other circuitry operating on the PCB. The signal at 5.5kHz offset is -129.4dB and the total jitter is 2.07ps.

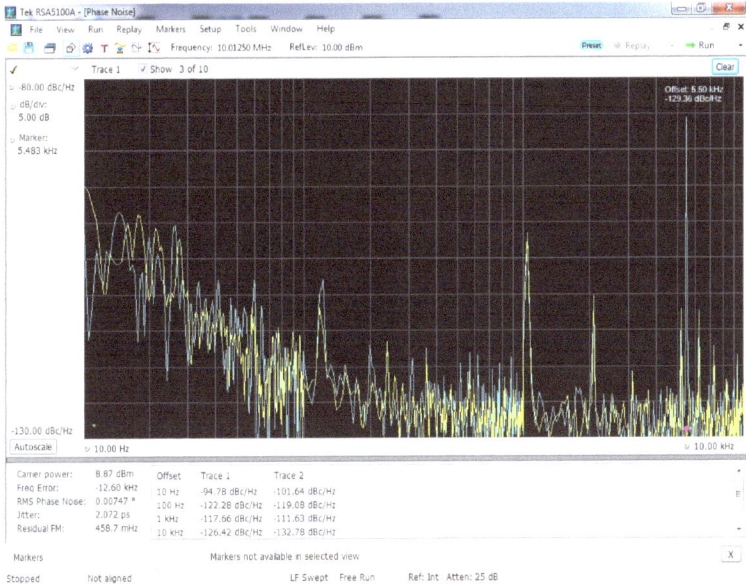

Baseline Phase Noise Measurement without Modulation (Yellow Trace) and with Modulation (Blue Trace)

Figure 10-21

The J2111A is then used to modulate the 3.3V POL regulator with a 50mApp sine wave at 5.5kHz. The result is shown in Figure 10-22 indicating a jitter of 8.1ps and a 5.5kHz level of -85.7dBc. The modulation at 5.5kHz significantly increased the jitter. The sensitivity of the clock is different at different frequencies, and so this measurement is generally repeated at many different modulation frequencies.

In Figure 10-22, the table in the modulated case shows the signal at 5.5kHz offset is -85.7dB and the total jitter is increased significantly to 8.1ps.

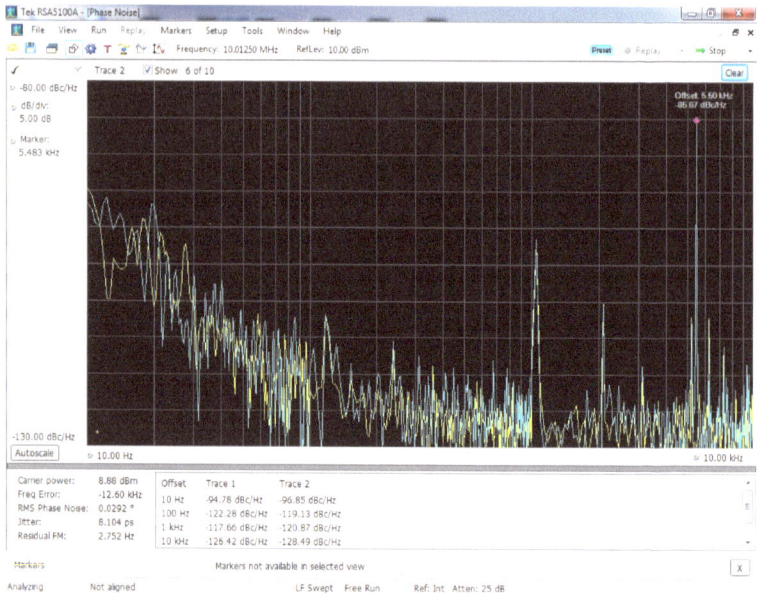

Phase Noise Measurement without Modulation (Yellow Trace) and with Modulation (Blue Trace)

Figure 10-22

Chapter References

1. F. Hämmerle & S.M. Sandler, *Reverse Transfer Measurement Using the Bode 100 and the Picotest J2111A Current Injector*, 2011,
www.picotest.com/articles/App Note Reverse Transfer Final.pdf

2. F. Hämmerle & S.M. Sandler, *Power Supply Crosstalk Measurement Using the Bode 100 and the Picotest J2111A Current Injector*, 2011,
http://www.omicron-lab.com/fileadmin/assets/application_notes/App_Note_Crosstalk_V1_0.pdf

3. F. Hämmerle & S.M. Sandler, *Power Supply Reverse Transfer Measurement Using the Bode 100 and the Picotest J2111A Current Injector*, 2011,
http://www.omicron-lab.com/fileadmin/assets/application_notes/App_Note_Reverse_Transfer_V1_0.pdf

4. *Voltage Regulator Test Standard for Agilent E5061B*, Ver. 1.0a, 2012,
https://www.picotest.com/articles/Agilent-App-Notes/Voltage%20Regulator%20Test%20Standard%20Manual%201.0d%20Agilent.pdf

5. *Voltage Regulator Test Standard for OMICRON Lab Bode 100*, Ver. 1.0d 2010,
https://www.picotest.com/downloads/VRTS01/Voltage%20Regulator%20Test%20Standard%20Manual%201.0d.pdf

Steven M. Sandler

Chapter 11

Measuring Step Load Response

THERE ARE SEVERAL reasons for step load testing, only one of which is to assess the stability of control loop.

Unlike a Bode plot, the assessment of stability via step load testing does not produce a quantitative phase margin result.

However, there is a direct relationship between the ringing seen in a step load response and the phase margin of the control loop.

A second reason for step load testing is to determine the power supply voltage excursions.

For example, a critical requirement for high-speed devices, such as FPGAs and CPUs, is that the voltage at the device must remain within the manufacturer's allowable operating range.

Additionally, a step load could also be used to assess the Q of passive filters, such as a ferrite bead and its associated decoupling capacitors, to assess the non-linearity of a circuit with respect to operating load current, to gain insight into the large signal behavior of a circuit or to determine the susceptibility of other circuits that are connected to the circuit under test.

Generating the Transient

In order to perform a step load response test, the step load must be created and applied to the DUT. Then the resulting voltage response, or other disturbance, must be monitored.

Current Injector vs. Electronic Load

Electronic loads regulate load current (or resistance), but generally do not allow control of the shape of the stimulus. Some can accept external modulation signals, allowing more complex load current profiles.

Current injectors are a class of instrument closely related to electronic loads though with some unique differences. They are small signal transconductance amplifiers (voltage to current converters) with much higher impedance, higher bandwidth and faster response times than traditional electronic loads. Current injectors are generally low current devices, therefore, they are mostly used for small signal testing, such as stability evaluation or testing of lower power devices including op-amps, voltage references and linear and switching regulators. The current injector accepts a modulation signal from the oscillator of an FRA/VNA, a function generator, pulse generator, or arbitrary wave generator allowing extensive control of the stimulus shape in both the time and frequency domains.

One of the more important considerations is that in order for the test equipment not to influence the performance of the DUT, the impedance of the load should ideally be at least an order of magnitude greater than that of the DUT at all frequencies of interest. This ratio is conservative; assuming that both the DUT and the load are resistive. In such a case the DUT sees the two impedances in parallel or a 9.1% reduction in loading impedance. The measured impedance presented by the Picotest J2111A and J2112A current injectors and a B&K Precision model 8540 electronic load are shown in Figure 11-1.

In Figure 11-1, the measurements are at 4V as noted. The open circuit is also included in orange.

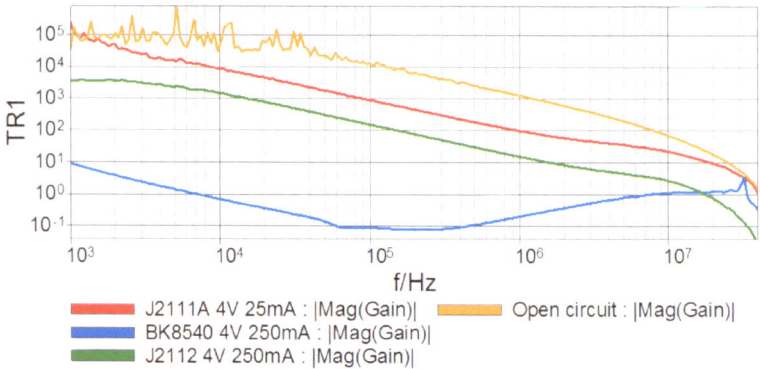

▬ J2111A 4V 25mA : |Mag(Gain)|	▬ Open circuit : |Mag(Gain)|
▬ BK8540 4V 250mA : |Mag(Gain)|	
▬ J2112 4V 250mA : |Mag(Gain)|	

Measured Impedance of the Picotest J2111A (Red), J2112A (Green) and B&K Model 8540 (Blue)

Figure 11-1

The J2111A is represented by a capacitance of approximately 1500pF while the J2112A is represented by a capacitance of 0.015µF and the B&K model 8540 is represented by 2.2µF. The substantially greater capacitance of the B&K load is a property of most electronic loads and this capacitance can easily resonate with the inductance of the interconnects.

While the impedance of the load may not be an issue for higher power VRMs and POLs, it is frequently an issue for lower power devices.

The 2.2µF capacitance of the B&K model 8540 electronic load can easily be much larger than the capacitance of many voltage references and lower power LDOs. The output capacitance of electronic loads is not generally specified, therefore, it is good practice to measure the impedance of the load to be determine whether it will influence the measurement.

When step load testing voltage references and op-amps, it is often necessary to generate current steps as small as 100uAmps. The current injector can provide this resolution, while electronic loads generally cannot.

The current injector is externally modulated. By connecting it to an AWG, almost any waveform or load current pattern can be generated. Several characteristics allow definition of the dynamic load current including the step amplitude, slew rate, repetition rate and the waveform or pattern.

Slew Rate

There are three limits of the slew rate. One is related to the interconnecting inductance between the electronic load and the DUT. This limit, neglecting saturation voltage or resistance, is calculated as:

$$Slew\ rate\ interconnect\ inductance = \frac{V_{DC}}{L} \qquad 11.1$$

This results in a rise time of:

$$T_r\ interconnect\ inductance = \frac{L \cdot I_{step}}{V_{DC}} \qquad 11.2$$

Most low power electronic loads use banana jacks for interconnecting to the DUT, while larger electronic loads use either studs or bus bars. Unfortunately, neither of these support low inductance connections to the DUT. Solving Equation 11.2 for inductance, the maximum interconnecting inductance for a given voltage and rise time is:

$$Interconnect\ inductance \leq \frac{V_{DC} \cdot T_r}{I_{step}} \qquad 11.3$$

Including the saturation voltage of the current injector or electronic load:

$$Interconnect\ inductance \leq \frac{(V_{DC} - V_{sat}) \cdot T_r}{I_{step}} \qquad 11.4$$

Assuming a desired 10ns rise time and 600mV saturation voltage, the maximum interconnecting inductance vs. DC voltage is shown in Figure 11-2 for several values of pulsed current.

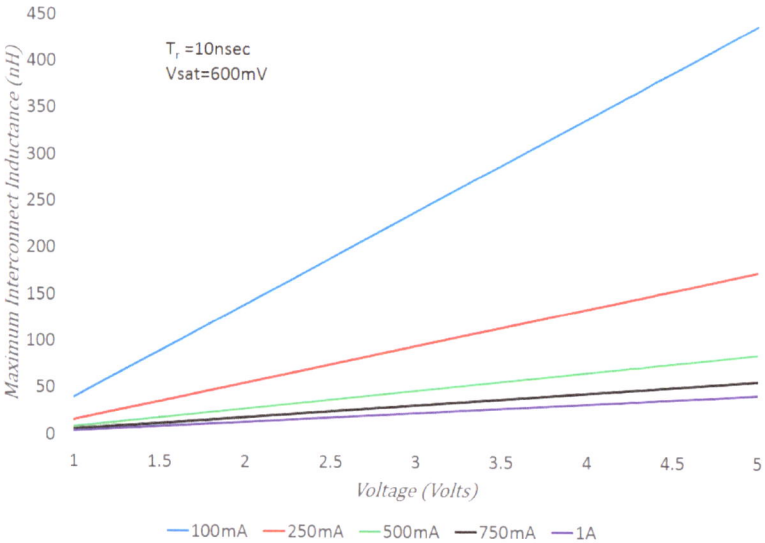

Maximum Interconnecting Inductance as a Function of V_{DC} for Several Values of Pulsed Current, assuming a 10nsec Rise Time and 600mV Saturation Voltage

Figure 11-2

In order to generate a 500mA pulse on the output of a 1.8V regulator with a 10ns rise time the maximum interconnecting inductance is 24nH, while a 100mA pulse for a 5V regulator allows 440nH. A short list of typical interconnect inductances are provided in Table 11.1.

Cable	Approximate Inductance
RG58	7nH/inch
Temp-Flex 10CX-04	1.5nH/inch
Twisted pair	17nH/inch
Banana lead pair	60nH/inch

Interconnect Inductance for Some Common Cables

Table 11.1

It is best to keep the connections as short as possible and in many cases low inductance cable, such as the Temp-Flex 10CX-04 in Table 11.1 is appropriate. In some cases it is possible to use SMA or BNC adapters to make very short connections between the current injector and the DUT.

Another slew rate limitation is set by the control loop bandwidth of the electronic load. Assuming a stable control loop in the electronic load, the rise time limit due to the bandwidth is:

$$Slew\ rate\ control\ loop = \frac{0.35}{BW} \qquad 11.5$$

This relationship is valid for assuming a single pole control loop and is derived in Chapter 13.

Since typical electronic loads have a bandwidth below 500kHz, the rise time of a typical electronic load ranges from a few hundred nanoseconds up to several microseconds. The bandwidth of the Picotest J2112A (up to 1A) current injector is approximately 40MHz with a rise time as fast as 6nS. The J2111A (up to 100mA) has a small signal bandwidth of approximately 15MHz for a rise time as fast as 25ns.

A third slew rate limit is the rise time of the AWG used to modulate the current injector, or electronic load, if externally modulated.

Current Modulation Waveform

If the repetition rate of the current step is much lower than the resonant frequencies of the regulator control loop and PDN resonances, the response of the circuit is the "naturally" decaying response. The natural response of a control loop with poor stability is a decaying ring, while for more stable loops it is generally an exponentially decaying waveform.

If the step repetition rate is equal to one or more of the resonant frequencies of the control regulator (bandwidth) and PDN, the result is the "forced" response, which can be much larger in amplitude than the natural response.

Using an Arbitrary Waveform Generator (AWG) to modulate the current injector, or electronic load, allows more effective waveforms to be applied. The image in Figure 11-3 shows both the natural and the forced response as a result of a pulse burst that occurs at the resonance frequency of a sample regulator attached to a PDN. The burst is created using an AWG to modulate a Picotest J2111A current injector.

An interesting observation is that the natural response differs slightly for each of the two current levels. The resonance is often a sensitive function of operating current, which is one reason to use the actual system impedance as the load.

In Figure 11-3, the burst shows the natural and forced responses to the same load current changes.

The yellow top trace is the output voltage, the middle blue trace is the input current and the lower green trace is the current signal.

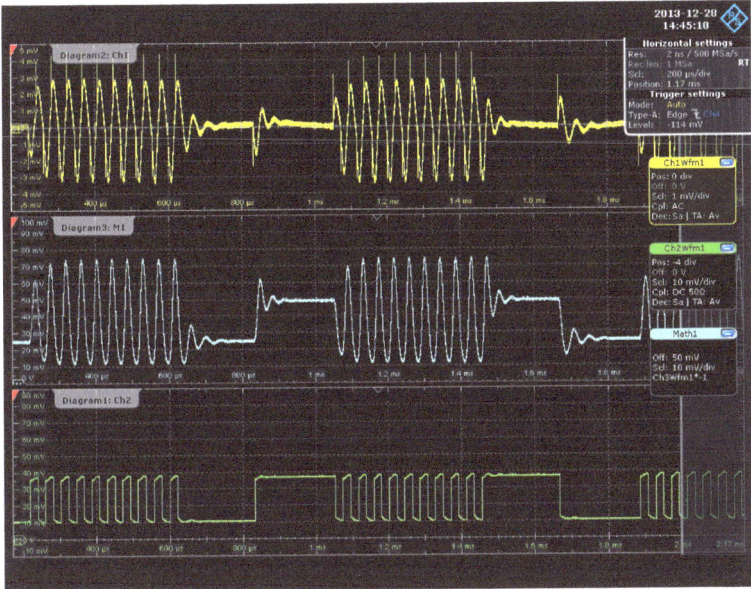

The Picotest J2111A Current Injector is Connected to a Picotest G5100A Arbitrary Waveform Generator to Produce a Burst-Current Profile

Figure 11-3

It is possible, if not likely, for a system to have multiple resonances. These resonances can be simultaneously excited by creating a current profile as seen in Figure 11-4.

In this case, the Agilent ADS simulator is used to determine the characteristics of the current profile that maximizes the output voltage excursion. This forced response is the result of a unique FPGA or ASIC data pattern, making it difficult to perform in a lab environment.

The transient voltage resulting from such an event can easily exceed the allowable operating voltage range of the connected integrated circuits, resulting in system malfunction or even permanent damage. This type of data dependent anomaly can be very difficult to troubleshoot or to artificially produce. It is best to make sure that the PDN is well-behaved, through impedance testing and simulation, before committing the design to production. if all of the resonances are simultaneously excited, as opposed to the excursion caused by a single resonance.

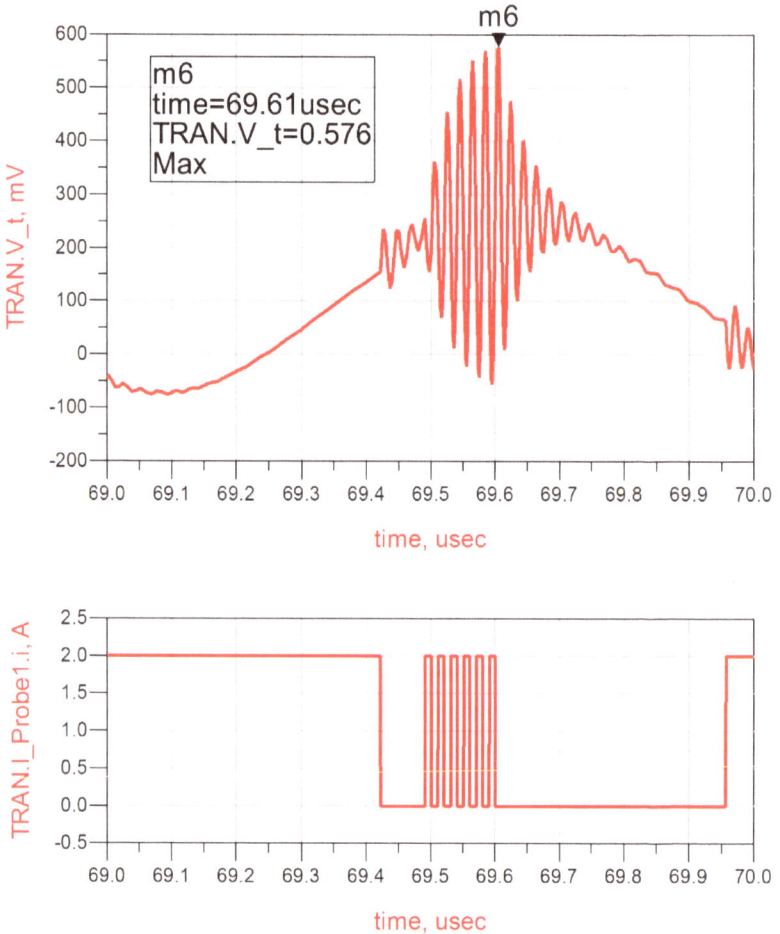

Much Larger Transient Excursions can Result in a System with Multiple Resonances

Figure 11-4

Measuring the Response(s)

The current wave shape, or profile, should be recorded along with the output voltage. The scaling should be selected such that the current signal rise time can be seen, as well as the current signal amplitude and waveform.

The technique for monitoring the voltage response is very similar to measuring ripple and noise, discussed in Chapter 12.

Large Signal vs. Small Signal

While step load testing is often performed to assess small signal stability, it is also possible for a voltage regulator to have nonlinear large signal responses necessitating the need for large signal step load testing. The measurement in Figure 11-5 shows a large signal effect on a regulator's positive voltage excursion. The use of color persistence highlights this effect. In this case, the step load pulse is not synchronous with the switching frequency of the POL regulator. Therefore, different responses result from reducing the current at different points in the switching cycle. This is common for POL regulators that operate at very low duty cycles. The dynamic range in this situation is very small, since the duty cycle starts very low and can only reduce to zero. A similar large signal effect appears in POL regulators that operate at very high duty cycles, such as with 3.3V input and 2.5V output regulators. In this case, the large signal effect occurs on the negative voltage excursion, since the dynamic range is limited by the high operating duty cycle. This can only increase to the PWM maximum duty cycle, which is often less than 100%.

In Figure 11-5, the upper trace is the voltage response and the lower trace is the current pulse. The color represents the occurrence frequency of a response, with blue being less frequent and red being most frequent.

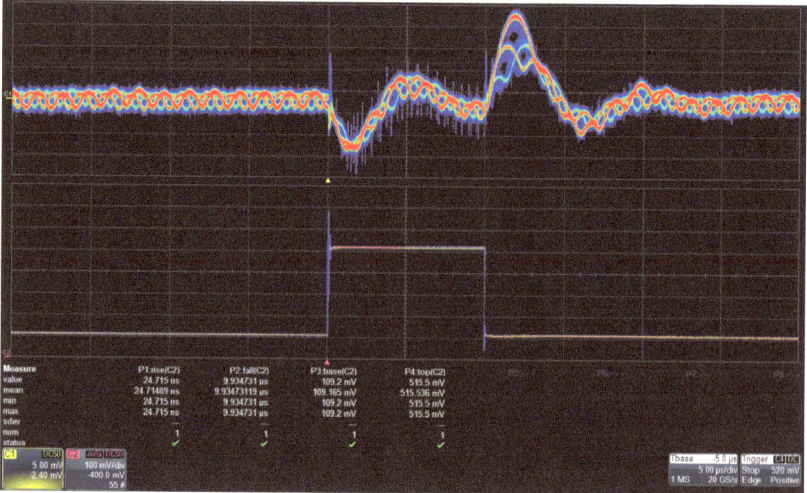

This POL Step Load Response shows Several Different Large-Signal Responses on the Positive Voltage Excursion

Figure 11-5

Notes about Averaging

The scope measurement in Figure 11-6 shows the step load response of a POL regulator. The top trace is the output voltage. The output ripple is clearly visible. It is possible to get a clearer image of the response by removing the output ripple from the measurement. This can be achieved in several ways, one of which is to bandwidth limit the oscilloscope. Another is to add an active filter if the oscilloscope supports that feature.

Trace averaging can also be used to remove the ripple from the measurement. Since the oscilloscope is triggered on the step load pulse, which is not synchronous with the POL switching frequency, the ripple voltage appears to be random with respect to the trigger. It averages to nearly zero as seen in Figure 11-7.

In Figures 11-6 and 11-7, the oscilloscope is triggered on the current step shown in the lower (red) trace.

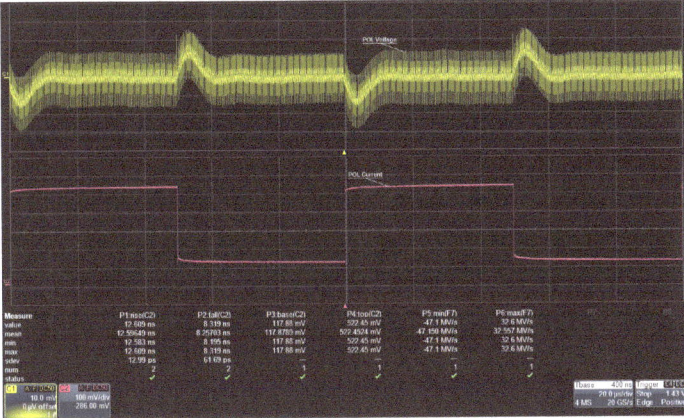

*Load Step Response of a POL Showing Switching
Frequency Ripple along with the Step Load Response
in the Upper (Yellow) Trace*

Figure 11-6

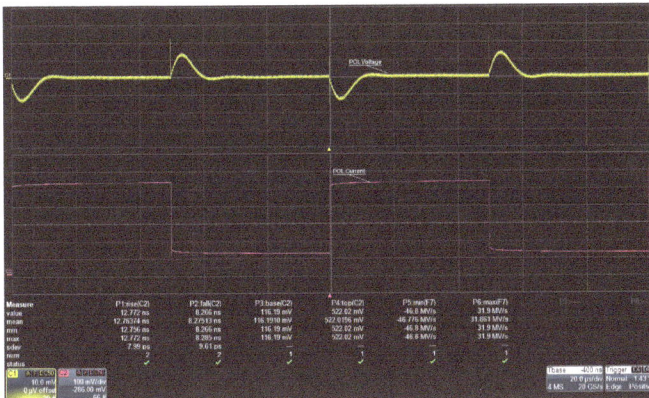

*Since the Switching Frequency is not Synchronous
with the Load Step, Trace Averaging Eliminates the
Ripple Seen in the Upper (Yellow) Trace—Note the
Large Voltage Spike at each Current Step Transition*

Figure 11-7

Averaging isn't always appropriate. As shown in Figure 11-8, the POL regulator is operating at a very high duty cycle, resulting in limited range.

A large signal effect, as discussed previously occurs on the negative voltage excursion. In this image, the negative voltage excursion is dependent on where in the switching cycle the current is increased.

Two different acquisitions are seen in the green and yellow traces, while the average result is seen in the blue trace. In this case, averaging reduces the large excursions which could be troublesome to system performance.

It is generally best to make such measurements using a high update rate with persistence color grading.

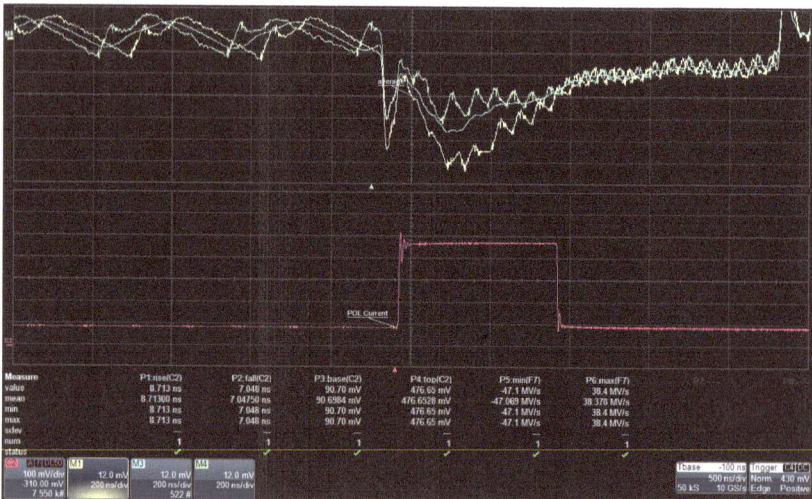

Two Different Acquisitions are Seen in the Green and Yellow Traces while the Blue Trace Shows the Result of Trace Averaging—the Current Step is Shown in the Red Trace

Figure 11-8

Sample Rate and Time Scale

In many cases, there is a very narrow spike at each current transition; as shown in Figure 11-7. Such spikes can be easily missed depending on the time scale and the sample rate of the oscilloscope. An example of this phenomenon is illustrated in Figure 11-9. Two different time base and sample rate settings are used to measure the same step load. In the measurement on the right the spike in the voltage excursion is evident, while in the measurement on the left it is not evident. Only the two oscilloscope settings were changed in these two captures.

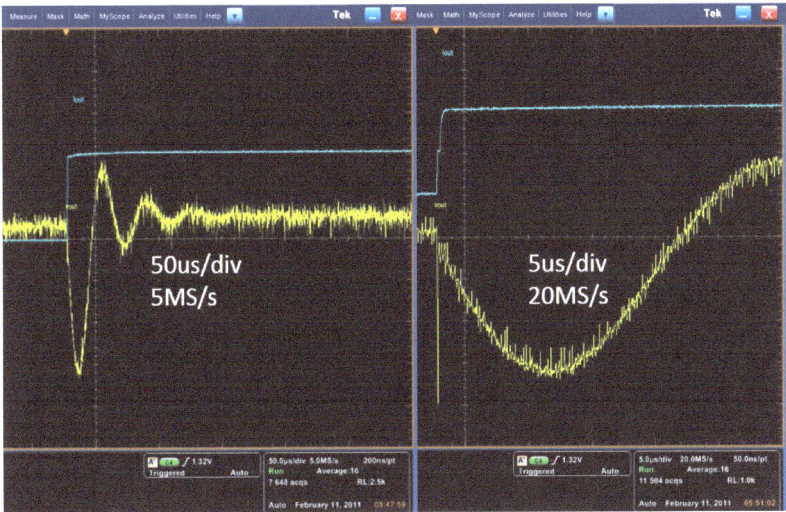

The Figure on the Left misses the Spike at 5MS/s and 50µs/div while the Figure on the Right Captures the Spike at 20MS/s and 5µs/div—in Both Cases, the Voltage Response is the Yellow Trace and the Current Step is the Blue Trace

Figure 11-9

Additional Examples

1. Voltage Reference 1mA Step Load

In this measurement of a voltage reference, the Picotest J2111A current injector is used to apply a 2mA step load. The sensitivity of the voltage reference stability can be seen by the different responses at the two current amplitudes. Both the ringing frequency and the Q are sensitive to the current amplitude. In this case, the voltage response is significantly worse at the higher load current. As this is a shunt voltage reference, the high load current represents a very low bias current. In Figure 11-10, The voltage response is the yellow trace and the current step is the green trace. The ringing frequency and Q are both shown to be sensitive to the voltage reference load current.

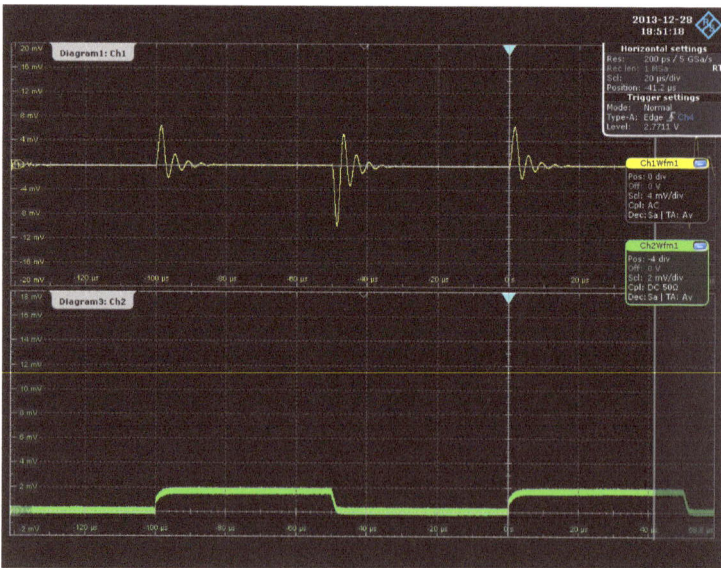

2mA Step Load Applied to a Shunt Voltage Reference

Figure 11-10

2. POL Switching Regulator using Multiport Probe

A multiport probe can be used to simplify the step load measurement. One of the 2 ports in the probe is connected to the Picotest J2111A current injector. The other port is connected to a 50Ω oscilloscope port, through a Picotest J2130A DC blocker, as discussed in the measuring *Ripple and Noise* chapter. The current injector is modulated by a Picotest G5100A arbitrary waveform generator. The setup for the measurement is shown in Figure 11-11.

Since the load step is not synchronous with the switching frequency and the oscilloscope is triggered by the load step, averaging can be used to remove the switching frequency ripple from the output voltage to make the response clearer.

The upper trace in this figure is the voltage response to the step without trace averaging. The middle trace is the same voltage with 50 averages.

While averaging can be used effectively to remove the switching ripple, caution is advised as there could be large signal effects. In Figure 11-11, note the voltage response without averaging (top trace) and with averaging (middle trace).

The lower trace is the pulse load current and the rise and fall time are measured as 25ns and 18ns, respectively.

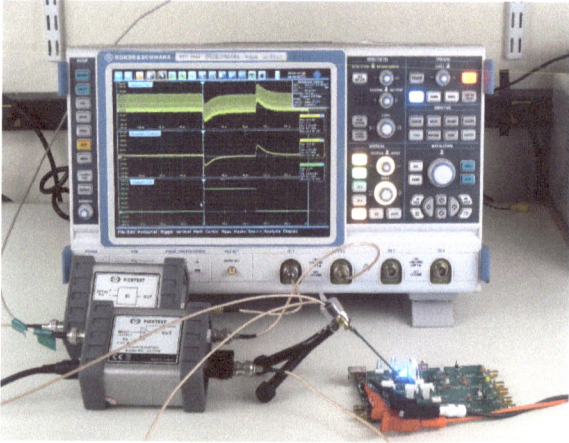

The Picotest 2-Port Probe is used to Simplify Small Signal Pulse Load Testing

Figure 11-11

Close-Up View Showing the Two-Cable Connections to the 2-Port Probe Held on the Output Capacitor using a Weighted Probe Holder as seen in Figure 11-11

Figure 11-12

3. Ferrite Bead with Capacitor

Ferrite beads are often used to keep RF frequency noise from leaking from the RF circuit onto the power supply bus.

If leaked, the noise can then exit the system as EMI or permeate the system with the RF and can have undesirable consequences.

Ferrite beads are generally very resistive at higher frequencies and much less resistive at lower frequencies. This allows the DC voltage to pass through to the RF circuit relatively unencumbered by the bead, while presenting high impedance to high frequencies.

When a ferrite bead is combined with a ceramic bypass capacitor, the resulting combination can have a very high Q. This creates its own noise at much lower frequencies.

An LI0805G201R-10 ferrite bead is used in this example. The bead is specified as having a resistance of $200 \pm 50\Omega$ at 100MHz and a DCR of 0.3Ω maximum.

An impedance measurement shows the ferrite bead to have a real series resistance of approximately $52\text{m}\Omega$ and inductance of 688nH, measured at 100kHz.

The bead is combined with a $1\mu\text{F}$ ceramic chip capacitor as shown in Figure 11-13.

Measurements of the capacitance and ESR of the $1\mu\text{F}$ capacitor are $0.88\mu\text{F}$ and $10\text{m}\Omega$, respectively.

The resonant frequency can be estimated as:

$$f_r = \frac{1}{2\pi\sqrt{688nH \cdot 0.88uF}} = 204kHz \qquad 11.6$$

And the circuit Q to be:

$$Q = \frac{\sqrt{\frac{688nH}{0.88uF}}}{52m\Omega + 13m\Omega} = 13.6 \qquad 11.7$$

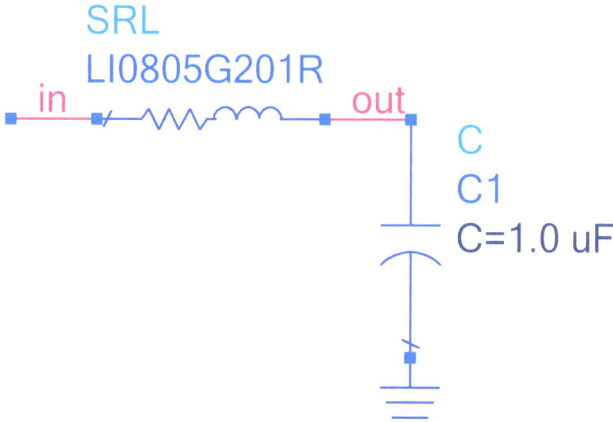

LI0805G201R Ferrite Bead Combined with a 1μF Ceramic Capacitor

Figure 11-13

The J2111A is used to create a fast 10mA edge step. This reveals the response of the circuit, which is consistent with the expected high Q and resonant frequency. The impedance of the J2111A is much higher than the 1μF capacitor and so the injector does not impact the measurement results.

It should be noted that the 2.2μF capacitance of the B&K model 8540, or similar electronic load, would significantly impact the measurement. In Figure 11-14, the lower trace is the 10mA current step (blue), while the upper trace is the voltage response at the 1μF capacitor (yellow).

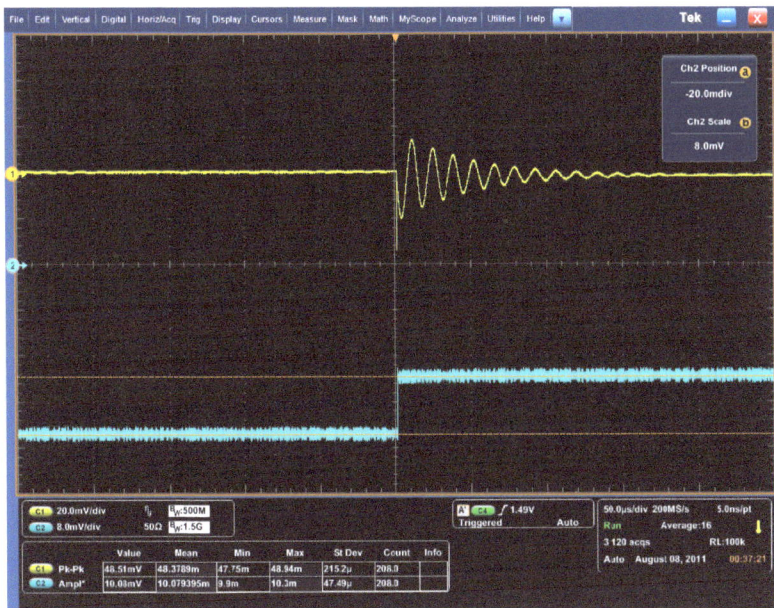

Step Load Response of the LI0805G201R Ferrite Bead coupled with a 1µF Ceramic Capacitor

Figure 11-14

4. Fast Response LDO

A high performance 5V LDO is connected for a 1 Amp peak step load, using a Picotest J2112A current injector.

The connection is made using a SMA Tee adapter and an SMA to BNC adapter to connect the current injector to the DUT as shown in Figure 11-15. A pulse generator is adjusted to provide an approximately 1Amp peak pulse with fall times of 5ns, 50ns and 100ns.

The measurement results are shown in Figure 11-16.

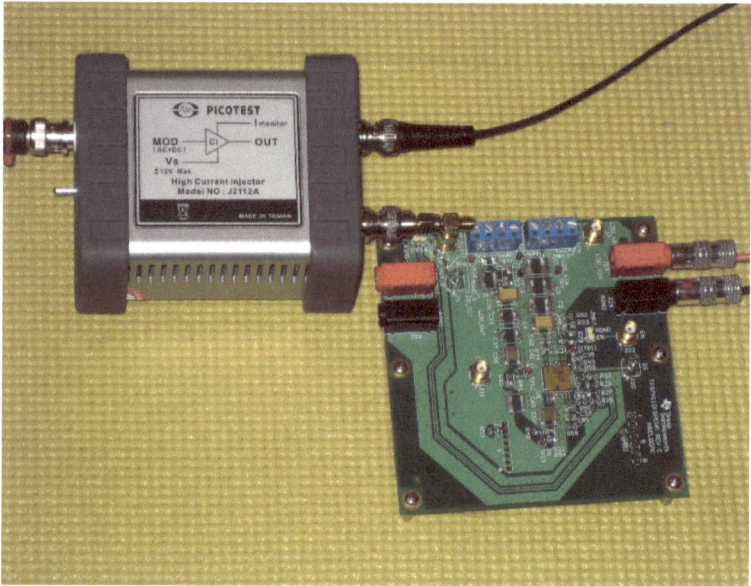

Picotest J2112A Direct-Connected to an LDO Demo Board using an SMA Tee Adapter and a SMA to BNC Adapter

Figure 11-15

While the current injector cannot produce a 5ns edge with this connection and a 5V output the fall time is close at 5.7ns.

The measurement result shows that with a 5ns edge there is significant 10MHz ringing, likely due to the PCB planes connecting the SMA connector to the ceramic decoupling capacitors on the LDO board, while at a 50ns edge it is greatly reduced and with a 100ns edge it is barely visible.

This demonstrates the versatility of the current injector with external modulation.

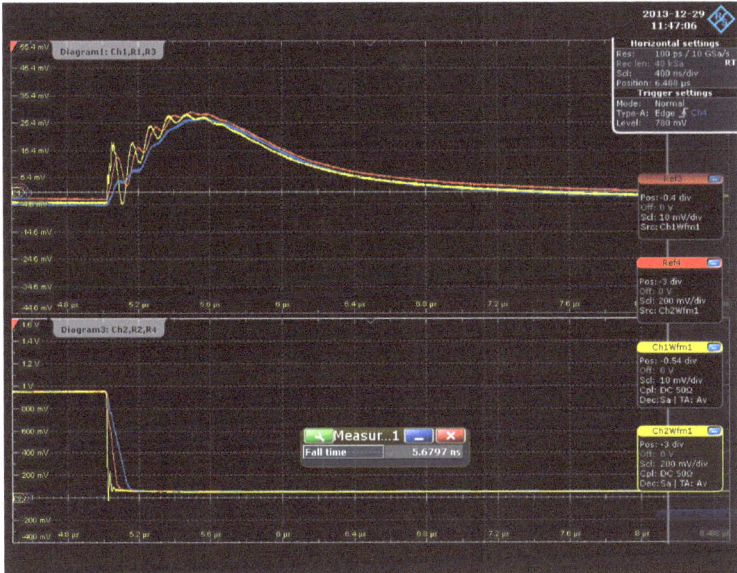

Picotest J2112A Direct-Connected to an LDO Demo Board using an SMA Tee Adapter. The Load is Approximately 1Amp Peak with a Rise Time of 5.7ns (Yellow) 50ns (Red) and 100ns (Blue)

Figure 11-16

Tips and Tricks

1. Measure the impedance of the load to be sure it does not influence the measurement. Ideally, the load impedance should be an order of magnitude greater than the impedance of the DUT.

2. The Picotest J2111A current injector can generally be connected using short banana leads, or 50Ω coaxial cable, as the current is low and the speed is moderate at 25ns. The Picotest J2112A can generate current pulses up to 1Amp peak and at a much faster rate and so requires much lower interconnect inductance. If a cable

must be used, Temp-Flex can minimize the connection inductance, but 50 Ω coaxial cable and banana leads are inappropriate for the J2112A.

3. Zoom traces can be used to see the edge in greater detail. This is often better than just changing the time scale, since both the detail and the entire signal can be seen at the same time.

4. The initial voltage spike resulting from the fast edge current pulse can be large and very narrow. Use the maximum sample rate that the scope supports to capture the pulse.

5. A probe such as the Picotest 2-port probe, is provides a simple and quick method for checking most voltage regulators, voltage references and op-amps.

6. Many designs have a large number of decoupling capacitors located close to each high-speed integrated circuit. The step load response should be measured at the decoupling capacitors closest to the integrated circuit pins and at the integrated circuit pins.

Chapter References

1. Steve Sandler *Target impedance based solutions for PDN may not provide a realistic assessment* EDN, May 1, 2013 http://www.edn.com/design/test-and-measurement/4413192/Target-impedance-based-solutions-for-PDN-may-not-provide-a-realistic-assessment

2. *New Injector Supports Testing POLs,* Picotest.com, Sept. 2, 2012 http://www.picotest.com/articles/New%20Injector%20Supports%20Testing%20of%20POLs.pdf

3. *Current Injector Outperforms Electronic Loads in Measuring POLs,* How2Power.com, October 2012 http://www.how2power.com/newsletters/1210/products/H2PToday1210_products_Picotest.pdf

4. Picotest.com, Nov 17, 2010 https://www.picotest.com/articles/Step%20Load.pdf

5. Temp-Flex http://www.tempflex.com/files/pdfs/coaxial-cable-low-inductance.pdf

6. Steve Sandler *Why a Little Ringing in Step Load Isn't Such a Little Problem,* EETimes.com Aug 14, 2013 www.eetimes.com/author.asp?section_id=36&doc_id=1319210

Steven M. Sandler

Chapter 12

Measuring Ripple and Noise

WHILE RIPPLE AND noise are generally considered interchangeable, they often represent different measurements, requiring different methods and techniques. For example, when we are discussing ripple and noise related to a switching regulator we usually mean the ripple signal that results after filtering the switched signal and the spike noise or ringing associated with the switching edges.

Most systems are complicated, with the ripple and noise being generated from several different sources. Since voltage regulators also have finite and non-zero output impedance, variations in the load current result in variation of the output voltage. A portion of the ripple at the input to the regulator also feeds through to the output attenuated by the power supply rejection ratio (PSRR) of the voltage regulator.

For the purposes of this chapter we'll divide ripple and noise into three distinct categories.

Ripple—A periodic variation in output voltage resulting from an expected event. For example, the switching action of a switching power supply is expected to result in an output ripple voltage. The rectification of the AC mains is expected to result in an input ripple signal. While the ripple may not be at a fixed or specific frequency, for example in a frequency modulated resonant converter, the ripple is an expected result.

Noise—Noise is a non-periodic random signal resulting from multiple sources. For instance, resistors contribute random noise. Many other semiconductors generate other types of noise, such as shot noise, popcorn noise and burst noise.

Spurious Signals—Spurious signals are fixed frequency signals other than the desired or tuned frequency. This can also include mixing products. For example, considering a 125MHz sine wave oscillator, any frequency other than 125MHz would be considered a spurious response.

Selecting a Measurement Method

In selecting the methodology for the measurement of ripple and noise, one decision is whether to measure in-system or removed from the system. Another choice is whether to make your measurements directly or indirectly. Yet another is which measurement domain (time or frequency) to use.

Interestingly, each of these choices is essentially independent of the others. For example, a measurement can often be made either in or out of the system and independently be either a direct or indirect measurement, made in either the time domain or the spectral domain.

In or Out of System

The in system vs. out of system selection is based, in part, on the goal of the measurement. As an example, if you are a manufacturer of switching power supplies, you have specified a certain ripple limit into a target, often generic load. This ripple is a measure of the power supply performance and measuring in-systems would include noise effects that are induced by the actual loading. Since the power supply specification is not inclusive of the actual load, it is best to measure ripple out of a system.

On the other hand, if the goal is to troubleshoot a system level noise issue, it is generally best to measure in-system and measuring out of the system is much less helpful.

In many cases, the power supply and the system are manufactured by separate companies, and so measuring in-system may not always be an option.

Direct or Indirect

A direct measurement is a measure of the ripple and/or noise as a measurement of the magnitude of the signal. This is an absolute measurement of the noise.

On the other hand, an indirect measurement is not a measure of the noise, but a measure of the impact of the noise.

A good example is an ADC clock.

The noise level of the ADC is strongly related to the clock jitter and the clock jitter has a strong relationship to the power supply noise. In many cases, it is more meaningful to assess the power supply noise by measuring jitter rather than the noise signal (This is one of those cases where it may be specified with a load—the load being the ADC clock).

One of the more significant considerations for direct vs. indirect measurement is whether we care about meeting a specified noise requirement or are we more concerned with the impact of the noise on the system performance.

Time or Spectral Domain

These two domains are quite different and provide different information. The time domain measurement, captured by an oscilloscope, is a wide bandwidth measurement. As such, it generally does not have the sensitivity, or noise floor, to measure low level noise. Though it is generally adequate for measuring switching power supply ripple and spike noise, the time domain measurement is not frequency selective, so it offers little information regarding the source of the noise.

The spectral domain, on the other hand, generally offers much better sensitivity and frequency selectivity, which helps the user in identifying the noise sources. The spectrum analyzer measurement, being a narrow band measurement, also has a much better dynamic range and much lower noise floor. Despite these distinctions, there are oscilloscopes with integrated FFT or spectrum analysis capability.

These can be used to measure the ripple and/or noise in either domain. The oscilloscope based spectrum analyzer still does not have the sensitivity or noise floor of a high performance spectrum analyzer, but is still useful in many cases.

Connecting the Equipment

Independent of the measurement method selected, the measurement setup is quite similar. The DUT is powered and the output is probed to monitor the noise. No external stimulus is required for this measurement.

The choices are in the probe selection and coupling of the probe to the equipment.

Passive Scope Probes

Pros

- Tend to be plentiful and are inexpensive.
- Allow connection of the equipment to higher voltages due to attenuation.

Cons

• Are generally 10X probes resulting in a 20dB reduction in sensitivity.

• A 1X probe often results in <6MHz bandwidth just for the probe.

• Limited in bandwidth (the best are < 1GHz).

• Can be fairly capacitive and, along with a ground clip or trace impedance, exhibits significant ringing due to the probe.

Tips and Tricks

1. If you do use a scope probe, try to measure as close to the point of interest as possible. This is generally an output or decoupling capacitor.
2. Use a spring clip for the ground or a scope tip to BNC adapter to minimize the length of the ground lead connection.
3. Always verify the setup by measuring a known value first.
4. The signal-to-noise and sensitivity can often be improved by using an external preamplifier.
5. Try to set the measurement to be as close to full scale as possible for best signal-to-noise ratio.

Active Scope Probes

Pros

• Are available in much wider bandwidth than passive probes.

• Have much lower tip capacitance and are much less sensitive to ground clip and PCB traces.

Cons

- Are expensive compared with passive probes.
- Are generally limited in operating voltage range.
- Are generally 10X, or greater, resulting in a significant reduction in sensitivity.

Tips and Tricks

1. Try to select the probe with the lowest attenuation that is compatible with the measurement.
2. Always verify the setup by measuring a known value first.
3. The signal-to-noise and sensitivity can often be improved by using an external preamplifier.
4. For the best signal-to-noise ratio, try to set the measurement to be as close to full scale as possible.

Direct 50Ω Terminated Connection

Pros

- Produces no attenuation, resulting in maximum sensitivity.
- Allows very wide bandwidth measurement.
- Allows the scope to operate at the maximum bandwidth and maximum sensitivity.
- Enables lowest capacitance measurement.

Cons

- Is a low impedance measurement and can load sensitive devices.
- Is limited to the voltage range of the test equipment,

typically < 5VDC.

Tips and Tricks

1. Insert a low frequency blocker, such as the J2130A, in front of the 50 Ohm equipment. This allows operation at higher voltages and isolates the 50 Ohms from the device being measured.
2. Always verify the setup by measuring a known value first.
3. The signal-to-noise and sensitivity can often be improved by using an external preamplifier.
4. A preamplifier can also be used to transform a high impedance signal or probe to the 50Ω impedance of test equipment.
5. Set the measurement as close to full scale as possible for the best signal-to-noise ratio.

CAUTION: In some oscilloscopes, increasing the sensitivity may reduce the measurement bandwidth.

Choosing the Equipment

Once the domain is selected, the specific equipment is chosen primarily based on the magnitude of the measurement.

While all of the equipment listed in Table 12.1 can measure from very low frequencies to beyond 1 GHz, some devices can display log frequency while others can only display linear frequency.

A linear frequency display is very limiting for wideband measurements.

Equipment	Sensitivity[*]	Noise Floor[*]	Measurement Methods
Agilent E5052A	10's-100's nV	-150 dBm	Direct and Indirect Spectrum/Jitter
Agilent N9020A	10's-100's nV	-150 dBm	Direct and Indirect Spectrum/Jitter
Tektronix RSA5106A	10's-100's nV	-150 dBm	Direct and Indirect Spectrum/Jitter
Tektronix MDO4104-6	10's μV	-85 dBm	Direct and Indirect Spectrum/Time-linear x axis
Tektronix MSO5204	xxx	xxx	Direct and Indirect Spectrum/Time-linear x axis/Jitter
Teledyne Lecroy WR640Zi	10's μV	-85 dBm	Direct and Indirect Spectrum/Time-linear x axis/Jitter

[*]can be improved with an external preamplifier

Equipment for Measuring Ripple and Noise
Table 12.1

The top three instruments listed in Table 12.1 can measure noise as spectral content and clock jitter or phase noise, though not time domain.

The bottom three instruments can measure time domain noise measurements with reduced sensitivity and a higher noise floor.

These three oscilloscopes also have the ability to measure clock jitter, though not with the same sensitivity as the top three devices listed.

Some spectrum analyzers display the results in dBm rather than Volts, though many are starting to offer various units. Converting between dBm and volts is simple.

Power is the RMS voltage squared divided by the load resistance, which for most test equipment is 50Ω, though there are other impedances. Video test equipment, such as cable TV, is generally 75Ω, but 50Ω is much more common and so it is used here as an example. The following equations can be used to convert RMS or peak voltages to dBm and convert dBm to RMS or peak voltages.

$$P = \frac{Vrms^2}{R} = \frac{Vrms^2}{50\Omega} \qquad 12.1$$

$$PdBm = 10 \cdot log\left(\frac{P}{1mW}\right) \qquad 12.2$$

$$PdBm(Vrms) = 4.3429 \cdot ln(Vrms^2) + 13 \qquad 12.3$$

$$PdBm(Vpk) = 4.3429 \cdot ln(Vrms^2) + 10 \qquad 12.4$$

$$(Vrms(PdBm) = \sqrt{e^{0.2303 \cdot PdBm - 2.9956}} \qquad 12.5$$

$$(Vpk(PdBm) = 0.3162\sqrt{e^{0.2303 \cdot PdBm}} \qquad 12.6$$

Examples

Spectral Domain Examples

1. Measuring a Known Signal

Prior to making any measurement, the setup should always be verified by measuring a precisely known value. In this case, a precision signal generator is used to apply a sine wave to the input of a WR640Zi oscilloscope operating in its spectrum measurement mode. The signal generator is set to produce a 224mVrms (0dBm) sine wave at 5kHz. Channel 1 of the oscilloscope is set for a 50Ω input termination and the instrument is set to measure the spectrum of channel 1 from 1Hz to 10kHz. The measurement setup is shown in Figure 12-2 and the 5kHz signal can be seen at the center of the display.

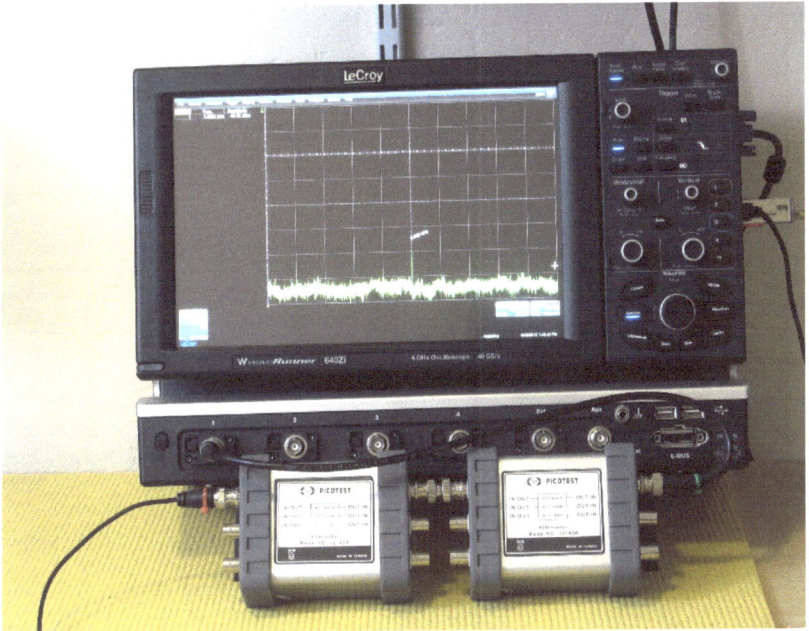

Setup Verification using a 224mVrms Signal fed through two 40dB Attenuators to a WR640Zi Oscilloscope operating in Spectrum Analysis Mode

Figure 12-1

The result, shown in Figure 12-2, correctly displays very close to 0dBm (0.007 dBm to be precise). Two J2140A cascadable attenuators are then inserted in series using the 40dB tap on each attenuator, for a total of 80db attenuation.

The display in Figure 12-3 correctly shows -80.43dBm with the 80dB attenuation inserted. Therefore, the measurement setup is verified.

Another observation from Figure 12-2 is that the tallest spurious responses in the display are approximately -75dBm, while the average noise level is closer to -95dBm. The spurious signals are -75dBm with respect to the signal or -75dBc. The results in Figure 12-3 show the attenuated signal to be -

80.43dBm, which is very close to the expected result. The tallest spurious signal in this case is 95dBm, or approximately -15dBc. If the signal level is further reduced, the 15dBc magnitude is also reduces, degrading the SNR of the measurement.

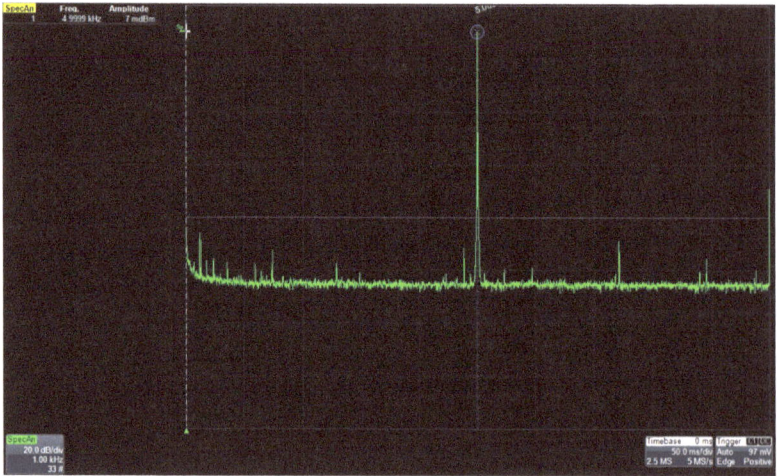

Measured Result of 224mVrms 5kHz Sine Wave displays 0.007dBm

Figure 12-2

An interesting note about the measurement in Figure 12-3 is that -80.43dBm represents 22μVrms or 30μVpk. This is far too small to see in the time domain signal, illustrating that the spectrum capability of the oscilloscope is much more sensitive and selective than the time domain measurements of the same oscilloscope.

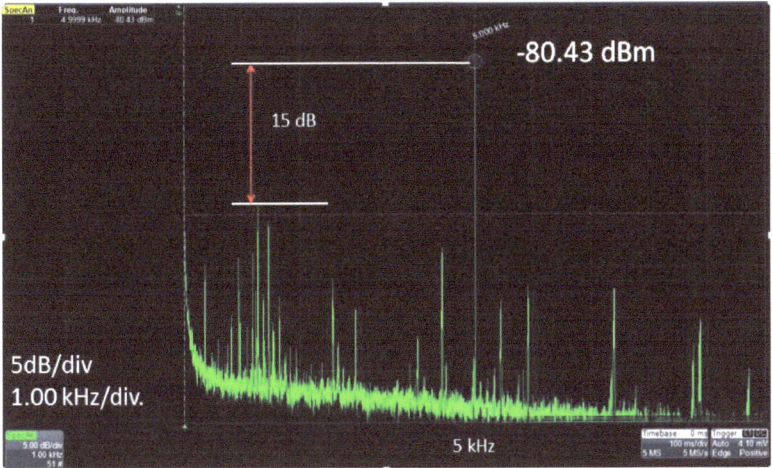

After inserting Two Cascaded 40dB Attenuators, the Measurement displays a Level of -80.43dBm for the Sine Wave being Measured

Figure 12-3

At these low levels, there are many noise sources that can interfere with the measurement. For this reason we always want to perform a measurement with the equipment we are measuring turned off in order to determine the noise floor of the measurement.

As equipment in a typical lab is turned on and off periodically and randomly, the noise floor measurement is best made immediately before or after the actual device measurement. Of course, as stated earlier, the addition of a low noise preamplifier can improve both the sensitivity of this measurement and the signal-to-noise ratio.

For comparison purposes, an N9020A spectrum analyzer is connected using a setup similar to the one described above and set to peak detect.

In this case the input signal is set to 70.7mVrms (-10dBm) at 5kHz and each attenuator is adjusted to 50dB, for a total attenuation of 100dB. The resulting display, shown in Figure 12-4 correctly displays very close to 1µVpk or 707nVrms (-110dBm), which is correct given the 100dB attenuation applied.

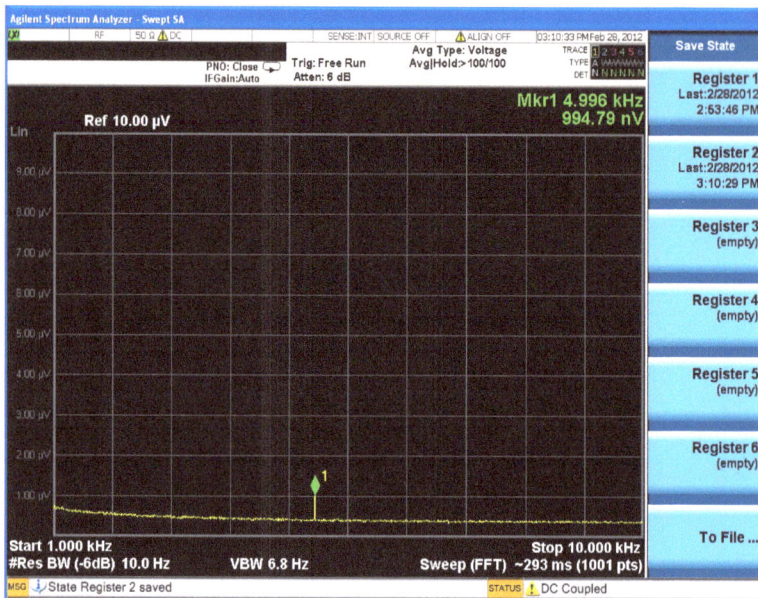

Taken using a Similar Setup to Figure 12-1, this Measurement shows the Much Lower Noise Floor and Higher Sensitivity of the N9020A Spectrum Analyzer

Figure 12-4

2. Measuring Linear Regulator and Voltage Reference Noise

A similar 50Ω direct connection to an RSA5106A real time spectrum analyzer is shown in Figure 12-5. In this setup, a J2130A DC blocker is added between the voltage regulator and the RSA5106A so that the 50Ω input of the test equipment does not load the voltage regulator being tested. Of course, the setup is first verified with a known signal, though not shown here.

A Picotest VRTS01 demonstration board is used for this measurement as it allows different regulators to be easily plugged into the demo board. This allows simple connections to the test equipment.

Three traces are shown in Figure 12-6.

The lowest level trace is the noise floor, measured with all connections in place, but no input power applied to the linear regulator.

The second green trace is the output of a specially made, ultra-low noise voltage regulator, discussed in Reference 2.

The third and most interesting trace is the blue trace, which is a measurement of the output of an LM317 linear voltage regulator.

It is interesting that in addition to the average noise, the LM317 has spurious responses at 1kHz and harmonics of 1kHz. It is also noteworthy that the measured spurious responses are on the order of 2.25μV and that the average noise floor is 10's of nV.

Since many engineers might be surprised by the spurious responses from a linear regulator, the measurement is repeated at a later time using the N9020A spectrum analyzer.

The measurement result seen in Figure 12-7 is quite similar to the prior measurement.

In Figure 12-5, note that the RSA5106A is connected via a J2130A DC Blocker.

In Figure 12-6, note the spurious responses of the LM317.

A DEMO Board with "Plug-In Regulators" is connected to the RSA5106A Spectrum Analyzer for Measuring Noise

Figure 12-5

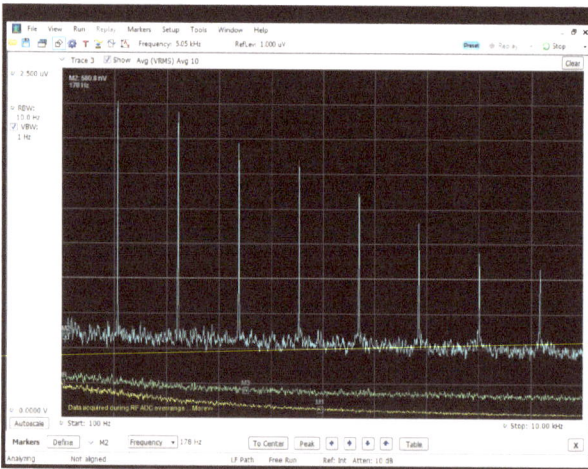

The Instrument's Noise Floor, the Output of a Custom-Designed Ultra-Low Noise Regulator and the Output of an LM317 Measured using the RSA5106A

Figure 12-6

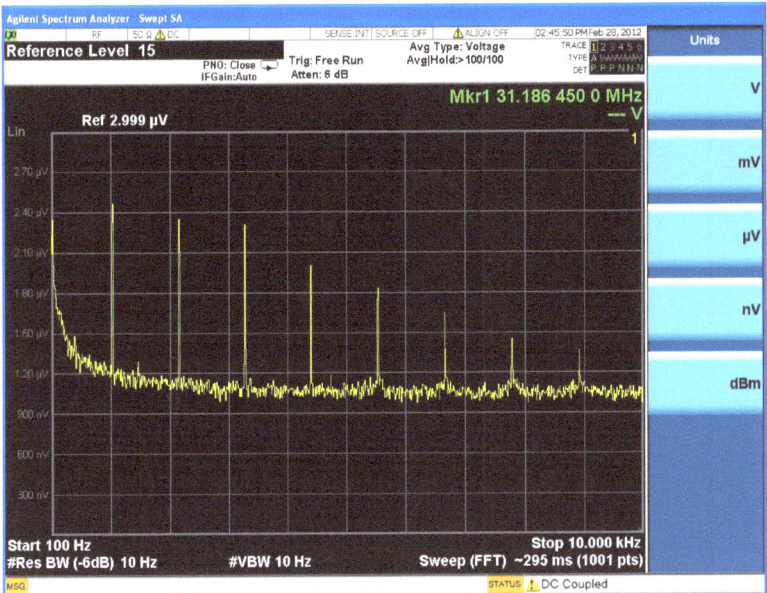

Repeating the LM317 Noise Measurement using the N9020A Verifies the Spurious Responses seen in the Previous Measurement of Figure 12-6

Figure 12-7

3. Measuring Low-Dropout Regulator (LDO) Noise

Since spurious responses can be a significant issue for analog and instrumentation equipment, where LDOs rather than standard linear regulators are commonly used, it is worthwhile measuring the noise of an LDO. The test setup is the same as in Example 2.

In this case, the amplitude scale is adjusted slightly as the LDO (TPS7A8001) noise is slightly greater than that of the LM317 noise. The result of the LDO noise, shown in Figure 12-8 also shows a spurious response, though in this device the fundamental spur is at 3.5kHz.

The second harmonic can also be seen in this image.

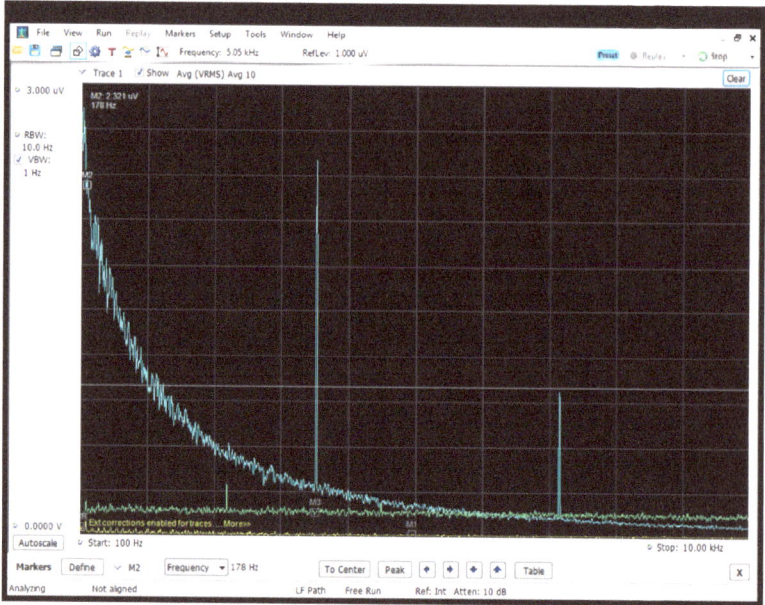

LDO Noise Measurement (Blue Trace) showing Spurious Response as well as Lower Frequency 1/f Noise—a Low-Noise Regulator is also Shown (Green Trace) with a Barely Visible Noise Floor (Yellow)

Figure 12-8

Time Domain Examples

4. Measuring Output Voltage Ripple of the TPS54225 Switching Regulator

This first time domain example is used to demonstrate the limitations of high input impedance oscilloscope terminations, as well as the drawbacks of using an attenuating probe.

This example also shows the signal-to-noise and sensitivity improvements obtained by inserting a preamplifier into the measurement.

The setup image in Figure 12-9 shows the connection of a Texas Instrument TPS54225 EVM board connected to an oscilloscope via a 2 port coaxial probe and the J2180A low-noise preamplifier. When the coaxial probe is not connected to the scope via the preamplifier, it is connected via the J2130A DC blocker seen in the foreground of Figure 12-9.

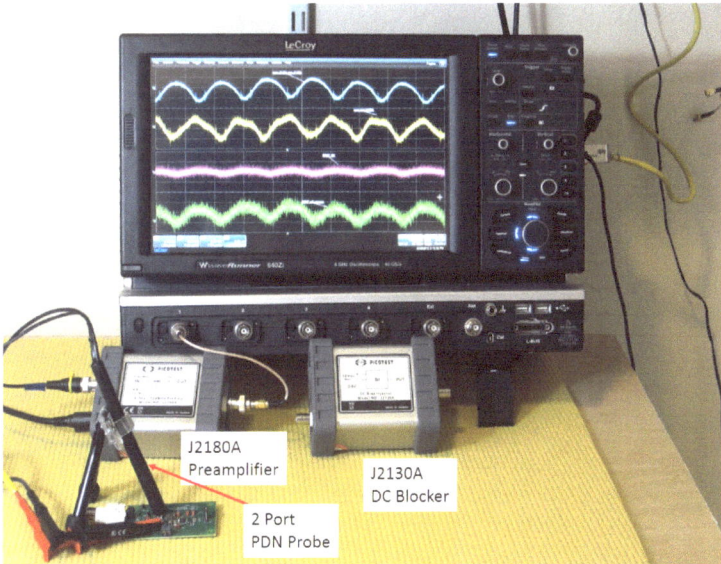

Switching Regulator Ripple Measurement Setup Showing a Multi-Port Coaxial Probe Connected on one end to the TPS54225 EVM Board and on the other end to the WR640Zi Oscilloscope via a J2180A Preamplifier

Figure 12-9

The measurement results, seen in Figure 12-10, show the lack of sensitivity of the oscilloscope with a 10X passive probe and without a DC blocker. Another limitation is that the oscilloscope sensitivity is reduced to 10mV/Div minimum in high input impedance mode. The sensitivity can be increased to 2mV/Div when the 50Ω termination is selected. The measurement using a standard 10X probe is barely visible at 10mV/Div as seen in the pink trace. The addition of the J2180A preamplifier increases the signal to where it is visible, though with very poor signal to noise, as seen in the green trace.

Using a 2 port PDN probe at 2mV/Div improves the fidelity of the measurement significantly, as seen in the yellow trace while adding the J2180A preamplifier significantly improves the fidelity and the signal to noise ratio.

In Figure 12-10, the waveforms are: multiport coax probe with J2180A preamp (blue), multiport coax probe without preamp and with J2130A DC Blocker (yellow), PP007 500MHz passive leadless probe without J2180A preamp (pink) and same probe with J2180A preamp (green).

The scale for all measurements is 2 mV/division, except for the pink trace, which is measured on a 10 mV/division scale.

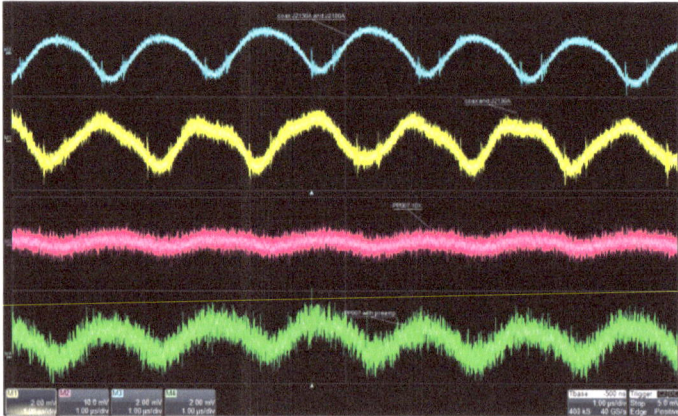

Switching Regulator Ripple Measurements taken on a TPS54225 EVM Board Illustrate the Effects of Using Different Probes

Figure 12-10

In Figure 12-11, the measurement of this EVM is repeated with the preamplifier and 2-port PDN probe. But this time, the display shows both the time domain response and the spectral response simultaneously. The spectral response allows identification of the correlated harmonics of the switching frequency, as well as signals that are unrelated to the switching action of the converter.

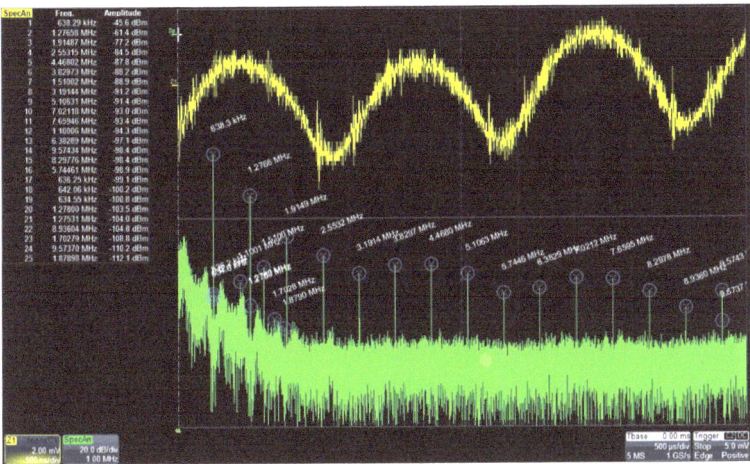

Comparing Time Domain and Spectral Domain Responses

Figure 12-11

One of the best features of this measurement technique, and the 2 port probe, is that since it is designed to work with a standard 50Ω input, the measurement can be applied to any type of test equipment from any manufacturer.

The ripple measurement of Figure 12-11 is shown on a Tektronix MSO5204 in Figure 12-12. Using the multiport probe, the measurement allows the signal to be connected to two different test instruments at the same time. All probe ports must be terminated into 50Ω.

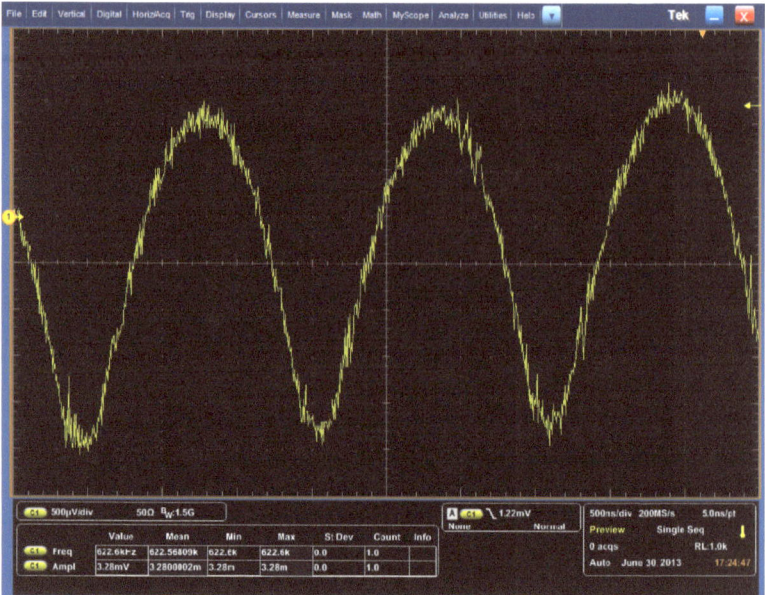

The Time Domain Measurement with the Multiport Probe uses a Standard 50Ω Input making it Compatible with all types of Test Equipment from any Manufacturer

Figure 12-12

5. **Measuring the Output Voltage Ripple of another Switching Regulator**

Using a 1 port probe combined with the preamplifier again, the ripple signal is measured at the output capacitor of a programmable DC/DC converter. The results of this measurement are shown in Figure 12-13. In this case, the ripple, observed at 2mV/Div, is very clean, indicating that there is little jitter in the measurement and that the high frequency signals are well attenuated. The measurement is clear enough to see the duty cycle of the DC/DC converter.

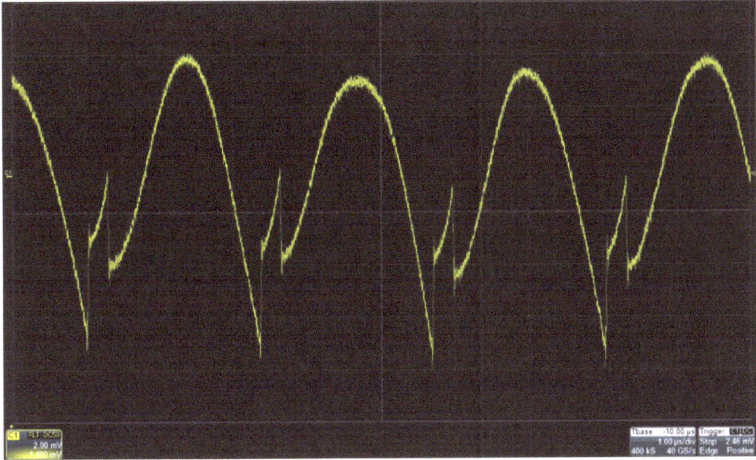

Comparing a Scope Probe Measurement with an AC-Coupled Coax Connection

Figure 12-13

Indirect Measurement Examples

6. Using Clock Jitter/Phase Noise as a Power Supply Noise Measurement

One of the more common indirect noise measurements is clock jitter either measured as a time jitter or as phase noise.

This example shows a phase noise measurement taken with an Agilent E5052B signal source analyzer. While this measurement is not a direct measurement of the power supply noise, it is a direct measurement of the impact of the noise on the performance of the clock.

Another advantage of this measurement is that the signal source analyzer has much better sensitivity than an oscilloscope. The same can be said for most instruments designed to make measurements in the frequency domain, including most spectrum analyzers.

Since the clock we are testing has a 50Ω output impedance, it can be directly connected to the test instrument, though it is generally preferred that a DC blocker be included in order to limit the DC into the sensitive front end of the instrument.

The oscillator is connected to the power supply being tested. In the setup shown in Figure 12-14, power comes into a voltage regulator through the red and black grabber clips in the top right of the image. The regulator output connects to the clock power supply input via J2 on a Picotest VRTS2 demo board. The output from a 125MHz clock on the board is connected directly to the test instrument input, via a DC blocking capacitor.

The phase noise measurement is shown in Figure 12-16 with each of the spurious responses highlighted. Each spurious signal is a representation of the ripple and noise from the voltage regulator.

Test Setup for Phase Noise Measurement

Figure 12-14

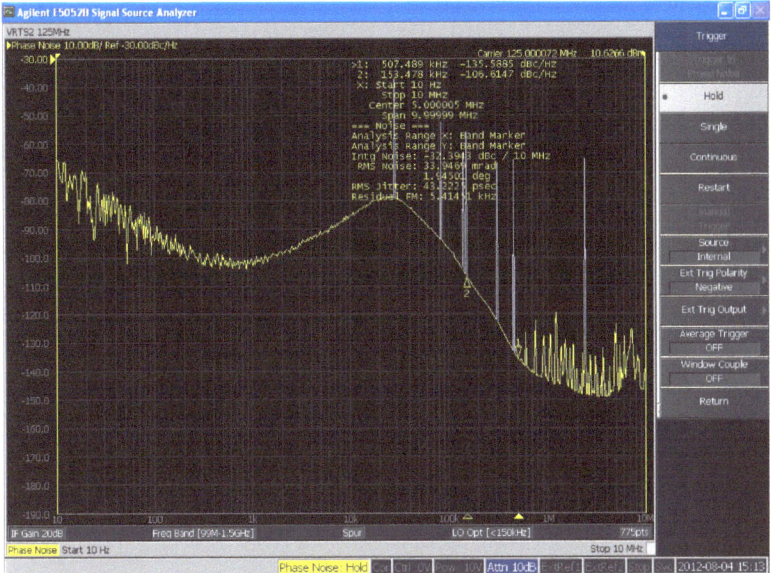

Phase Noise Measurement for 125-MHz Clock Showing the Spurious Responses from the Voltage Regulator

Figure 12-15

While the measurement in Figure 12-15 provides the clock noise, the measurement does not tell us the impact of noise voltage from the regulator on the clock performance.

If we wish to determine the sensitivity of the clock to the voltage regulator noise, we can modulate the power supply via a known amplitude and measure the corresponding spurious response of the clock.

A modified version of the phase noise measurement setup in Figure 12-14 and shown in Figure 12-16 allows the clock power supply voltage to be modulated using an external signal generator. For instance, the Picotest J2111A or similar current injector, as shown, or a line injector, such as the Picotest J2120A.

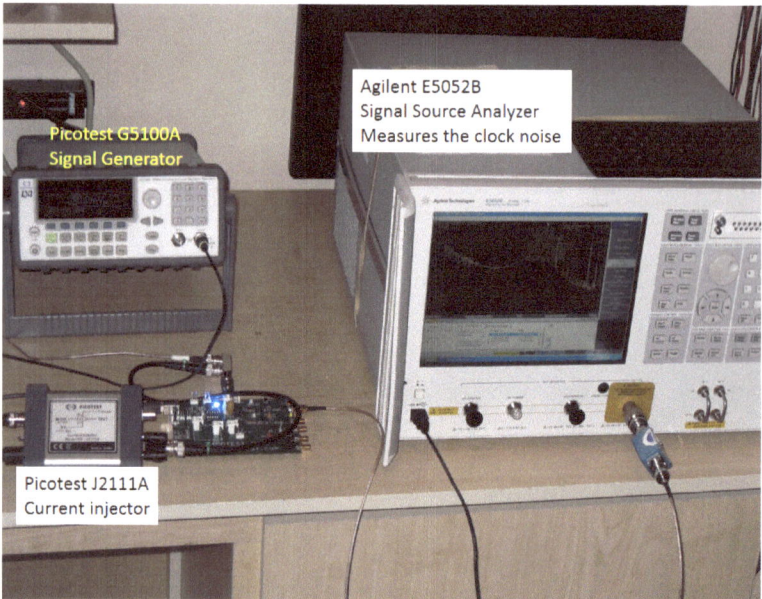

*Setup Showing the J2111A Current Injector used to
Modulate the Power Supply while the E5052B
Measures the Phase Noise*

Figure 12-16

The clock sensitivity can then be determined as phase
noise/Volt. An example of a clock sensitivity measurement is
shown in Figure 12-17, using a Tektronix RSA5106A analyzer
with a 1kHz 10mVpp modulation signal into the voltage
regulator. This measurement is then repeated at many
frequencies. With the sensitivity known at each frequency, the
power supply noise can be calculated as in Equation 12.7.

$$Vnoise = \frac{Clock\_spur}{Clock\ sensitivity} \qquad 12.7$$

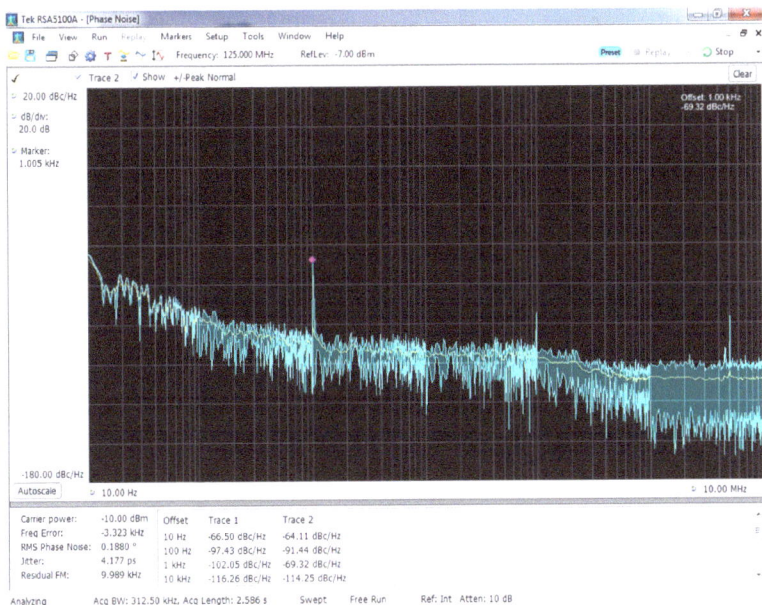

Using a Setup Similar to the One Shown in Figure 12-16, the Tektronix RSA5106A Measures the Phase Noise of the 125-MHz Clock

Figure 12-17

In Figure 12-17, a 10mV 1kHz signal modulates the clock power supply and the resulting spurious response is indicated by the marker.

In Figure 12-18, the same setup is applied again, this time using the N9020A spectrum analyzer.

In this case, a 10mV, 250kHz modulation signal is applied to the power supply—it modulates the clock power supply and the resulting spurious response observed in the measurement.

As illustrated here in these three examples, the phase-noise measurement is quite versatile, working with all spectrum type test instruments.

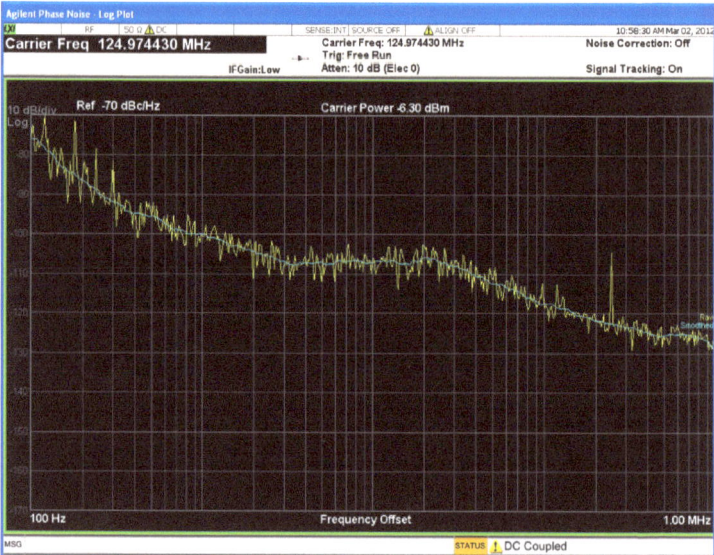

Using a Setup Similar to Figure 12-16, the Agilent N9020A Measures the Phase Noise of a Clock Powered by a Modulated Power Supply

Figure 12-18

In-Circuit Examples

The same demo board used in Figure 12-16 is now operated with the onboard 2.8MHz, 3.3V switching regulator rather than the external power supply. In addition to the 125-MHz clock previously noted, this demo board also contains a 10MHz clock. Both of these clocks are powered by the on-board switching regulator. In addition, the demo board provides an SMA connector attached directly to the switching regulator output. So using a coax cable, the output of the regulator can be connected through a Picotest J2130A DC blocking capacitor to the input of an oscilloscope as shown in Figure 12-19.

The result, shown in Figure 12-20, demonstrates that in

circuit measurement allows us to observe the total noise on the power supply port. It also shows that the noise is not necessarily generated by the power supply. We can clearly see the 2.88MHz spur and the harmonics of this frequency, but we can also see the 10MHz clock and its harmonics, as well as the 125MHz signal.

The result of this measurement is more meaningful in terms of determining the ripple voltage during operation, while measurement of the voltage regulator alone is a better indication of the ripple caused solely by the switching action of the regulator.

Demo Board with 10-MHz and 125-MHz Clocks Powered by an On-Board 2.8MHz Switching Regulator for "In-Circuit" Measurement of Power Supply Ripple

Figure 12-19

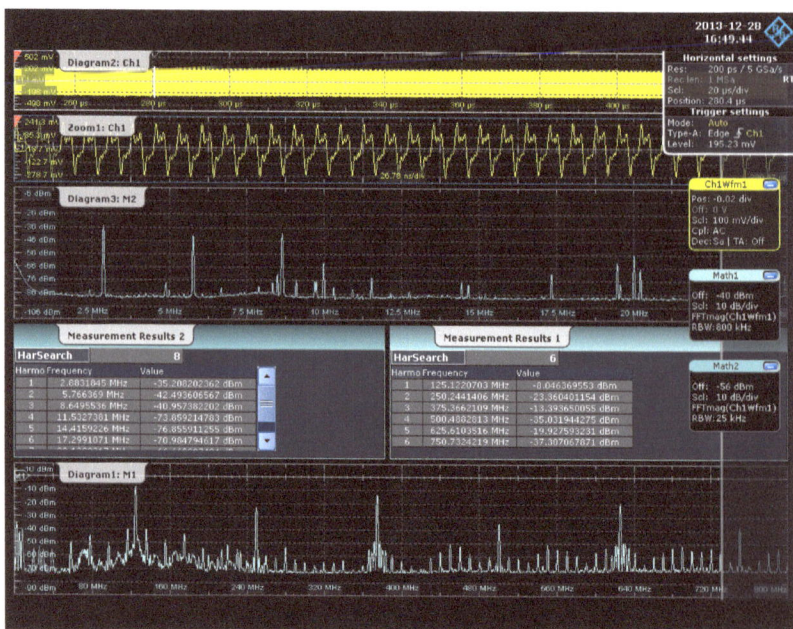

In-Circuit Measurement of Ripple Voltage Showing both the Time Domain and Spectral Domain Responses

Figure 12-20

In Figure 12-20, the top spectrum plot shows the 2.8MHz POL harmonic components and the lower spectrum plot shows the 125MHz harmonic components. The two spectrum views are used to capture different frequency ranges.

Averaging and Filtering

Using the same setup as in Figure 12-19, the ripple is measured using natural sampling and with various triggers and filters. The results, shown in Figure 12-21, show various effects. The top trace shows sampled data without any trigger (auto run). Trace M1 shows the impact of averaging the sampled data. Since there are many uncorrelated frequencies, the averaged waveform amplitude is near zero. If the oscilloscope is triggered and averaged, the result is mostly the trigger frequency as shown in trace M2. Triggering on either the 10MHz clock or the 2.88MHz switching regulator with bandpass filtering at the center frequency allows each signal to be displayed independently in the time domain. It is easy to lose the data of interest as a result of trace averaging, so be very selective in using this feature.

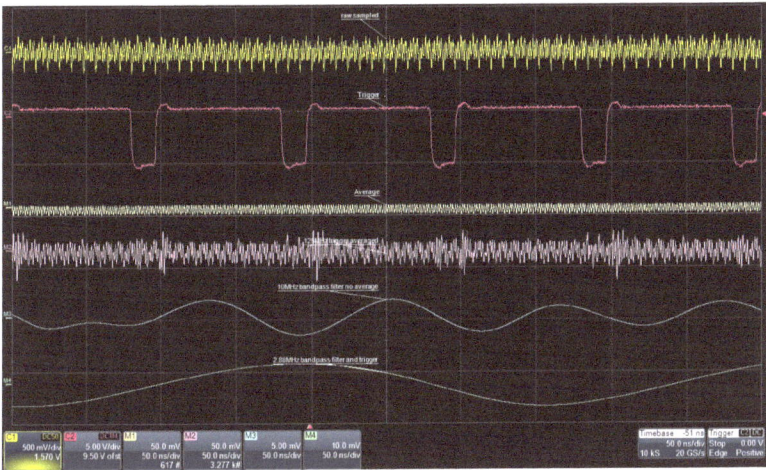

Effects of Triggering, Averaging and Filtering on Ripple and Noise Measurements

Figure 12-21

Chapter References

1. http://www.evaluationengineering.com/special-reports/201307/instruments/Resolving-Finer-Detail.html
2. Steve Sandler, *Unconditionally Stable Linear Voltage Regulator Power Electronics*, Mar. 23, 2012, http://powerelectronics.com/power-electronics-systems/unconditionally-stable-linear-voltage-regulator

Chapter 13

Measuring Edges

THERE ARE SEVERAL different reasons why you might need to measure signal edges.

In a switching power supply or POL regulator, you might want to measure the switching related loss in the MOSFET, as it is a significant term in the overall efficiency. This switching loss can also be used to optimize the gate drive design or the selection of the switching frequency.

In high-speed circuits, the interest is more often related to timing relationships, eye diagrams for signal integrity purposes, or for determining time jitter for an ADC circuit. Edge speeds also tell us approximately how high in frequency the signal harmonics extend.

The present state-of-the-art for POL switch edge speeds is on the order of 1nS (rise time). HEMPT technologies, including eGaN, GaN and GaAs devices, will significantly increase these speeds. Typical high-speed CMOS logic edge speeds are presently in the range of 350pS, while very high-speed logic and Time Domain Reflectometry (TDR) edges can be as fast as 7pS.

One positive factor related to edge measurement is that external stimulus of the circuit is not required.

Despite this benefit, measuring fast edges requires consideration of many characteristics such as the bandwidth required for the measurement system. This includes the probe and the oscilloscope as well and possibly the design of the printed circuit board being measured.

Other significant characteristics include the oscilloscope sampling rate, impedance of the probe, and in 50Ω applications, even the selection of coaxial cable.

All of these can impact the accuracy of the results.

Additional discussions related to the probes and coaxial cables can be found in Chapter 5 *Interface Cables and Probes.*

Relating Bandwidth and Rise Time

Oscilloscopes are single order systems in order to provide flat response and no overshoot.

This makes deriving the relationship between the oscilloscope bandwidth and rise time simple. The single order system is modeled as a resistor and capacitor with an RC product of τ.

With that definition, we can calculate the percentage charge of the capacitor in Equation 13.1.

$$\% \, charge = 1 - e^{-\frac{t}{\tau}} \qquad\qquad 13.1$$

Solving for t as a function of the percentage charge results in Equation 13.2.

$$t = -\tau \cdot \ln \left(1 - \% \, charge\right) \qquad\qquad 13.2$$

The time difference between any two charge differences is shown in Equation 13.2.

$$T_{rise} = \tau \cdot \ln(1 - \% \, charge1) - \tau$$
$$\cdot \ln(1 - \% \, charge2)$$
$$= \tau \cdot \ln\left(\frac{V2}{V1}\right)$$

13.3

The rise time from 10 to 90% is, therefore, 2.19722 time constants as calculated in Equation 13.4.

13.4

$$T_{rise_{10/90}} = \tau \cdot \ln\left(\frac{90\%}{10\%}\right) = 2.19722 \cdot \tau$$

While the rise time from 20% to 80% is 1.386 time constants as calculated in Equation 13.5.

13.5

$$T_{rise_{20/80}} = \tau \cdot \ln\left(\frac{80\%}{20\%}\right) = 1.38629 \cdot \tau$$

Solving Equation 13.5 for the time constant, τ, results in Equation 13.6, in this case solved for the 10% to 90% rise time.

13.6

$$\tau = \frac{T_{rise_{10/90}}}{\ln\left(\frac{90\%}{10\%}\right)} = \frac{T_{rise_{10/90}}}{2.19722} = 0.45517 \cdot T_{rise_{10/90}}$$

We can also calculate the bandwidth of the same RC circuit in Equation 13.7.

$$BW = \frac{1}{2\pi \cdot \tau}$$

13.7

Solving Equation 13.7 for the time constant, τ, results in Equation 13.8.

$$\tau = \frac{1}{2\pi \cdot BW} \qquad\qquad 13.8$$

Since we have solved both the rise time and the BW for τ we can equate Equations 13.6 and 13.8.

$$0.45517 \cdot T_{rise_{\frac{10}{90}}} = \frac{1}{2\pi \cdot BW} \qquad\qquad 13.9$$

And now we can calculate either rise time from BW or BW from rise time as in Equations 13.10 and 13.11.

$$T_{rise_{\frac{10}{90}}}(BW) = \frac{0.34966}{BW} \qquad\qquad 13.10$$

$$BW(T_{rise_{\frac{10}{90}}}) = \frac{0.34966}{T_{rise_{\frac{10}{90}}}} \qquad\qquad 13.11$$

The bandwidth of a 4GHz oscilloscope is measured using an Agilent 8257D signal source connected to the oscilloscope using a high quality 50Ω coaxial cable.

The signal source is swept while the oscilloscope spectrum analyzer captures the response to determine the approximate 3dB BW in Figure 13-1. The spectrum analyzer is in maximum hold mode. The BW of the oscillscope measures approximately 4.1GHz. The rise time using Equation 13.10 results in a 10%-90% rise time of approximately 85pS, which is confirmed in Figure 13-2.

The rise time of the oscilloscope is measured using the same high quality 50Ω coaxial cable and a 7pS rise time signal.

The signal is obtained using an 80E10B TDR module in a Tektronix DSA8300 sampling oscilloscope.

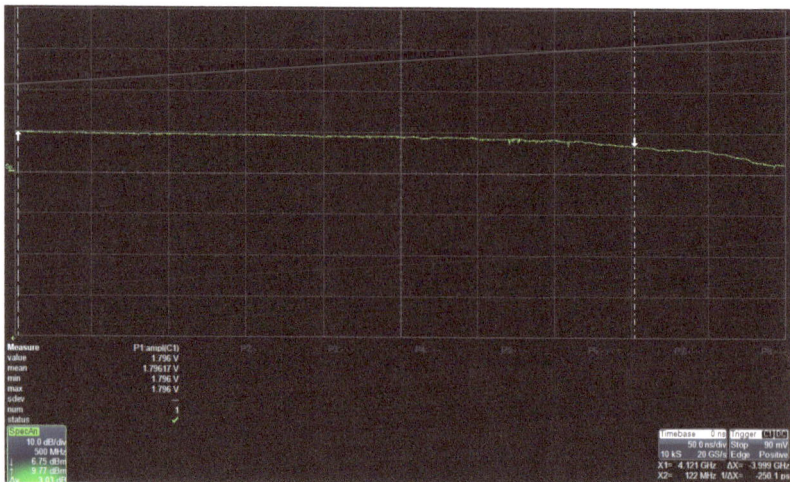

Measurement Showing the 3dB BW at Approximately 4.1GHz

Figure 13-1

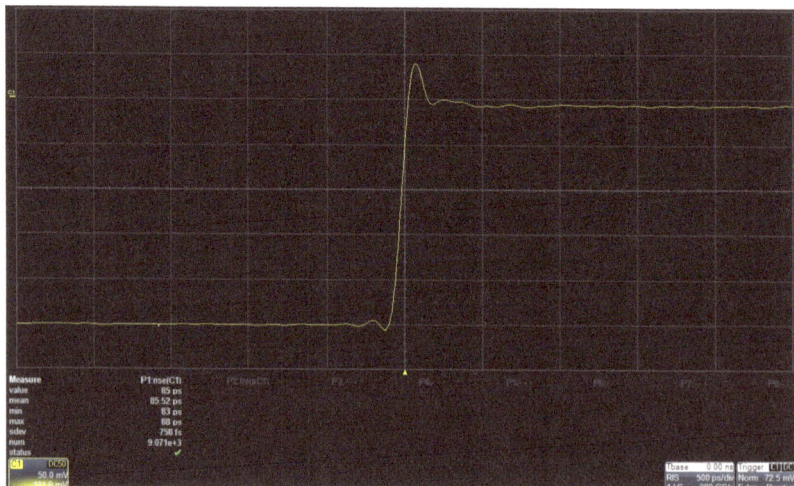

Rise Time Measurement

Figure 13-2

A similar measurement, shown in Figure 3, is performed with a 2GHz oscilloscope. The rise time of the oscilloscope, calculated using Equation 13.10 along with the measured BW of 2.079 GHz results in a rise time of 168pS.

In Figure 13-3, at 1MHz, the 500mV signal measures 499.8mV and at 2.079GHz the signal measures 336mV. This is close to the -3dB bandwidth.

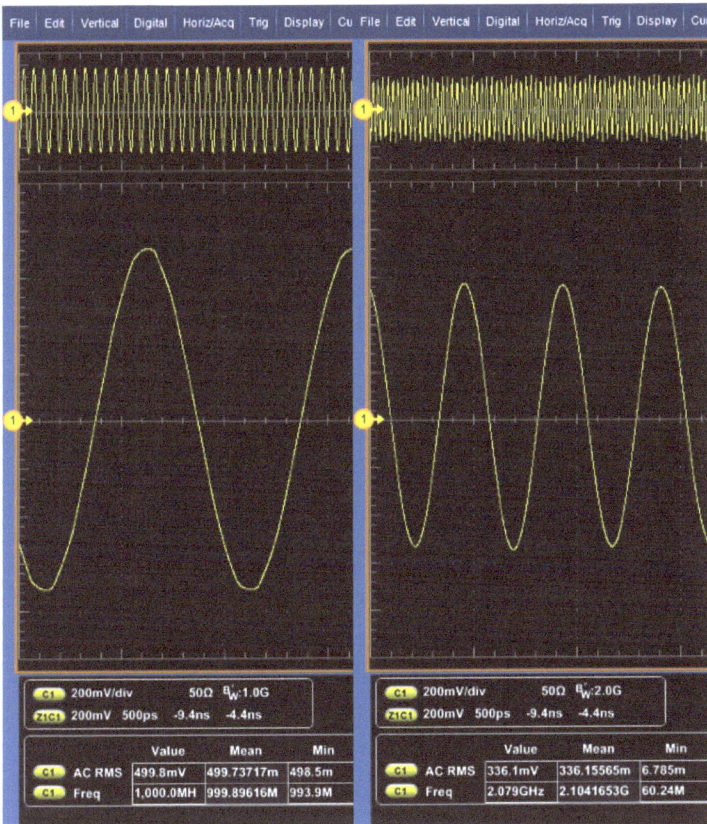

Similar Measurement with a 2GHz Oscilloscope

Figure 13-3

Cascading Rise Times

Cascading two independent rise times, such as the rise time of the signal being measured and the oscilloscope rise time, or the rise time of a probe and the rise time of the oscilloscope, is a root sum square function.

$$measurement\ rise\ time = \sqrt{T_{rise1}^2 + T_{rise2}^2}$$
13.12

We can confirm this using a simple simulation. The simulation schematic shows a simple single order RC network with $\tau=1$.

In one case, the rise time of the pulse is very fast compared to the time constant, while in the other the pulse rise time is set to present a 10% to 90% rise time equal to the rise time of the RC network.

We showed in Equation 13.4 that the rise time of the RC network is 2.91722 τ.

The pulse rise time can therefore be set to:

13.13

$$Pulse\ rise\ time = \frac{2.19722 \cdot \tau}{90\% - 10\%} = 2.747 sec$$

$$Cascaded\ rise\ time = \sqrt{2.19722^2 + 2.19722^2}$$
$$= 3.107\ sec$$
13.14

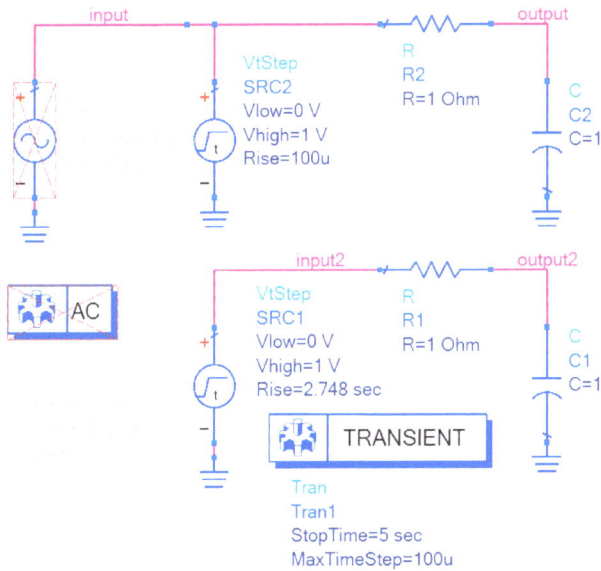

ADS Simulation Schematic for Transient and AC Simulations

Figure 13-4

The simulation results in Figure 13-5 show that the rise time of the RC network is 2.195 seconds, very close to the 2.1972 seconds calculated in Equation 13.4.

The results in Figure 13-6 show 3.125 seconds, again very close to the 3.107 seconds calculated using Equation 13.14.

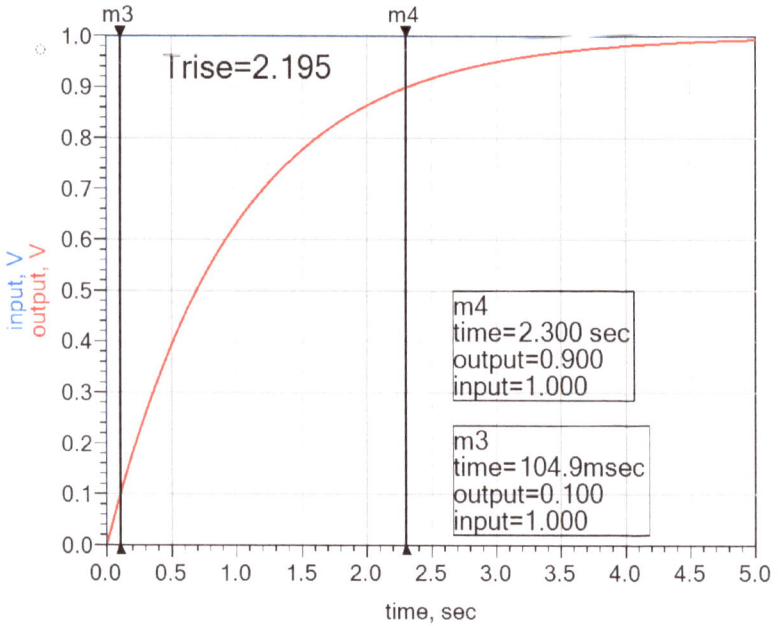

Simulated Rise Time of a Single-Order Stage

Figure 13-5

Calculating the expected BW from this rise time results in Equation 13.15.

$$BW\left(T_{rise_{\frac{10}{90}}}\right) = \frac{0.34966}{2.195} = 159 \; mHz \qquad 13.15$$

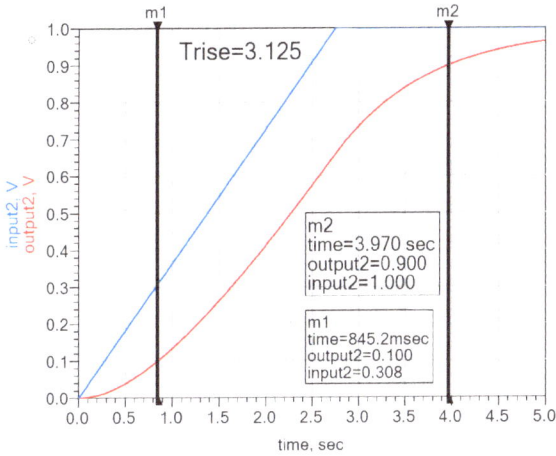

Simulated Rise Time of Two Cascaded Single-Order Stages

Figure 13-6

The AC simulation of the circuit shows the -3dB BW to be 158.5mHz, confirming Equations 13.11 and 13.16.

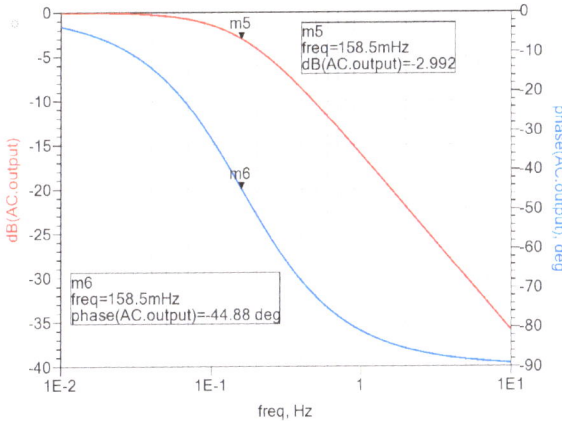

Simulated BW for the Single-Order Stage

Figure 13-7

387

Impact of Filters and Bandwidth Limiting

The impact of bandwidth limiting, or filtering, follows the same bandwidth relationships as the scope itself, with one exception. If the bandwidth limiting is single order, then Equations 13.10 and 13.11 are applicable.

If the filtering is of a higher order, the bandwidth will still impact the rise time, but to a much more severe extent. The rise time measurements of a ≈200pS edge are displayed in Figure 13-8 at three different BW limits, as well as with no BW limiting.

The four traces are skewed in time for clarity.

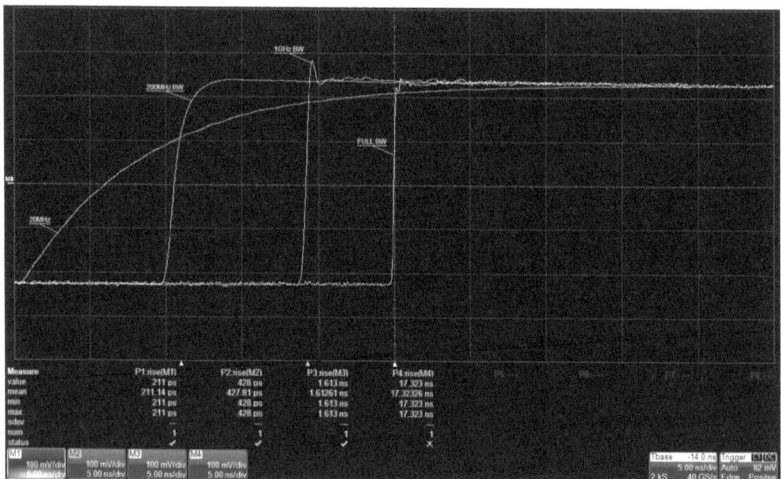

Rise Time Measurements for 20MHz, 200MHz, 1GHz and FULL BW

Figure 13-8

Applying Equations 13.10 and 13.12 to each of the BW limited measurements, assuming that each bandwidth limit is precisely the state value, you can determine the most probable rise time of the pulse being measured using a least squares solver as shown in Figure 13-9.

Given

$$211 \cdot 10^{-12} = \sqrt{Tpulse^2 + \left(\frac{0.34966}{4.1 \cdot 10^9} \right)^2}$$

$$426 \cdot 10^{-12} = \sqrt{Tpulse^2 + \left(\frac{0.34966}{1 \cdot 10^9} \right)^2}$$

$$1.63 \cdot 10^{-9} = \sqrt{Tpulse^2 + \left(\frac{0.34966}{2 \cdot 10^8} \right)^2}$$

$$17.32 \cdot 10^{-9} = \sqrt{Tpulse^2 + \left(\frac{0.34966}{2 \cdot 10^7} \right)^2}$$

$$minerr(Tpulse) = 1.902 \times 10^{-10}$$

Minimum Error Solver used to find the Most Probable Signal Edge Speed

Figure 13-9

The applied pulse edge is likely 190pSec. This estimate can be further improved by measuring the actual BW for each limiter setting.

$$Measured\ rise\ time\ =\ \sqrt{T_{pulse}^{2} + \left(\frac{0.34966}{4.1GHz}\right)^{2}} \qquad 13.16$$

Combining this pulse edge with the 4.1GHz measured oscilloscope bandwidth, the expected measurement rise time is:

$$4GHz\ scope\ rise\ time \qquad 13.17$$

$$= \sqrt{190pS^{2} + \left(\frac{0.34966}{4.1GHz}\right)^{2}} = 208.3pSec$$

The measured value, shown in Figure 13-12, is very close to this result. Performing the same calculation for the 2GHz oscilloscope:

$$2GHz\ scope\ rise\ time \qquad 13.18$$

$$= \sqrt{199pS^{2} + \left(\frac{0.34966}{2.097GHz}\right)^{2}} = 253pSec$$

The measured value, shown in Figure 13-13 is also very close to this expected result.

The relationship between rise time and BW also extends to the harmonic content related to edge speed. Two pulses are simulated in Figure 13-10.

Each signal is a 10MHz square wave. One simulation uses a 5pS edge, while the other uses a 5nS edge.

The harmonic content for the two signals is shown in Figure 13-11. It illustrates the impact of the 5nS rise time on the harmonic content.

Using Figure 13-15 to calculate the BW associated with a 5nS rise time, results in a BW of 70MHz as shown in Figure 13-18.

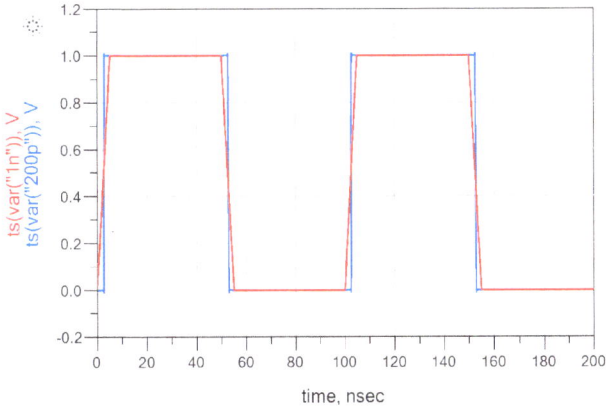

Simulation Pulse with 5Ps (Blue) and 5nS (Red) Edges

Figure 13-10

$$BW\left(T_{rise_{\frac{10}{90}}}\right) = \frac{0.34966}{5nS} = 70MHz \qquad 13.19$$

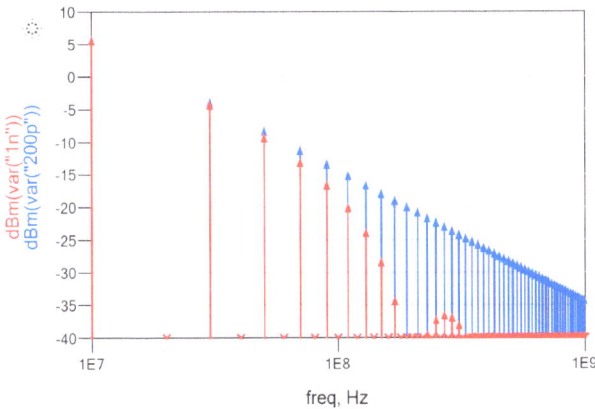

Spectrum Resulting from the 5pS (Blue) and 5nS(Red) Edges

Figure 13-11

Sampling Rate and Interleaved Sampling

The Nyquist criteria states that a signal needs to be sampled more than two times per cycle to avoid aliasing. When it comes to measuring edges, the sample rate needs to be much higher in order to provide reasonable fidelity. The pulse edge in Figure 13-12 is the same 211pS edge used in Figure 13-8.

While acceptable fidelity is obtained at 20GS/s, the fidelity is greatly improved at 40GS/s with 8 samples per edge or 4X oversampling.

Higher sample rates are possible with many oscilloscopes using Random Interleaved Sampling (RIS) or Equivalent time to produce higher "effective" sample rates. Whereas a Real Time oscilloscope records an entire waveform in a single sweep, interleaved sampling records data from many sweeps allowing datapoints to be obtained between the Real Time samples.

One point that may not be obvious is that for the oscilloscope to create an image from a number of samples, it is essential that the signal remain perfectly still.

Since the data is taken from many signal samples, each from a different trigger, it means that each cycle must be identical to every other cycle or the resulting image will be distorted. When measuring fast edges it is possible that there will be enough signal jitter to distort the end result.

Therefore, one should be cautious about using these features, though used properly they can be very helpful.

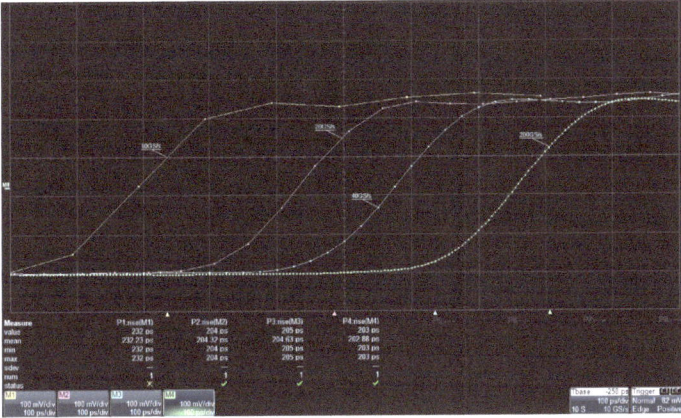

Impact of Sampling Rate on Edge Fidelity 10GS/s, 20GS/s, 40GS/s and Random Interleaved Sampling (RIS) at 200GS/s

Figure 13-12

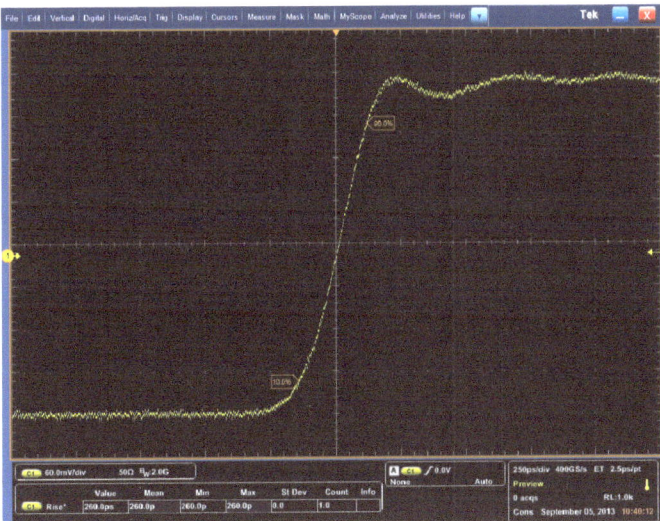

Equivalent Time (ET) Sampling

Figure 13-13

Interpolation

The oscilloscope measures a single amplitude for each sample. The result is a measurement "dot" at each sample. These "dots" are highlighted in Figure 13-14.

The dots are then connected to create the final waveform. Linear interpolation connects these dots is use straight lines between each dot. At low sample rates, the edge can appear quite jagged and the measurement result may be inaccurate.

Another method of interpolation supported by most oscilloscopes is Sinx/x interpolation.

This method draws a smoother line and at low sample rates may provide a more accurate rise time measurement. The measured rise times for 10GS/s using linear and Sinx/x interpolation are shown in Figure 13-14 along with a 40GS/s linear interpolation. The results are as follows:

10GS/s Linear	10GS/s Sinx/x	40GS/s Linear
229pS	208pS	198pS

As shown in Figure 13-14, at 10GS/s the linear interpolation is quite angular and the rise time is approximately 15% higher than the 40GS/s linear interpolation.

The Sinx/x interpolation at 10GS/s is much closer to correct than the 10GS/s linear interpolation.

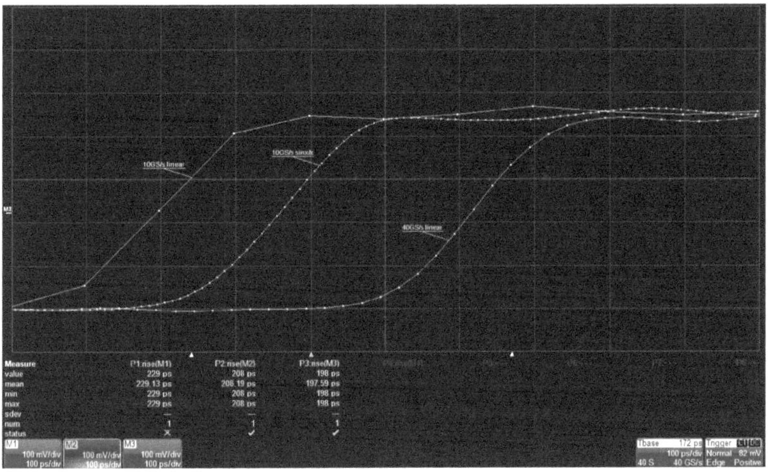

Sinx/x vs Linear Interpolation

Figure 13-14

The measurement should ideally be made at the highest possible sample rate, however, if the sample rate must be low, the Sinx/x often generates better images. Keep in mind that these are both approximations of the data between dots, and so neither is absolute in its measurement accuracy.

Coaxial Cables

The major benefits of using coaxial cables are that if they are properly matched to a 50Ω input, the bandwidth can be very high.

Some cables can even exceed 40GHz.

The coaxial cable presents a controlled impedance and effective shielding. Not all coaxial cables are equal. The variable parameters are impedance, shielding effectiveness, and attenuation. Some cables have double braided shields, while higher frequency cables often have a foil shield, as well as a braided shield in order to improve the high frequency performance.

When measuring high speed edges, these cable improvements can often produce higher fidelity measurements. While a good coaxial cable may be expensive, it is a good investment and with proper care will last many years.

Effects of High Frequency Losses

The loss characteristics of a coaxial cable can play a very significant role in rise time measurements and can clearly illustrate the benefits of a high quality, low loss cable.

The two traces in Figure 13-15 show the rise time of a 200pS edge using two different 2 meter long cables. The signal is measured at 20%-80% due to the leading edge ripple that can disrupt the rise time. Significant impact due to high frequency losses is evident. The 50% rise time is not nearly as impacted by the lossy cable as the 50% to 80% time rise time is. This effect is sometimes referred to as the dribble effect.

An excellent paper from Lawrence Livermore Laboratory is included in the reference section of this chapter and discusses this effect in great detail.

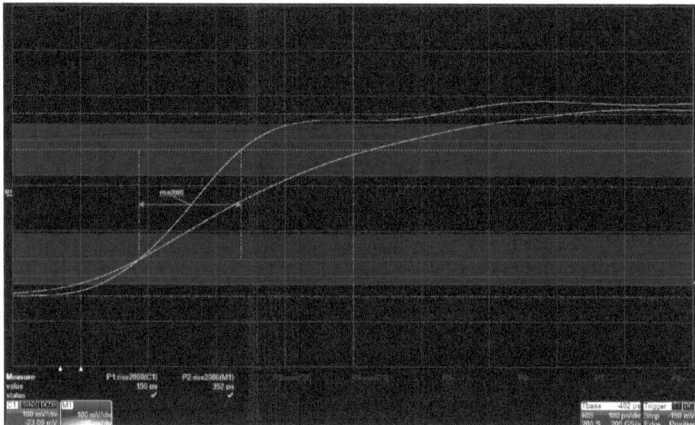

Impact of Frequency Dependent Losses on Pulse Edge

Figure 13-15

The dribble effect is a function of the cable length and the frequency spectrum. Longer cables are more affected and higher frequencies are more affected than lower frequencies.

Velocity Factor and Timing Delay

The speed of light is $2.99792458 \cdot 10^8$ Meters/second. At this velocity the rate it takes for a signal to propagate through an ideal cable is 3.336nS/meter.

Coaxial cables are not ideal and the propagation is reduced by the Velocity Factor (VF). This value ranges from approximately 0.66 to 0.85 for 50Ω coaxial cables.

$$Signal\ Delay = \frac{3.336nS}{m \cdot VF} = \frac{84.73pS}{in \cdot VF} \qquad 13.20$$

Depending on the cable, the delay is typically 99.8-128.4pS/in.

A port splitter is used to simultaneously connect a fast pulse edge to two channels of the oscilloscope.

The measurement in Figure 13-16 shows the time delay using equal length RG174 coax cables.

Both cables are from the same manufacturer and from the same wire lot, so they are very well matched.

The signal on CH2 is deskewed to achieve close to zero delay for the equal length cables as indicated by the 2fS delay. One of the equal length RG-174 cables is replaced with another RG174 cable from the same lot, but 3.25 inches longer than. The delay is then measured and reported by the oscilloscope as 419pS.

The velocity factor for RG174 cable is specified to be 66.0% and, therefore, the expected delay for a 3.25 inch cable length difference is:

13.21

397

$$Signal\ Delay = \frac{84.73pS}{in \cdot 66\%} \cdot 3.25in = 417.23ns$$

The result is very close to the expected value. If you are using edges for timing purposes, it is essential that the cables are well matched and/or the oscilloscope channels are properly deskewed.

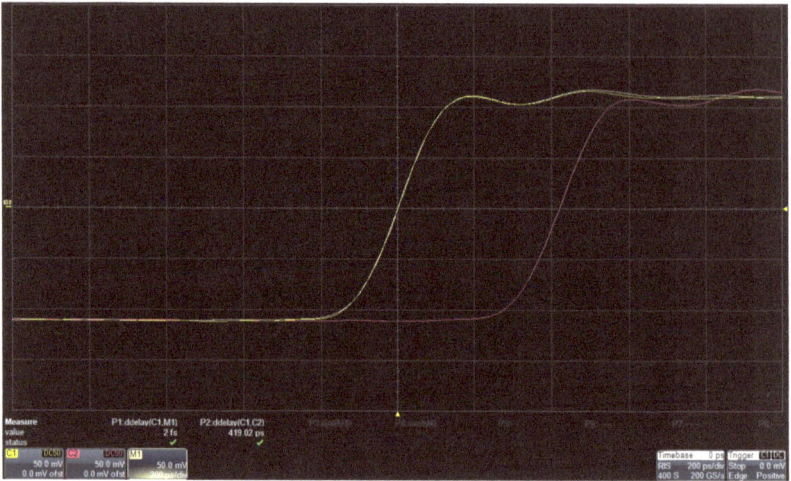

Signal Delay Between one of the Two Equal Length RG174 Coax Cables and the 3.25 Inch Longer Coax Cable

Figure 13-16

The Criticality of the Probe Connection

When measuring high-speed signals, everything in the measurement connection path is critical. All probes and cables have some capacitance, all connections have series resistance and inductance, and there are no perfect 50Ω cables.

In order to demonstrate the sensitivity of these

connections, the high-speed edge signal, terminated into a a 50Ω coaxial thru terminator is measured using a 2.5GHz active scope probe as shown in Figure 13-17.

In this measurement, the ground lead of the probe is connected using a foil ground plate that is included with the probe.

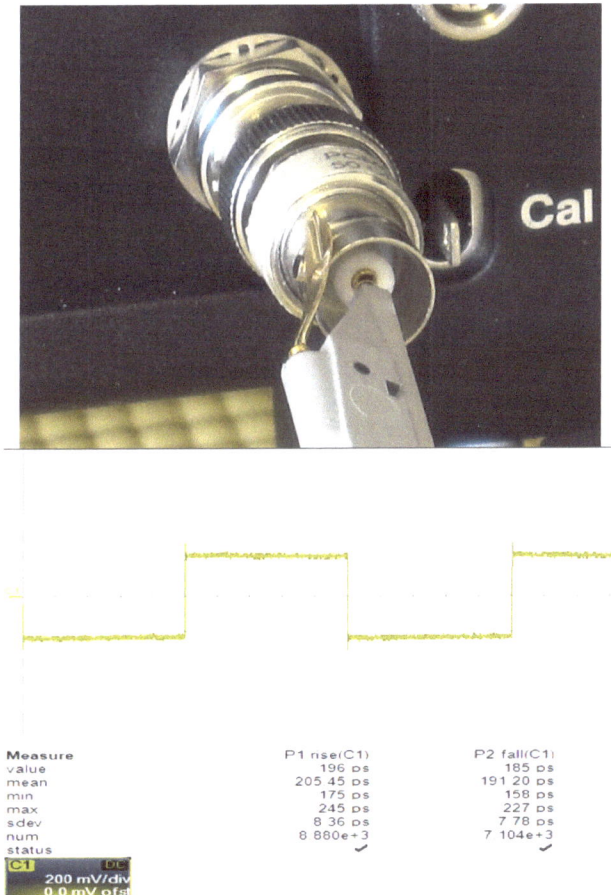

Measure	P1 rise(C1)	P2 fall(C1)
value	196 ps	185 ps
mean	205 45 ps	191 20 ps
min	175 ps	158 ps
max	245 ps	227 ps
sdev	8 36 ps	7 78 ps
num	8 880e+3	7 104e+3
status	✓	✓

C1 DC
200 mV/div
0.0 mV ofst

2.5GHz Active Probe Connected using a Ground Foil

Figure 13-17

The foil ground is replaced in Figure 13-18 with a ground clip that is also included with the probe. The rise time measurement has more than doubled as a result of replacing the short foil ground with the ground clip.

It is essential that you validate your equipment settings, and the potential impact of your interconnection, by measuring a known edge speed signal.

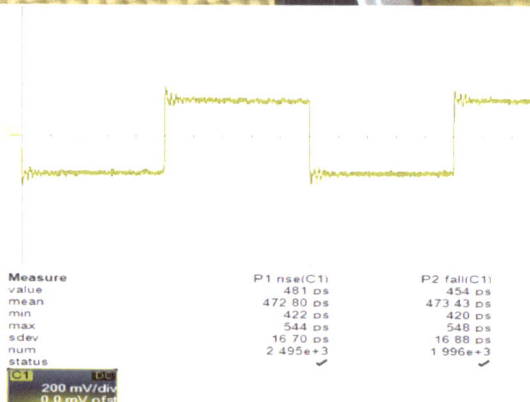

Measure	P1 nse(C1)	P2 fall(C1)
value	481 ps	454 ps
mean	472.80 ps	473.43 ps
min	422 ps	420 ps
max	544 ps	548 ps
sdev	16.70 ps	16.88 ps
num	2.495e+3	1.996e+3
status	✓	✓

C1 DC
200 mV/div
0.0 mV ofst

2.5GHz Active Probe Connected using a Ground Wire and Clip

Figure 13-18

The inductance of a flat wire with length, width and thicknes of l, W and T, respectively, are defined by:

$$L(nH) = 2 \cdot l(cm) \left[ln\left(\frac{2 \cdot l(cm)}{W(cm) + T(cm)} \right) + 0.5 \right.$$
$$\left. + 0.2235 \left(\frac{W(cm) + T(cm))}{l(cm)} \right) \right] \qquad 13.22$$

At very narrow foil widths, the inductance decreases with foil thickness.

This is also true of wire gauges. As wire diameter reduces, the inductance per length increases. The inductance of a one inch long foil is shown in Figure 13-19 for thicknesses of 0.01, 0.05, and 0.1cm and widths from nearly zero to 3cm.

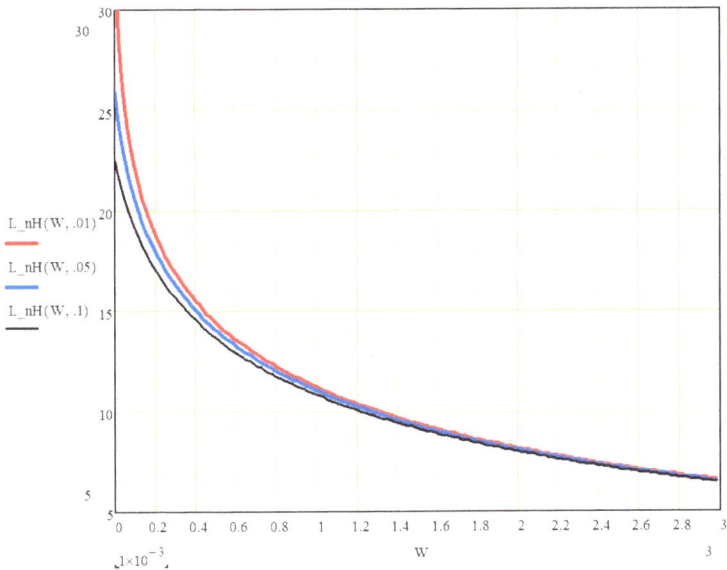

Inductance per Inch of a Copper Foil versus Width for Three Thicknesses

Figure 13-19

Comparatively, the inductance of a round wire with length and diameter in cm is:

$$L(nH) = 2 \cdot l(cm)\left[ln\left(\frac{4 \cdot l(cm)}{dia(cm)}\right) - 0.75\right] \qquad 13.23$$

The inductance of a one inch long round wire with the conductor diameter from nearly 0 to 2cm is shown in Figure 13-20.

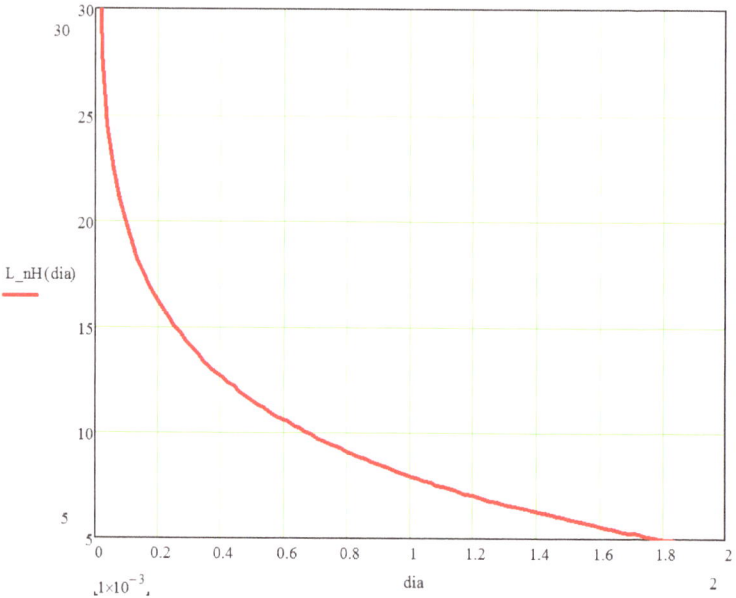

Inductance per Inch of Round Wire versus Conductor Diameter

Figure 13-20

While the ground wire has an inductance of approximately 25nH per inch, the flat foil can be much lower, depending on the foil width. As a rule, copper foil is better than a round wire and should be as wide as possible. The thickness is much less important, though thicker is better. Many hobby supplies offer .016" and .032" thick copper bars in various widths.

Printed Circuit Board Issues

Printed circuit board traces also have inductance defined by Equation 13.22.

A short trace added to bring a switching node to a test point can easily add a few nH of inductance, while adding a header to the circuit board for measurement purposes adds capacitance.

A "BNC" connector can also add 5pF of capacitance, so avoid them unless you are terminating them into 50 Ohms or using SMA connectors.

Probes

While probes are discussed in detail in Chapter 5, it is worth showing just a few comparisons that specifically relate to measuring edges.

The measurement in Figure 13-21 shows the 50 Ohm connection is still best with a 420pS rise time. The 2.5GHz active probe comes close at 470pS, which increased as a result of the reduced measurement bandwidth. A 500MHz passive probe significantly reduces the measurement BW as indicated by the significant increase in rise time to 747pS.

The addition of even a spring ground, which approximately 6nH of inductance, results in ringing, though with improved rise time due to the undercompensated response. The use of a 6" ground clip packaged with the probe results in a 1.73nS rise time.

Example 1: Measuring an Edge with Active and Passive Probes

The measurements in Figure 13-23 compare the rise time of a 50Ω connection, a 2.5GHz active probe, and a 4GHz active probe using a short center pin with a short ground foil (as depicted in Figure 13-17.

The same measurements are repeated in Figure 13-23 except this time the 4GHz probe uses a 2-inch ground wire and clip (as depicted in Figure 13-18).

The results of these measurements are summarized in Table 13.1.

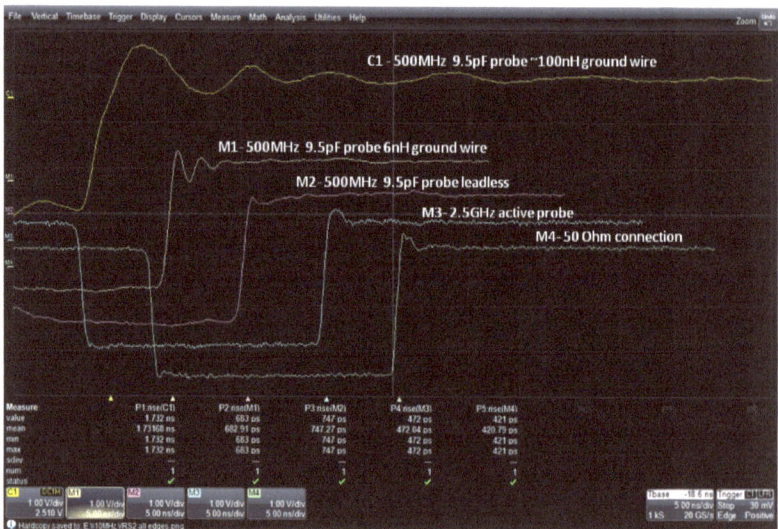

Comparing Rise Time Measurments using 5 Different Probes

Figure 13-21

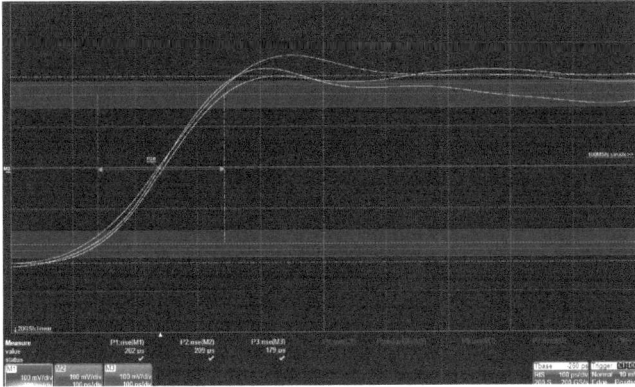

Comparing Rise Time Measurements using a 50 Ohm Connection, 2.5GHz Active Probe and 4GHz Active Probe with a Short Foil Ground as depicted in Figure 13-17

Figure 13-22

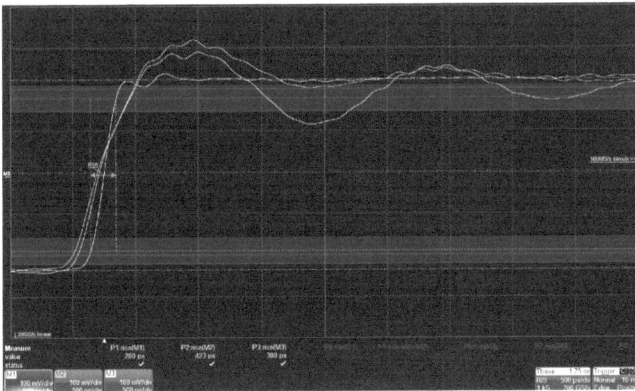

Comparing Rise Time Measurements using a 50 Ohm Connection, 2.5GHz Active Probe and 4GHz Active Probe with a 2-inch Ground Wire and Clip as depicted in Figure 13-18

Figure 13-23

	50Ω	2.5GHz active	4GHz active
Ground foil	202pS	209pS	179pS
2 inch ground/clip	200pS	423pS	380pS

Comparison of results from Figure 13-22 and Figure 13-23

Table 13.1

Example 2: Rise Time of a High-Voltage Probe

This example determines the rise time and bandwidth of a 800MHz high voltage probe, capable of measuring 2500Vpk with 1.8pF capacitance.

The probe is stated to have a rise time of 525pS maximum.

The measurement in Figure 13-24 is the measurement of a 208pS rise time signal.

The oscilloscope is already displaying a reduced oscilloscope bandwidth limit of 800MHz at this scale. If the sensitivity is increased to allow the signal to be nearly full screen, the oscilloscope bandwidth limit will be reduced further.

The oscilloscope rise time measurement indicates 625pS.

Using Equation 13.12, and accounting for three rise time terms, the oscilloscope BW limit, the scope probe rise time, and the source signal rise time, results in:

$$Measured\ rise\ time \qquad 13.24$$

$$= \sqrt{T_{pulse}^2 + \left(\frac{0.34966}{Scope\_BW}\right)^2 + T_{source}^2}$$

Solving for the probe rise time:

$$T_{probe} = \sqrt{T_{measured}^2 - \left(\frac{0.34966}{Scope_{BW}}\right)^2 - T_{source}^2}$$

13.25

$$T_{probe} = \sqrt{625pS^2 - \left(\frac{0.34966}{800MHz}\right)^2 - 208pS^2}$$
$$= 395pS$$

13.26

And using Equation 13.11, we can compute the probe bandwidth as:

$$BW\left(T_{rise_{\frac{10}{90}}}\right) = \frac{0.34966}{T_{rise_{\frac{10}{90}}}} = \frac{0.34966}{395pS}$$
$$= 886MHz$$

13.27

This represents a reasonable margin above the 800MHz probe specification.

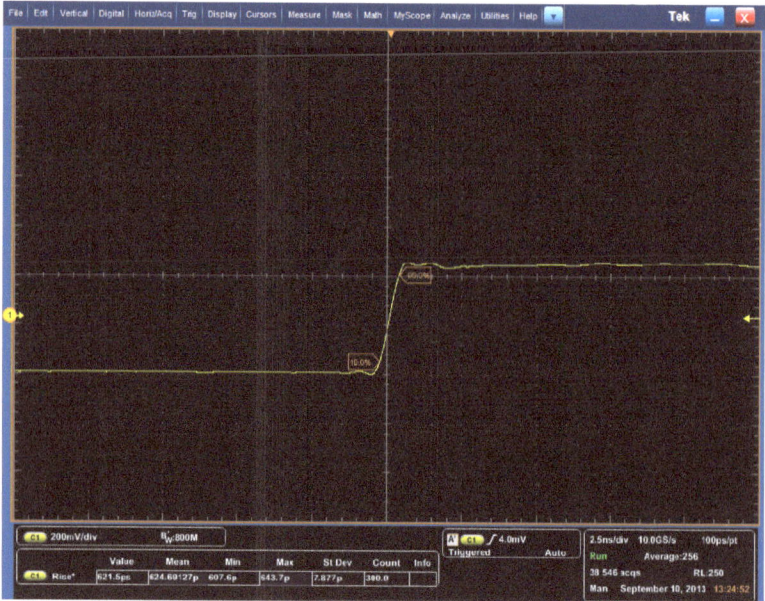

Rise Time of a High-Voltage Probe used to Measure High-Voltage GaN Devices

Figure 13-24

Tips and Tricks

1. Always verify the test setup, equipment, and interconnections by measuring a known value.
2. Be aware that some oscilloscopes reduce the scope bandwidth when using certain probes or high sensitivity ranges.
3. Use a spring clip for the ground or a scope tip to BNC adapter to minimize the length of the ground lead connection.
4. Try to set the measurement to be close to full scale for the best signal to noise ratio.

Chapter References

1. Q. Kerns, F. Kirsten, C. Winningstad Lawrence Livermore Laboratory, University of California, Berkley Feb, 12, 1964, *Pulse Response of Coaxial Cables* http://lss.final.gov/archive/other/lbl_cc_1b.pdf
2. Edward B. Rosa, The Self and Mutual Inductances of Linear Conductors, U.S. Dept. of Commerce and Labor, Bureau of Standards : 1908

Chapter 14

Troubleshooting with Near Field Probes

EMI COMPLIANCE TESTING consists of a far field emissions measurement.

There is no definitive relationship between near and far field measurements. Though near field measurements generally don't directly correlate with far field measurements, the near field measurement is still an excellent troubleshooting tool.

The signal strength in the near field is related to the square of the distance from the source. By getting in close, a particular noise source can be pinpointed thanks to the strong sensitivity to distance from the source. Near field measurements can help in identifying ground plane gaps as well as locating leaks in enclosures, connectors and EMI seals.

Near field measurements can also be used to identify source characteristics that can be used to determine the source of the emissions in a subsequent far field measurement.

The Basics of Emissions

All electrical signals produce electromagnetic emissions.

The emissions are a combination of electric (E) fields and magnetic (H) fields. E fields are generated by voltage more so than current. They present a high impedance in the near field, while H fields are low impedance and generated by current more than voltage. The E and H fields are orthogonally oriented.

In the far-field they converge to a finite impedance defined by the ratio of the fields. The impedance in free space is computed as the product of the speed of light, c, and the permeability of free space, μ, or approximately 377Ω as shown in Equation 14.1.

$$Z = \mu \cdot c = 4\pi \cdot 10^{-7} \cdot 2.99792458 \cdot 10^8 \qquad 14.1$$
$$= 376.73\Omega$$

The E field impedance, Z_E, and the H field impedance, Z_H, are related to the free space impedance and the distance from the source as represented in Equations 14.2 and 14.3.

$$Z_E = Z \cdot \frac{\sqrt{d^2 + 1}}{d} \ \Omega \qquad 14.2$$

$$Z_H = Z \cdot \frac{d}{\sqrt{d^2 + 1}} \ \Omega \qquad 14.3$$

The emission wavelength, λ is calculated in air as a function of the speed of light in Equation 14.4.

$$\lambda = \frac{c}{f} = \frac{2.99792458 \cdot 10^8}{f} \ meters \qquad 14.4$$

In order to determine whether we are measuring the near field of far field, it is important to define the position of the far field boundary. The distance from the source to the far field boundary is generally accepted to be defined as shown in Equation 14.5.

$$Far\ Field = \frac{\lambda}{2\pi} = \frac{2.99792458 \cdot 10^8}{2\pi f}$$
$$= \frac{4.77 \cdot 10^7}{f}\ meters$$
$$= \frac{1.878 \cdot 10^9}{f}\ in$$

14.5

The E field and H field impedance curves are calculated from Equations 14.2 and 14.3 and are presented graphically in Figure 14-1. This figure also shows the convergence to the free space far field impedance of 377Ω. Both the E and H field impedance are approximately 3dB from the convergent impedance at a relative distance of 1.

In the far field, E fields and H Fields converge to377Ω.

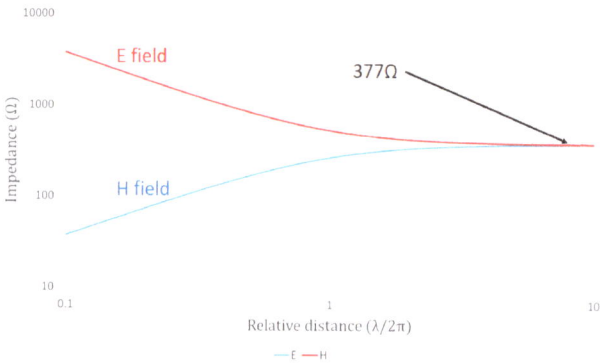

E Field (Red) and H Field (Blue) Impedance as a Function of Distance from the Source

Figure 14-1

While the definitions of the regions are a little fuzzy, the near field is often considered to be less than 10% of the far field distance. The transition region is often considered to be from 10% to 80% of the far field distance. The relative distance in inches from the source $(\lambda/2\pi)$ vs. frequency is shown graphically in Figure 14-2. The far field boundary at 100MHz is approximately 19 inches and approximately 3.8 inches at 500MHz. We can be relatively certain that we are in the near field if our probe is within a half inch from the source.

In Figure 14-2, the far field distance is greater than 2 inches up to approximately 1GHz

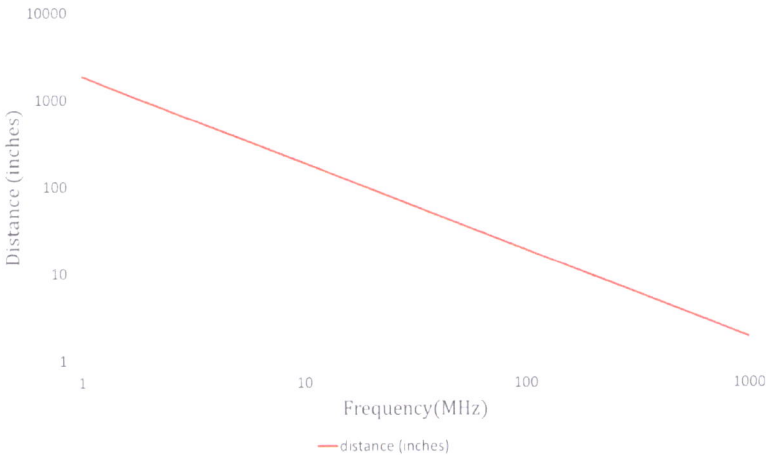

The Far Field Distance in Inches as a Function of Frequency

Figure 14-2

Steven M. Sandler

Near Field Probes

There are many articles that show how to make H field probes at home. While it is certainly possible to construct these probes, there are benefits to purchasing a set from a one of the several commercial suppliers.

Commercial probes are calibrated and offer better symmetry (for shielding and common mode rejection). They can generally be purchased as a complete set, which is usually less expensive than buying the probes individually.

H field probes are symmetrical shielded wire loops, with a very small gap in the shielding at the center of the loop. Different size loops provide a balance between sensitivity and selectivity. The larger probes are more sensitive and, therefore, provide a larger signal. The tradeoffs for the increased sensitivity are reduced selectivity, offering less information about the specific location of the source and reduced high frequency bandwidth.

For this reason, you will want to have several sizes of H field probes.

E field probes are also available with different sensitivity and selectivity choices. The E field probe is generally a "stub" rather than a loop. Similar to the H field probes, the larger the E field probe is the greater the sensitivity. The smaller it is, the more selective it is. Larger E field probes are often spheres, providing a much greater surface area than just an exposed wire tip.

An EMI bundle, which includes a complete set of near field probes, along with a 20dB 0.1Hz-100MHz preamplifier, as shown in Figure 14-3, is available from Picotest. The set includes three different sizes of H field probes, with the largest loop being the most sensitive and the smallest loop being the most selective and operating to the highest frequency.

This set also includes a short stub E field probe.

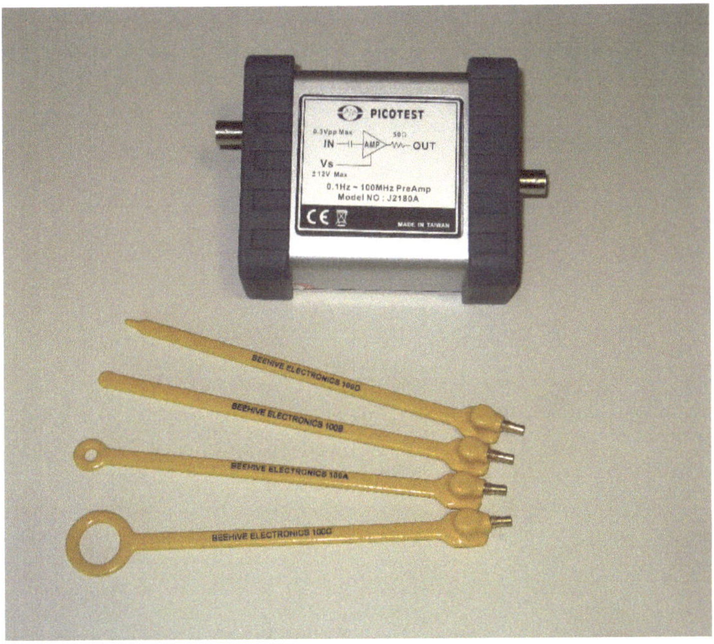

Picotest offers an EMI Bundle that Includes a Complete E and H Field Probe Set and a Low-Noise Preamplifier

Figure 14-3

Probe and Orientation

The sensitivity of the probe is also related to the orientation of the probe with respect to the signal path. This is more so in the H field probes than E field probes, though E field probes are also impacted by orientation. The response of the probe, up to 5GHz or more, is published for the majority of near field probes, though often using a linear scale. In order to show the low frequency performance better, a small fixture was constructed.

As shown in Figure 14-4 it has BNC and SMA connectors with wires between the connectors and a 1Ω shunt resistor.

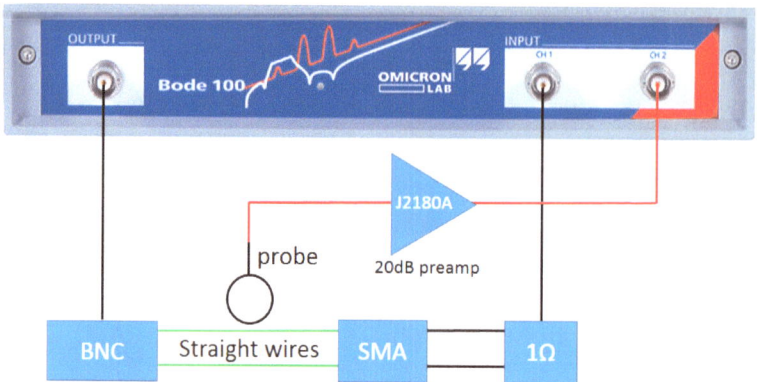

Test Setup for Measuring the Low-Frequency Probe Response Characteristics using a VNA

Figure 14-4

The straight wires provide a straight signal path where current is forced to flow through the 1Ω resistor.

The measurement of the voltage across the resistor is equal to the current in the straight wires. The probe is connected through a 20dB low noise wideband preamp to improve the measurement sensitivity. A picture of the measurement setup is shown in Figure 14-5.

In this image, the 1cm H field probe is positioned on top of and in parallel with the straight wire.

The Picotest J2180A preamplifier—as shown in Figure 14-5—adds 20dB gain to improve the sensitivity of the measurement.

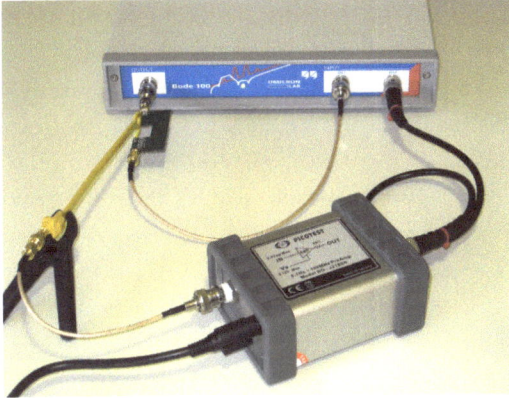

The 1cm H Field Probe is shown Aligned Parallel with and Directly on Top of the Current-Carrying Trace

Figure 14-5

This setup is used to measure the 6cm H field probe and also the E field spherical probe. The 6cm probe is also measured with the probe perpendicular to the wire as shown in Figure 14-6.

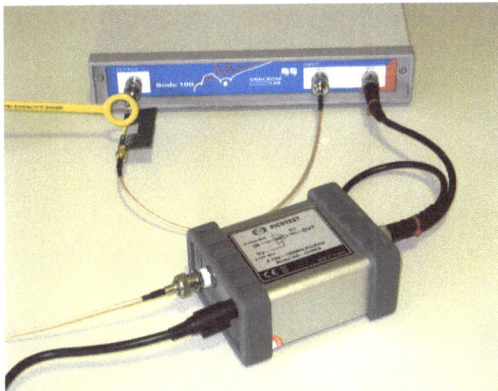

Using the Same Setup the 6cm H Field Probe is Shown Aligned Perpendicular to the Current-Carrying Trace

Figure 14-6

The results of all of the measurements are shown in Figure 14-7.

This plot shows many features of the near field probes. First, the plot shows that at low frequency the probe response is proportional to the signal frequency, limiting the low frequency use to 100kHz or more. Second, this plot shows the sensitivity that is related to the probe size. Third, it shows that the sensitivity is highest when the probe is in parallel with the signal path.

This is an important characteristic as this sensitivity can also help pinpoint a particular trace or wire as a noise source. The measurement is certainly a low impedance measurement, H field dominant.

This too is shown in the figure as the large spherical probe has a much lower sensitivity to H fields than E files. This particular spherical probe is specified to have at least 30dB rejection to H fields.

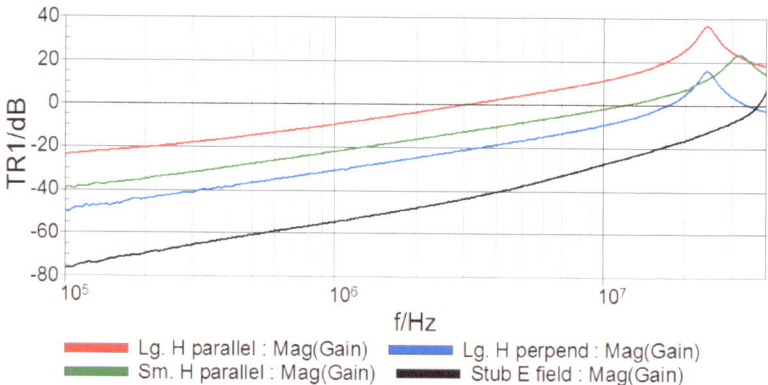

The Transfer Functions for Various Probes and Orientations—the Measured Signal Represents the Transimpedance, or Signal Voltage, as a Function of the Current in the Wire

Figure 14-7

The signal level at the 1Ω resistor is also measured to verify that it is flat over the measurement range. As shown in Figure 14-8, the current is uniform within 1.9dB over the entire measurement range.

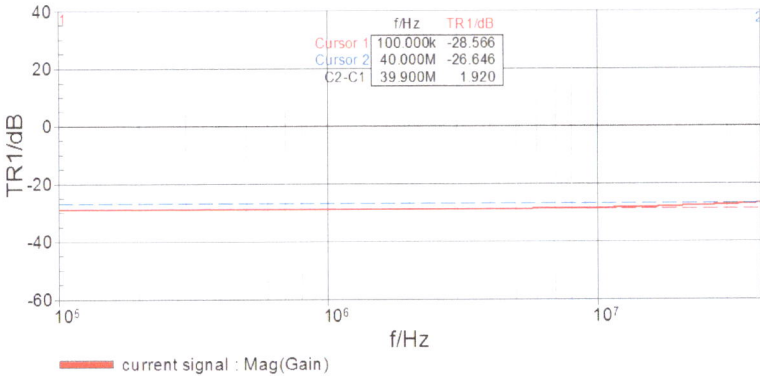

current signal : Mag(Gain)

The Signal Level Measured at the 1Ω Resistor is Flat within 1.9dB over the Measured Frequency Range

Figure 14-8

The Measurement Instrument

The most sensitive instrument for measuring near field signals is the spectrum analyzer, which can typically measure signals of 10's of nV or less.

In many cases, an oscilloscope with spectrum analysis functionality has usable sensitivity, especially with the addition of the preamplifier, which is one reason that it is included in the bundle.

One of the major drawbacks of using the oscilloscope is that they generally have linear time scaling. The limitations of linear time scales are discussed in detail in Chapter 3. One way to circumvent this limitation is to make multiple spectrum measurements with different frequency spans.

The Rohde & Schwarz RTO allows multiple FFTs so that several spans can be viewed simultaneously.

Spectrum Gating

It is helpful when troubleshooting EMI issues to determine the exact source of the noise. The near field probes can be used to isolate the physical location of the noise, but not the association to a particular event. In order to associate the spectrum with an event, the spectrum and time traces must be time correlated. In some oscilloscopes, such as the Tektronix MDO and the Rohde & Schwarz RTO1044 the time waveform and spectrum waveforms can be time correlated. This is often referred to as gating, where the spectrum is obtained from the time measurement within the gate window.

An example of time-gating is shown in Figure 14-9.

A square wave FM modulated 20MHz signal is shown in the upper yellow trace. The upper pink trace is a frequency measurement tracking function of the yellow trace. We can clearly see the square wave modulation. Two gating windows are placed, one during the upper frequency time slice and one during the lower frequency time slice.

An ungated spectrum measurement shows both frequencies at the same time, while the two gated measurements were able to separate the signals.

In Figure 14-9, the blue trace shows the FFT of the entire window, showing both frequencies.

The two gated windows effectively isolated the two frequencies.

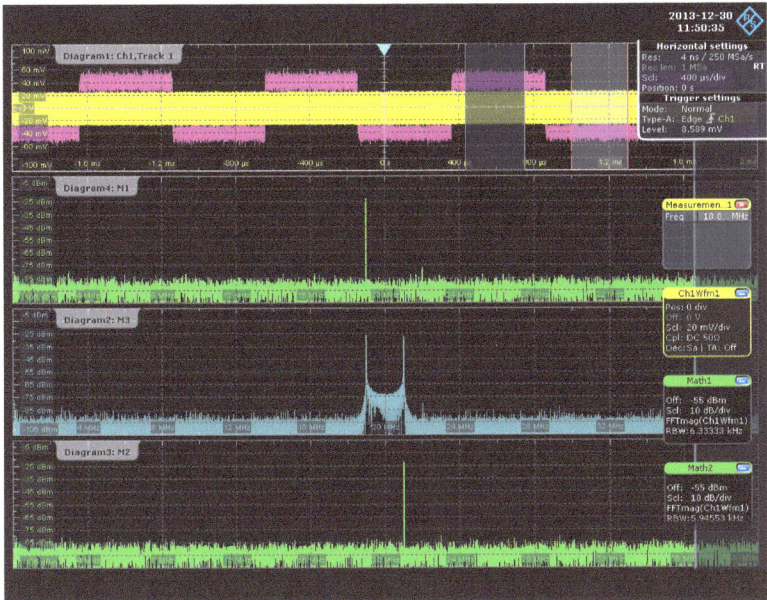

The Upper Trace (Yellow) is a 20MHz FM Signal and the Pink Trace shows the Frequency vs Time

Figure 14-9

In the same way, it is possible to demodulate and AM signal. The scope image in Figure 14-10 shows a 97% modulated 1MHz sine wave with two gating window; one during the high amplitude time slice and one during the low amplitude time slice. The pink trace shows the modulated spectrum without gating. The lower two traces show the spectrum of each of the gating windows. The gating function separated the two amplitudes.

Using this technique it is possible to search the time window to find the troublesome signal in the spectrum view.

In Figure 14-10, the two gated windows effectively isolated the two amplitudes. The green trace in the upper window is the trigger for the modulation.

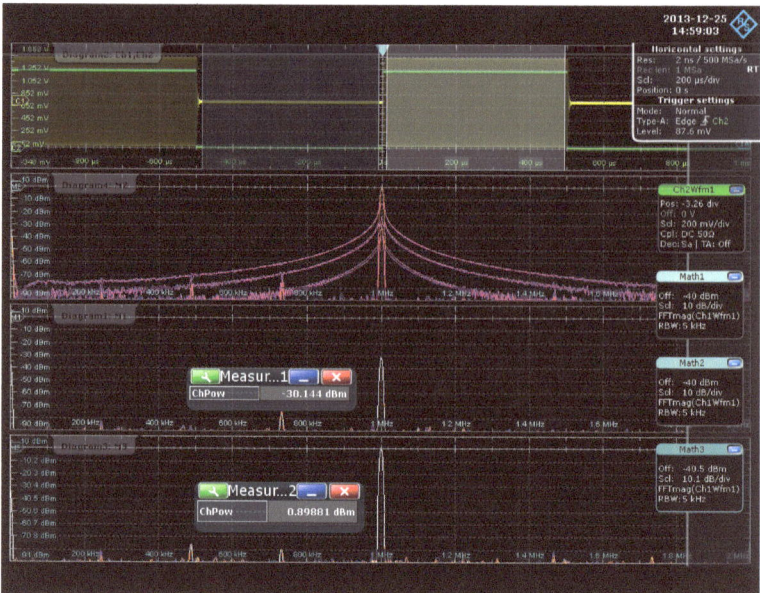

The Upper Trace (Yellow) is a 1MHz AM Signal and the Pink Trace shows the FFT of the Entire Window, Showing all Amplitudes

Figure 14-10

Examples

1. Locating Signal Sources

A near-field probe measurement of a DC/DC converter evaluation board, using a 6cm H field probe, is shown in Figure 14-11. The largest peak is at approximately 30MHz and the second largest is at 160MHz. The 160MHz is easily located near the switch node using a 1cm H field probe. Looking at the switch node shows the leading edge ringing to be at 163MHz and so that is the source of the second largest signal. Improved PCB layout or damping could reduce the emissions from this source.

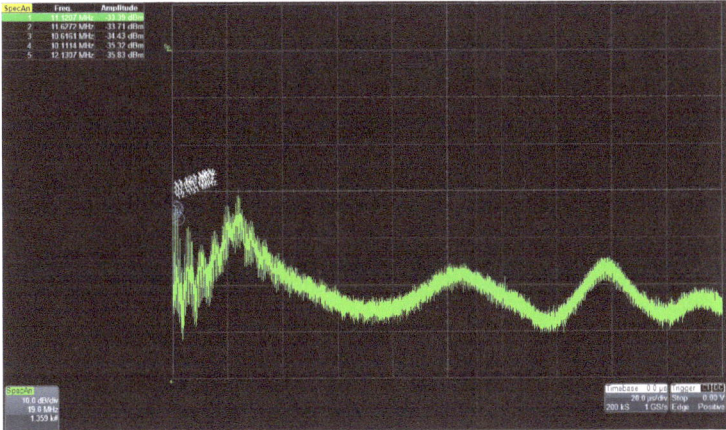

Using a 6cm H Field Probe the Spectrum shows the Strongest Emissions at 30MHz Followed by 160MHz

Figure 14-11

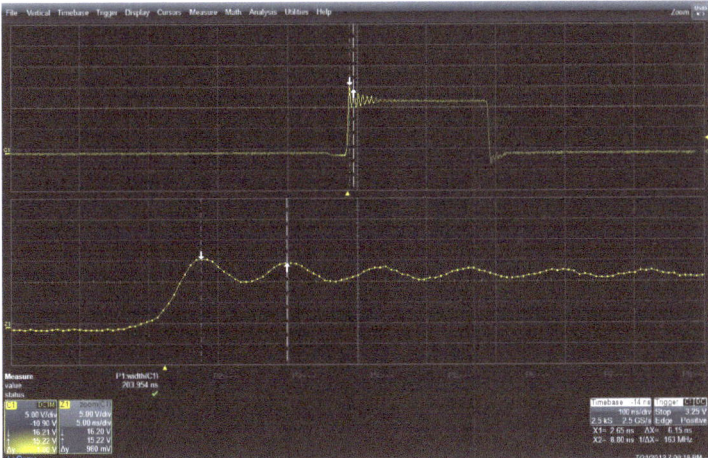

The Ringing at the Leading Edge of the Switch Node is at 163MHz and Clearly the Source of the 160MHz Emissions in Figure 14-11

Figure 14-12

Continuing to search the board with the small near field probe, a very strong 30MHz signal was found near the ceramic input capacitors.

With the board unpowered, the impedance was measured using a single port measurement at the input capacitor. The measurement result is shown in Figure 14-13.

The ceramic capacitor is resonating with the PCB. This resonance could be eliminated by damping this peak or a PCB redesign.

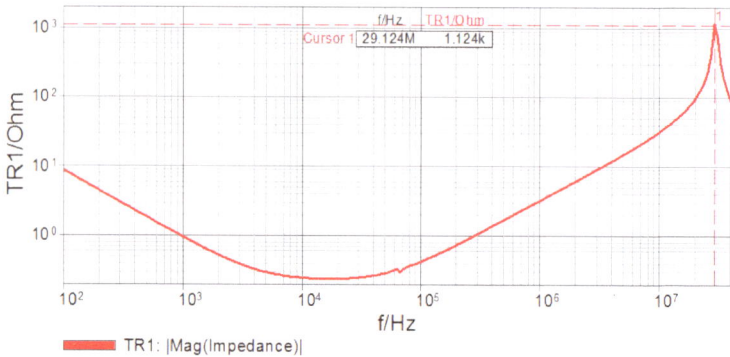

Unpowered Single Port Impedance Measurement at the Ceramic Input Capacitor

Figure 14-13

The measurement in Figure 14-11 also shows a signal at approximately 40MHz. Using the H field probes, the signal is isolated to the area of the switching MOSFET and the PWM controller.

The H field measurement is shown in Figure 14-14.

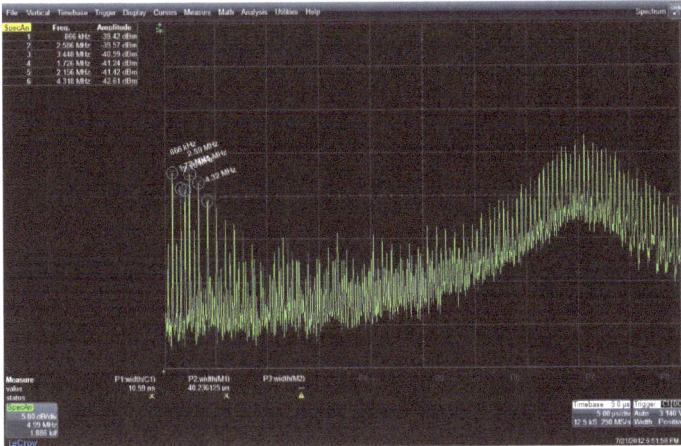

The Measurement using the H Field Probe near the Controller and Switching MOSFET indicates a Broadband Response at Approximately 40MHz

Figure 14-14

If this signal is from the PWM controller, the signal will be mostly current and, therefore, low impedance and predominantly H field.

On the other hand, if the signal is from the MOSFET and not due to current in the MOSFET's connecting traces, then the source of the signal is most likely be the drain voltage, which is high impedance and, therefore, predominantly E field.

Repeating the measurement using the spherical E field probe shows a much larger signal level as shown in Figure 14-15.

Since this is clearly more E field than H field, the signal source is the drain voltage.

One other clue about the source is that the individual signals are all multiples of the 432kHz switching frequency indicating that it is related to the switch period.

In Figure 14-14, the signal is much larger with the E field probe than with the H field probe indicating the signal is mostly E field, likely from the MOSFET drain voltage.

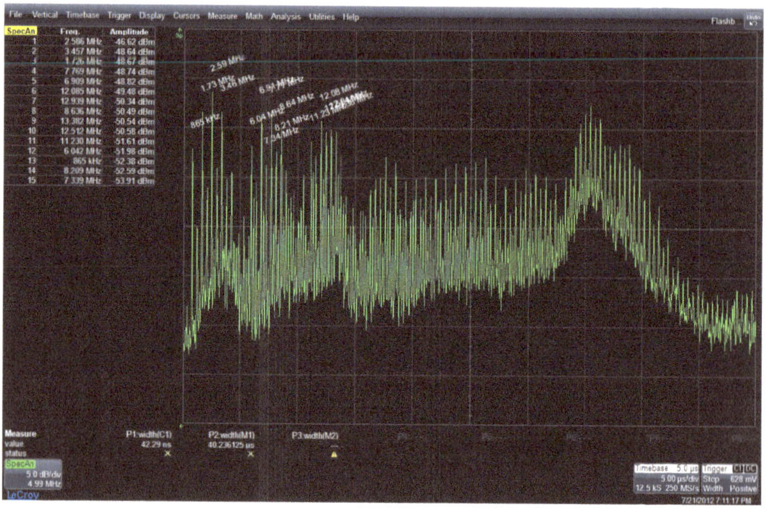

The Measurement from Figure 14-14 Repeated using the Spherical E Field Probe near the Controller and Switching MOSFET

Figure 14-15

2. Identifying Source Characteristics of a Wireless Battery Charger Model

A Qi wireless charger evaluation set is used to demonstrate the identification of characteristics that can be used to identify sources in far field measurements.

A picture of the Qi charger module is shown in Figure 14-16.

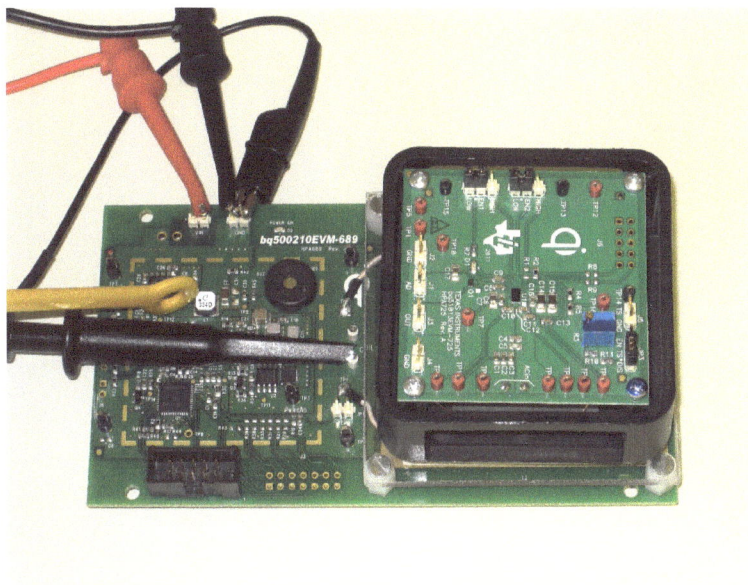

Qi Wireless Charger Evaluation Hardware with a Medium Size H Field Probe in the Vicinity of the 592kHz Buck Regulator

Figure 14-16

As in the prior example, we start with the largest H field probe and scan the unit looking for emission "hot spots."

Depending on the measurement instrument you might want to add a preamplifier, such as the Picotest J2180A used here. The measurement results in Figure 14-17 show the measured signal without the preamp, as well as with the preamp.

This set of measurements clearly shows the benefit adding a preamplifier to the measurement.

The measurement also shows that there are harmonically related and harmonically unrelated signals.

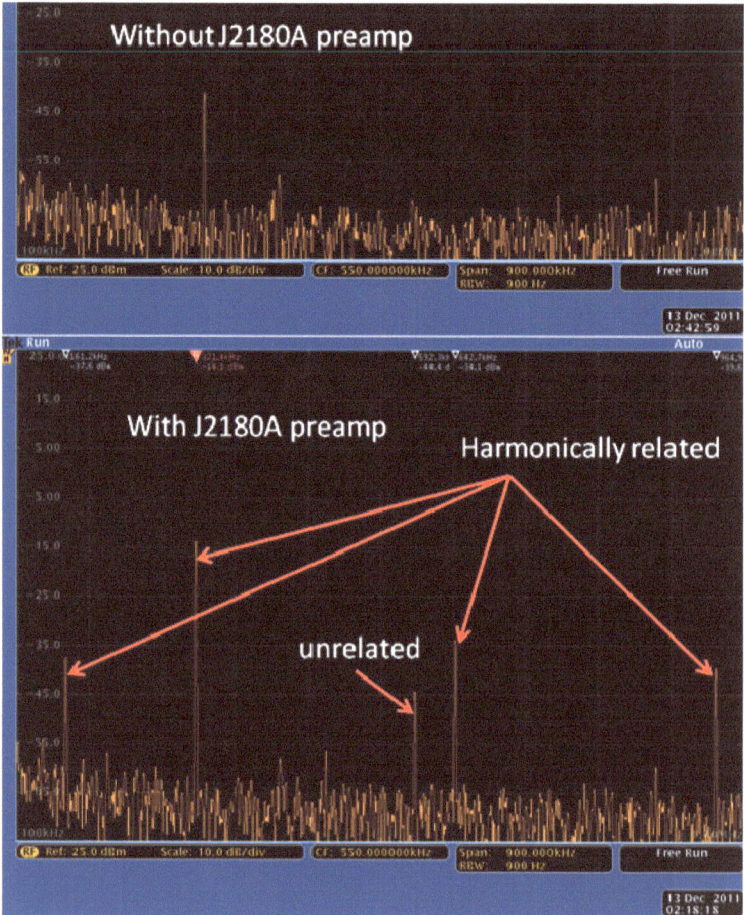

*Scan Results using a 6cm H Field Probe with and
without the Picotest J2180A Preamplifier*

Figure 14-17

Switching from the 6cm probe to the 1cm probe, and scanning the circuit board, three hot spots are identified and they are annotated in Figure 14-18.

Each of these hot spots had particular characteristics that help identify them in a far field measurement.

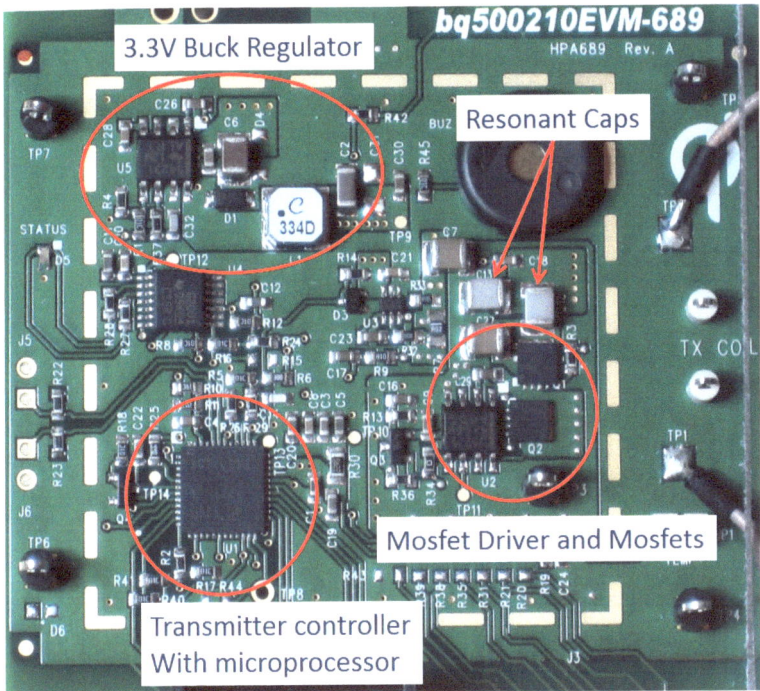

Qi Wireless Power Transmitter identifying Noise Source Locations

Figure 14-18

While an oscilloscope is sufficient in many cases, a spectrum analyzer has much greater sensitivity and a considerably lower noise floor as seen in Figure 14-19.

This measurement is made with the 1cm H field probe in the region of the buck regulator.

In Figure 14-19, note there are signals harmonically related to 140kHz—and those that are not.

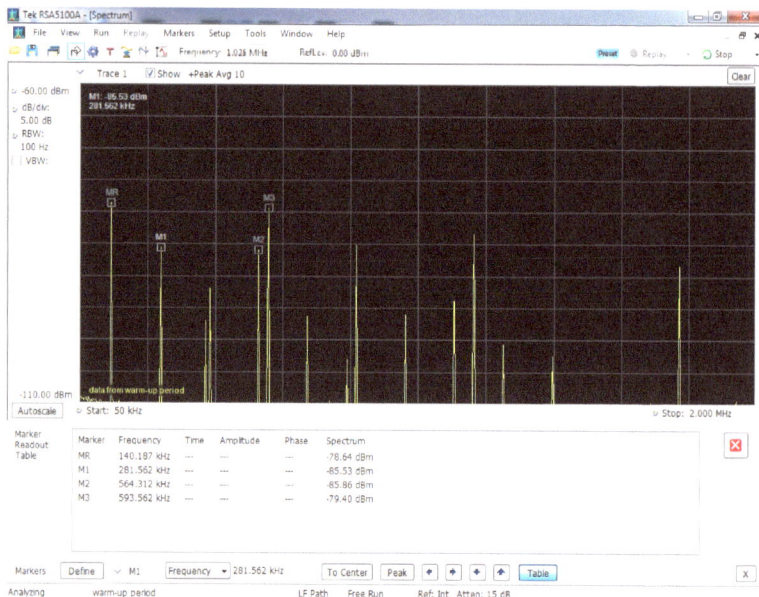

A Spectrum Analyzer has a Much Lower Noise Floor and Much Greater Sensitivity than an Oscilloscope

Figure 14-19

The source of the 140kHz is strongest in the region of the MOSFET and driver.

It is also noted that the 140kHz varies with input voltage, load current, and alignment of the transmitter and receiver. This is a valuable characteristic to note, since in the far field measurement we can easily adjust any of these three parameters to see if it has an effect.

The 592kHz is identified as the buck regulator switching frequency along with all of its harmonics. This signal is not sensitive to input voltage, load current, or alignment of the transmitter and receiver.

The measurement in Figure 14-20 uses the same probe near the microprocessor. While the microprocessor frequency is not specified, it is clearly at 31MHz. We can see the 31MHz and the second harmonic in this image. There is no signal at 15.5MHz, so the fundamental is 31MHz. Any signals identified in the far field measurements that are harmonics of this clock frequency can be associated with the microprocessor clock.

A broadband noise signal is also identified in the region of 25MHz.

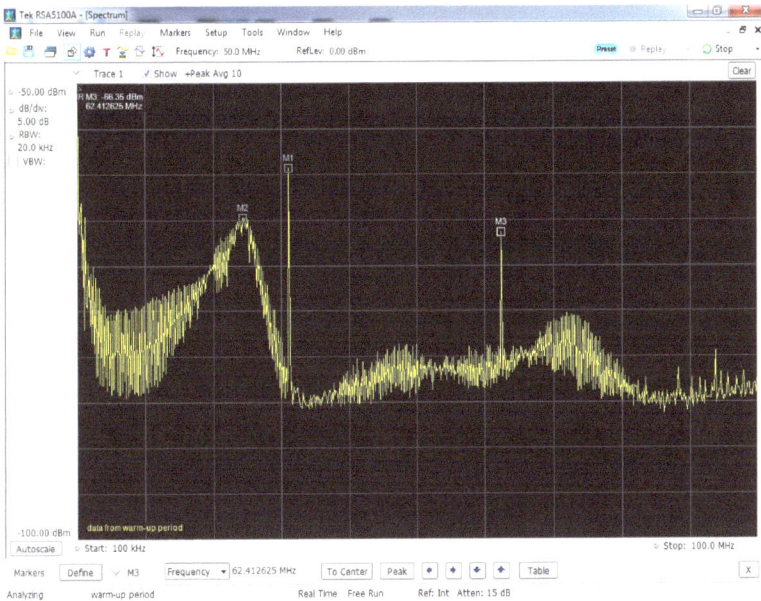

Marker M1 indicates the 31MHz Clock Fundamental and Marker M3 indicates the Second Harmonic of the 31MHz Clock—Marker M2 indicates 25MHz Broadband Emissions

Figure 14-20

The hotspot is identified as being from the MOSFET and driver by scanning the board with both the 1cm H field probe and the spherical E field probe.

The use of both E field and H field probes can help to isolate signals that are the result of current transients and those that are the result of voltage transients.

The 25MHz is predominantly from the MOSFET or driver current, since it is predominantly H field, while the 40MHz is more likely from the MOSFET voltage as it is predominantly E field.

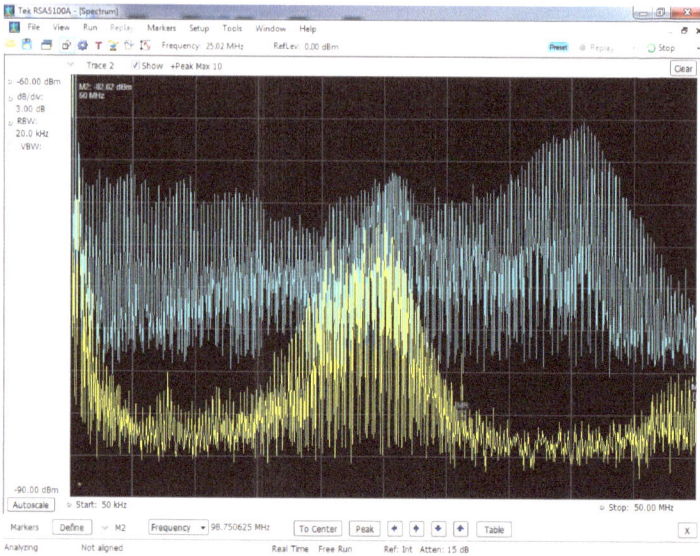

Measuring near the MOSFET and Driver with both the 1cm H Field Probe (Yellow) and Spherical E Field Probe (Blue) shows that the 25MHz is Predominantly from the MOSFET Current while the 40MHz is Predominantly from the MOSFET Voltage

Figure 14-21

These identifying characteristics can be used to identify the source of various signals in subsequent far field measurements as being related to the buck regulator, the MOSFET and driver or the microcontroller.

Emissions that change in frequency with changes in input voltage, load current or transmitter and receiver alignment are from the MOSFET driver area, while the 25MHz broadband is from the MOSFET driver current and 40MHz broadband is form the MOSFET voltage. Signals that are a harmonics of 31MHz are generated by the microprocessor clock. Signals that are harmonically related to 592kHz are generated by the buck regulator.

Since the microprocessor clock and buck regulator frequencies have tolerances, it is a good idea to make these near field measurements on the test unit prior to far field testing in order to determine the exact frequencies of the test sample.

3. Emissions from a High-Speed CMOS Logic Gate

An NC7SZ04 high-speed CMOS inverter is used to buffer a 10MHz clock. Relatively long traces are used to connect the decoupling capacitor to the logic gate through a 0.2Ω resistor. The resistor allows the power supply current feeding the gate to be measured directly. The circuit is shown in Figure 14-22. During each switching edge of the logic gate there is a current spike in its power supply.

This current transient is the result of charging the capacitance of the logic gate and PCB traces and also any logic gate shoot thru current during its switching transitions.

In Figure 14-22, resistor R20 and capacitor C16 provide an AC-coupled 50Ω impedance match for the interconnect of the 10MHz buffer to the test equipment. No connections are made from J5 in these measurements. The logic gate output is capacitively coupled to a 50Ω oscilloscope input. The voltage across the logic gate power supply is measured using a 4GHz 0.6pF active probe. The measurements are shown in Figure 14-23.

A Close-Up View of the CMOS Gate, Decoupling Capacitor and 0.2Ohm Resistor

Figure 14-22

In Figure 14-23, the AC coupled voltage at the Vcc pins (blue) is shown in the lower red trace. This measurement was made with a 10X transmission line probe.

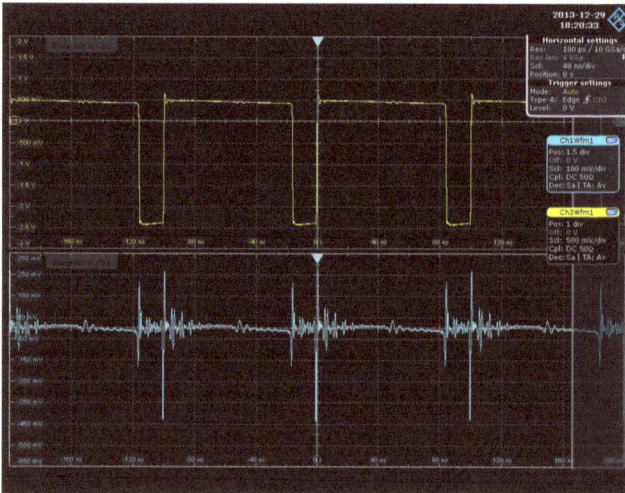

The Upper Trace (Yellow) shows the 10MHz clock Buffer Output

Figure 14-23

The very narrow current signal, during the logic gate transitions, results in a voltage spike greater than 1V. This current is carried through the traces connecting the decoupling capacitor (C14) to the 0.2Ω resistor (R18) and through the trace connecting R18 to the logic gate supply pin. The current flow results in H field emissions.

In order to measure these emissions, the 1cm H field probe is placed directly over the traces connecting the decoupling capacitor (C14) to the logic gate power supply through the 0.2Ω resistor (R18) as seen in Figure 14-24. The measurement is also performed with the spherical E field probe in the same location.

In Figure 14-24, the circuitry in the upper left hand corner is a 2.8MHz switching regulator.

Demo Board Showing the Position of the 1cm H Field Probe over the Traces Connecting the Decoupling Capacitor (C14) to the 0.2Ω Resistor (R18)

Figure 14-24

The emissions are shown in Figure 14-25 with the fundamental at 10MHz. Due to the narrow pulse width both odd and even harmonics are present.

As shown in Figure 14-7, the probe transfer function is +20dB/decade at low frequency up to a peak at 30MHz and the probe sensitivity at 20MHz is approximately the same as the sensitivity at 40MHz. The 10MHz fundamental measurement indicates an amplitude that is 3dB lower than the 20MHz second harmonic. Adjusting for the 6dB lower sensitivity at 10MHz results in an amplitude that is 3dB larger at the fundamental than at the 20MHz second harmonic. A smaller signal at 8.4MHz can also be seen on the far left of the screen without a marker. This signal is the 3rd harmonic of a switching regulator also powered on this board.

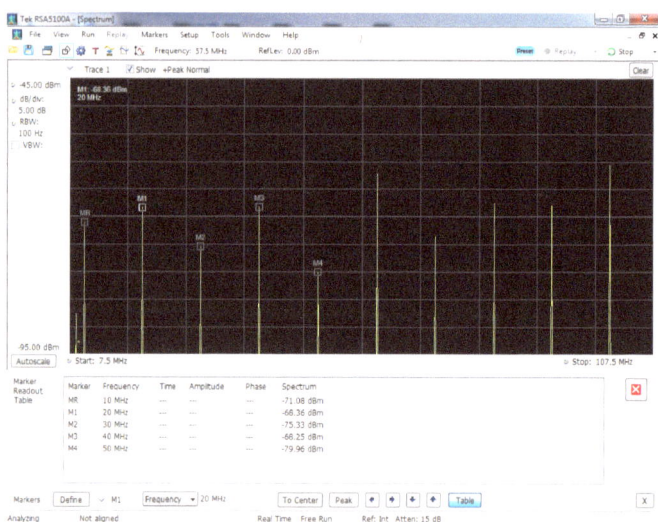

Harmonic Content from the 1cm H Field Probe Located Directly over the Traces Connecting the Decoupling Capacitor to the 0.2Ω Resistor

Figure 14-25

In Figure 14-25, the amplitudes must be adjusted to reflect the probe sensitivity shown in Figure 14-7.

The results of both the E field and H field probes are shown for the frequency range of 5MHz to 3GHz in Figure 14-26. Interestingly, in this case, both E and H fields are present, though not always at the same frequency.

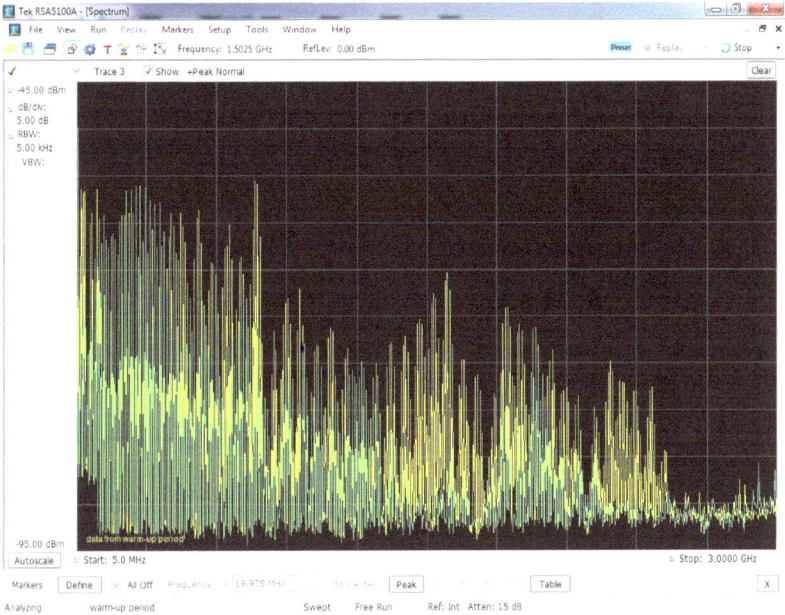

E Field (Yellow) and H Field (Green) Measurement Results

Figure 14-26

While the oscilloscope doesn't generally have log scaling in the spectrum view, some oscilloscopes allow multiple spectrum views with different frequency spans.

The image in Figure 14-27 uses two spectrum views to show the 10MHz and harmonics in addition to the wideband noise to 4GHz.

In Figure 14-27, the blue traces are the near field probe signal and a zoom view of the near field probe signal. Two spectrum traces (green) are used.

The upper spectrum shows the 10MHz and harmonics while the lower spectrum shows the response to 4GHz.

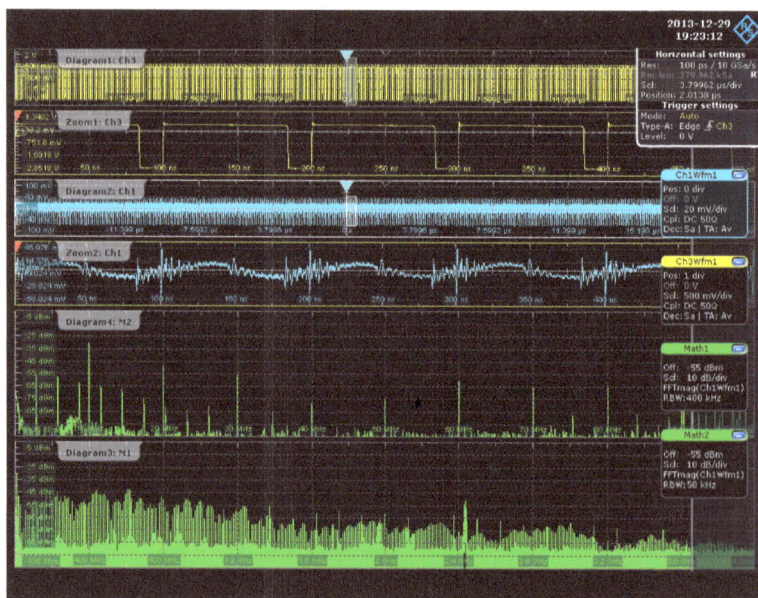

The Top Traces (Yellow) are the 10MHz Clock and a Zoom View of the Clock

Figure 14-27

The position of C14 and the trace routing on this board were created so that this characteristic could be demonstrated.

The emissions can be significantly reduced by relocating C14 close to the logic gate power supply pins and using very short connecting traces.

Tips and Tricks

1. Start with the largest size probe and the widest frequency bandwidth. Once you have an idea where the signal is physically located and the frequency range of interest you can reduce the probe size and the measurement span to get better selectivity in order to pinpoint the source.
2. The largest signals occur when the H field probe is aligned parallel with the radiating conductor.
3. Measure using both E and H field probes. This will help to determine the likely source.
4. Reducing the resolution bandwidth improves the noise floor and dynamic range of the instrument at the expense of a slower sweep.
5. When using a handheld near field probe it is helpful to have the fastest possible sweep so that there is a real time connection between your placement of the probe and the image displayed.
6. Balance the frequency span and resolution bandwidth to obtain a reasonably fast response, as well as a tolerable dynamic range and noise floor.
7. While an oscilloscope can be sufficient in many cases, the use of a spectrum analyzer provides higher resolution. In addition the sweep speed is generally faster.
8. When using an oscilloscope spurious responses may appear due to the sampling rate. It is also possible that there are spurious responses from other equipment operating nearby. Always measure with the power off so that any stray signals can be identified.
9. Not all analyzers contain full wave detectors and not all probes are symmetrical. Rotate the probe completely to be sure that you do not miss a potentially problematic signal.

10. Make these near field measurements on the test unit prior to far field testing in order to determine the exact frequencies of the test sample. Tolerances change these frequencies from unit to unit.
11. It is important to account for the probe transfer function in order to evaluate the measured spectral amplitudes.
12. The signal level of the near field probe below the probe resonance is generally proportional to the signal frequency. For this reason, the fundamental signal is not always the largest, which can be misleading.

Chapter References

1. Charles Capps, *Near Field or Far Field?*, EDN Aug 16, 2001
   http://www.edn.com/design/communications-networking/4340588/Near-field-or-far-field-
2. *Measuring Wireless Power Charging Systems for Portable Electronics*, Application note, Tektronix, 48W_28034__1_MR_letter.pdf
   http://www.tek.com/document/application-note/measuring-wireless-power-charging-systems-portable-electronics
3. Picotest, *Troubleshooting EMI: Use Versatile Instrument And Preamp To Search For Embedded Noise*, www.picotest.com/blog/?p=908
4. Picotest, *Effectively Using the EM-6992 Near Field Probe Kit to Troubleshoot EMI Issues*
   https://www.picotest.com/blog/?p=936
5. Signal and Noise Measurement Techniques
6. Douglas C. Smith, *Using Magnetic Field Probes*, IEEE 1999 EMC Symposium Proceedings
   http://www.emcesd.com/pdf/emc99-w.pdf
7. Vladimir Kraz, *Near-Field Methods of Locating EMI Sources*, Compliance Engineering May/June 1995
   http://www.onfilter.com/library/NEARFIELD.pdf

Chapter 15

Higher Frequency Impedance Measurements

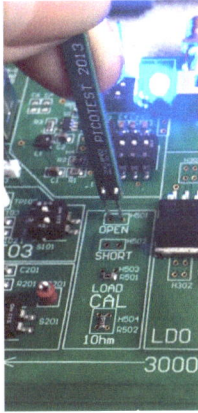

HIGH-SPEED CIRCUITS depend on proper impedance matching to assure signal and power integrity. In the case of high-speed FPGAs, it is often necessary to maintain a controlled power supply impedance to 10GHz or more.

High-speed digital and RF signals require carefully controlled impedance paths between drivers and receivers to assure high fidelity data signals. Inadequate impedance control can result in degraded circuit performance, failure of the circuit to function, and in extreme cases, can result in permanent damage to the sensitive high-speed circuits.

Measuring these impedance paths is often necessary in order to troubleshoot a new design, as well as to verify the final design. High frequency measurements can be made using time domain measurement or using S-parameters. These two measurement methods are often combined in a single instrument.

Time Domain

Time domain impedance measurements are performed using either time domain reflectometry (TDR) or time domain transmissometry (TDT). TDR is a single port measurement based on the measurement of the incident and reflected signals of the port, while TDT is a two-port measurement based on the incident signal measured at the receiver port. The time domain measurement is helpful for locating particular issues in a transmission path, as well as measuring the transmission line impedance.

It takes a finite time for an electrical signal to travel through a transmission line. The speed of light in a vacuum is $2.99792458 \cdot 10^8$ m/s; slower when traveling through other mediums. The ratio of the speed in a dielectric to the speed in air is the velocity factor (sometimes referred to as fractional velocity). The velocity factor is related to the dielectric constant of the medium as:

$$F_{velocity} = \frac{1}{\sqrt{\varepsilon_r}} \qquad \text{15.1}$$

The dielectric constant (ε_r) of a transmission line is dependent on the insulator material, but ranges from a low of approximately 1.3 for foam polyethylene to 2.3 for polyethylene while a typical FR4 printed circuit board is approximately 3.6-4.5.

The dielectric constant of some common materials and their respective velocity factors are shown in Table 15.1.

The signal travels through the transmission line at a velocity of:

$$Signal_{velocity} = \frac{1}{\sqrt{\varepsilon_r}} \cdot 11.8 \; in/ns \qquad \text{15.2}$$

The TDR measurement uses a single port so the signal must travel the length of the transmission path twice, once to go from the TDR instrument to the end of the measurement and once for the signal to return to the instrument from the end of the measurement.

The TDT uses two ports so the signal only travels the length of the transmission path once.

Insulator	Dielectric Constant (ε_r)	$F_{velocity}$	Signal velocity (in/ns)
Foam polyethylene	1.3	0.88	10.38
Polyethylene	2.3	0.66	7.79
FR4	3.6-4.5	0.53	6.25
Air	1	1	11.80

A Few Common Dielectric Constants and the Resulting Signal Velocities

Table 15.1

TDR

The reflection coefficient (ρ or Rho) is the ratio of the reflected voltage to the incident voltage and is related to the measurement instrument port impedance (Z_o) and the impedance of the DUT (Z_{dut}).

$$\rho = \frac{V_{reflected}}{V_{incident}} = \frac{Z_{dut} - Z_o}{Z_{dut} + Z_o} \qquad 15.3$$

The value of ρ ranges from -1 for $Z_{dut}=0$ to +1 for $Z_{dut}=\infty$. The value of ρ if $Z_{dut}=Z_o$ is zero meaning that there is no reflected signal.

The TDR instrument measures the reflection coefficient, Γ as shown in Equation 15.4.

$$\Gamma = \frac{1 + \rho}{1 - \rho} \qquad\qquad 15.4$$

Most TDR instruments can also display impedance directly, which is calculated from Γ as shown in Equation 15.5.

$$Z_{dut} = \Gamma \cdot R_{ref} \qquad\qquad 15.5$$

Calibration

It is essential to properly calibrate the instrument prior to making any measurements or the results can be significantly incorrect.

In the case of reflectometry, the calibration required is the Short Open Load calibration and for the transmissometry the calibration is the thru calibration. The calibration capabilities and procedures vary significantly from instrument to instrument. The Tektronix DSA8300 has a simple process for a complete Short Open Load and THRU (SOLT) calibration. In order to assure that the calibration and measurement setup are acceptable it is best to measure a known quantity before making new measurements.

After calibration, the instrument can easily measure a 10mΩ resistor, as shown in Figure 15-1. The Agilent E5071C has SOLT calibration for the VNA mode of operation, but only deskew and loss compensation for the TDR instrument. The Lecroy SPARQ can de-embed cables and fixtures using a 2nd tier calibration, though the process is somewhat more complicated. It is important that the calibration fixture be exactly the same as the DUT in order to completely de-embed the cables and fixtures.

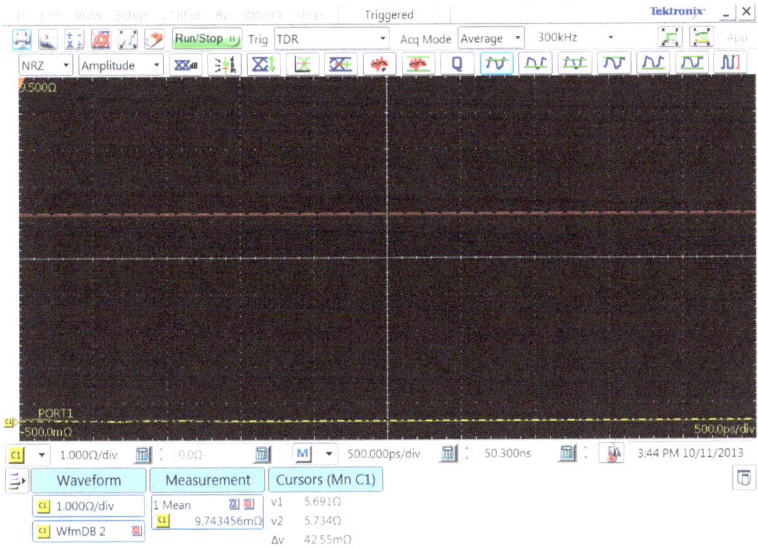

This Measurement Made with a Tektronix DSA8300 using an 80E10B Sampling Module confirms the Measurement of a 10mΩ Resistor after Performing Short-Open-Load Calibration

Figure 15-1

Reference Plane

The measurement reference plane is the point at which the instrument time=0. The reference plane can be relocated to the DUT input by performing a deskew adjustment with the transmission cables connected up to, but not including the DUT.

In some instruments, such as the Agilent E5071C, this deskew is performed automatically.

The image in Figure 15-3 shows the resulting measurement after the instrument was deskewed at the front panel with nothing connected to the instrument. The deskew calibration results in time being defined to be zero at the front panel connector.

In Figure 15-2, the very small artifact just before t=0 is the internal connection of the instrument to the front panel connector.

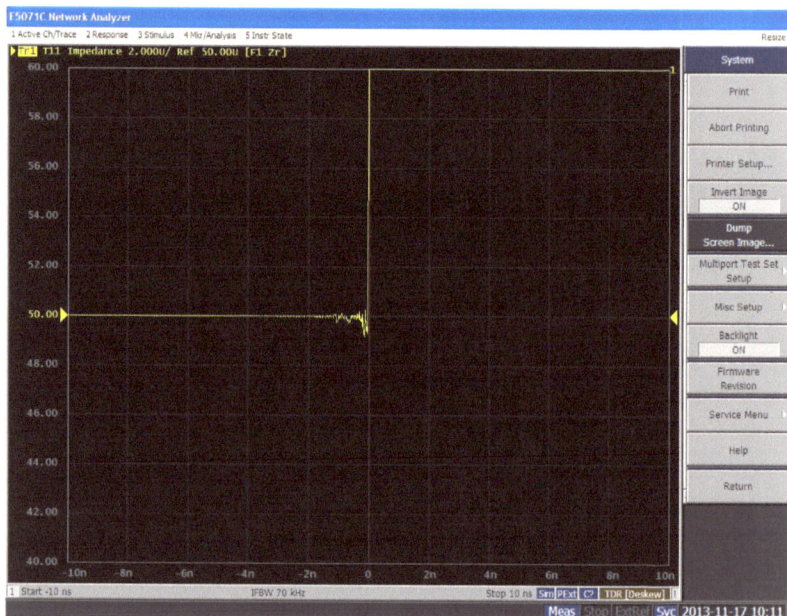

Measurement after Performing Deskew with Nothing Attached to the Front Panel

Figure 15-2

A high quality 50Ω, 18GHz, low-loss coaxial cable is then attached to the instrument and the TDR measurement is performed. The small artifact at t=0 is due to the interface between the connector of the coaxial cable and the connector of the instrument.

The impedance of the coaxial cable is directly displayed, and in this case, the impedance of the cable is approximately 49Ω. The cable impedance is relatively flat, which is representative of a low loss cable. Once the signal reaches the end of the cable, the reflection is sent back to the instrument port. In this measurement, it took the signal 3.8ns to travel from the instrument port to the end of the cable and back to the instrument. Since the cable is open at the far end, the signal impedance instantly increases at the end of the cable.

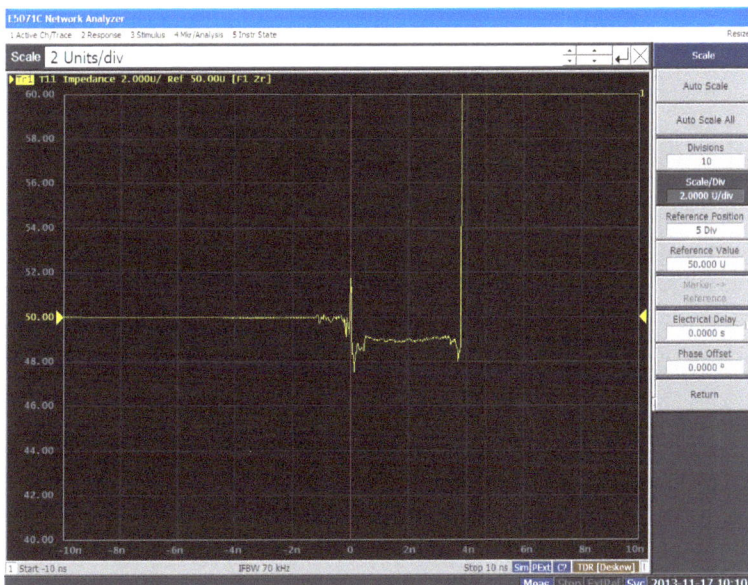

TDR Measurement with a High-Quality 18GHz Semi-Rigid Cable Connected to the Deskewed Instrument

Figure 15-3

In Figure 15-3, the deskewed reference is set at t=0ns and the small artifact at this point is due to the connection between the instrument and the high quality cable. The signal velocity is such that the signal traveled the length of the cable twice in approximately 3.8ns. Also note the flatness of the cable which is representative of the cable loss.

The measurement in Figure 15-4 shows two interfaces. The instrument is connected to a high quality semi-rigid coaxial cable and then connected to an FR4 PCB stripline via a low quality edge launch SMA connector. The instrument is deskewed with the semi rigid coaxial cable connected to the instrument, but not to the stripline.

This sets the time reference to zero at the end of the semi-rigid coaxial cable. The small artifact at approximately -4.6ns is due to the connectors between the instrument and the semi-rigid coaxial cable. The measurement from this point to the reference plane shows the impedance of the coaxial cable, which is very close to, and just above, 50Ω.

The slight impedance slope of the semi-rigid cable is the result of the cable loss.

The low quality SMA connector at PCB stripline results in the large artifact at t=0.

In Figure 15-4, the reference plane (t=0) is set to the end of the semi-rigid cable.

The end of the cable is then connected to an FR4 PCB stripline via a low quality edge launch SMA connector.

Note the instrument to semi rigid interface at approximately -4.6ns, the large artifact from the low quality cable and the slope of the stripline impedance indicating the higher loss than the semi rigid cable, though very close to 50Ω.

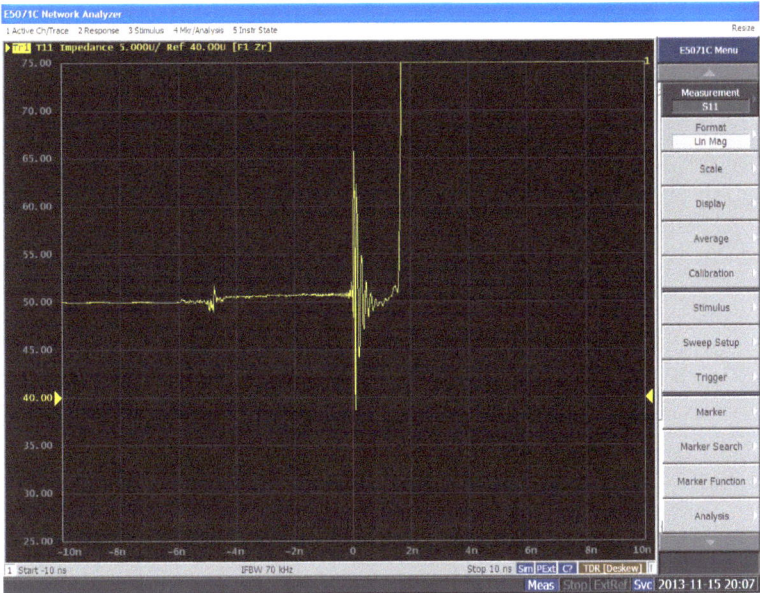

Instrument Connected to a High-Quality Semi-Rigid Cable and then Deskewed

Figure 15-4

The image in Figure 15-5 shows an assembly of the TDR instrument connected to a complete transmission path consisting of an SMA adapter, SMA coupler, PCB trace, SMA coupler and Short, Open and Load terminators.

These elements are added individually to illustrate how the impedance changes with each addition to the transmission path and the impact of each connection.

The corresponding measurements are shown in Figure 15-6 where the yellow trace shows the relatively constant 50 Ohm impedance path when the 50Ω terminator is added.

The poor quality PCB connector is clear in this trace as well. The poor quality PCB connector is evident from the ringing that occurs at the point that the PCB is inserted.

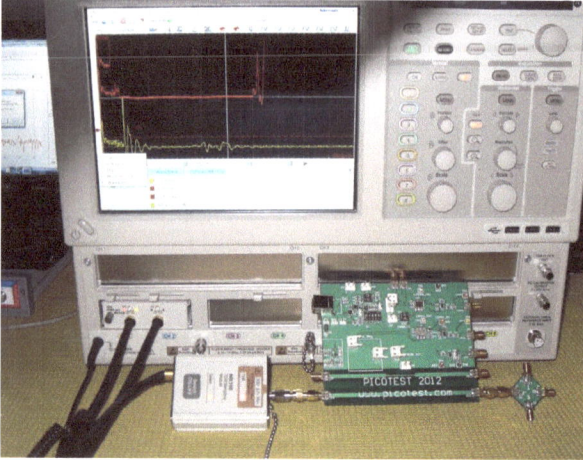

Setup Showing the TDR Instrument and all Transmission Path Elements Connected

Figure 15-5

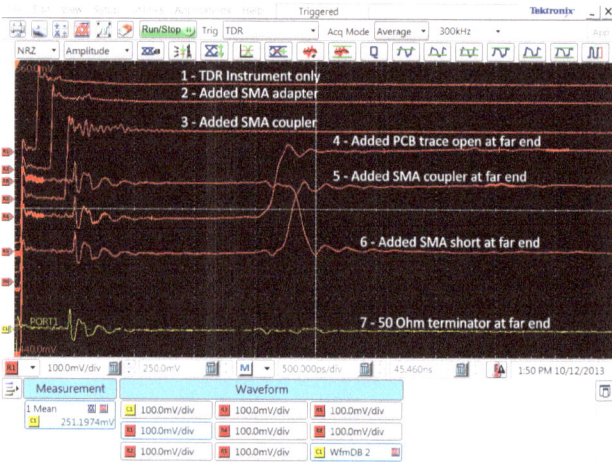

This Measurement shows Reference Lane Changing with Each Length added to the Transmission Path

Figure 15-6

Setting TDR Pulse Rise Time

The resolution of the TDR measurement is related to the rise time of the TDR step. This does not mean that we need to always use the highest rise time setting. Most instruments allow the rise time to be adjusted. Reducing the rise time reduces many of the reflections that result from pre-step discontinuities and post-step ringing making the signal we are interested in easier to see.

A reasonable guideline for setting an appropriate rise time is to set it to twice as fast as the fastest expected signal edges in circuit. For example, ultra high speed CMOS logic gates have a typical rise time of approximately 350ps and so the rise time can be set to 175ps. The coaxial cables connected to the instrument slow down the rise time of the step, so it is best to use only high quality, low-loss cables in order to keep the cables as short as possible.

The measurements are shown in Figure 15-7 and illustrate the effect of the step rise time. The 50Ω trace on the demo board from Figure 15-4 (and shown pictorially in Figure 5-15) is connected to the instrument using only an SMA coupler. The instrument is deskewed using only the SMA coupler, setting the reference plane to the end of the coupler.

The upper trace is measured with a 22.3ps rise time, while the middle trace is measured with a 150ps rise time. For comparison, a high quality coax cable is attached to the instrument to show the further improvement resulting from the high quality connector. While the impedance varies wildly (and unacceptably) with a 22.3ps rise time, the response is significantly better at 150ps, which is sufficient for high speed CMOS. Using a high quality cable and connector, terminated into a 50Ω terminator, the artifact is almost imperceptible at 150ps. In Figure 15-7, The response is significantly better at 150ps, which is sufficient for high speed CMOS, while not usable for higher speed devices. A significant improvement is seen in the connection of a high quality coaxial cable, terminated into 50Ω

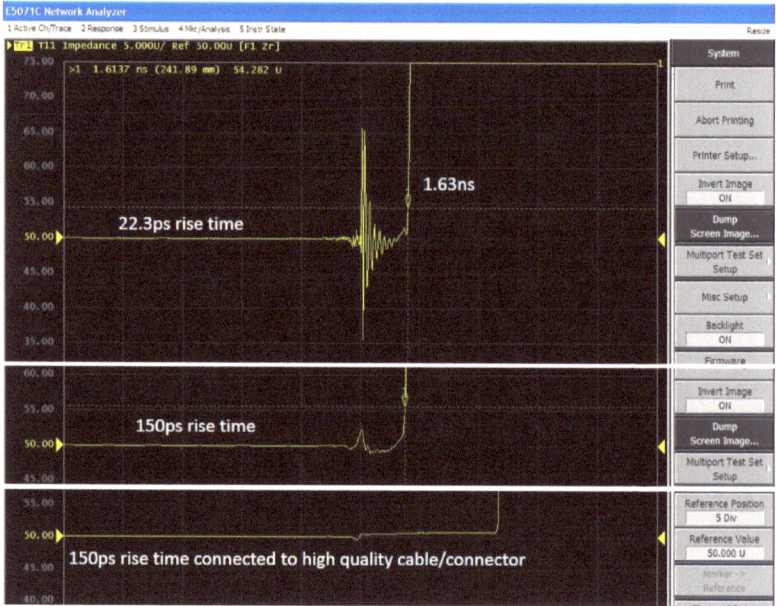

FR4 PCB Stripline at 22.3ps Rise Time and 150ps Rise Time

Figure 15-7

Interpreting TDR Measurements

A simple simulation model is used to perform a TDR assessment, Simulation allows simple and precise control over all of the elements of the measurement, while measurements do not offer such control. The simulation model shown in Figure 15-8 includes a 20ps TDR pulse through a 50Ω source (reference) resistance, internal to the TDR instrument. Two ideal 1.5ns transmission lines, one slightly below the reference resistance and one slightly above the nominal reference resistance are connected in series. The circuit is then terminated into an ideal 50Ω termination resistance.

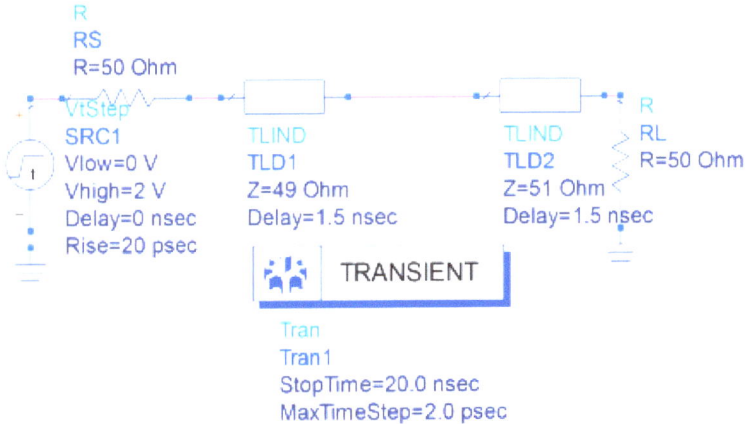

R
RS
R=50 Ohm

VtStep
SRC1
Vlow=0 V
Vhigh=2 V
Delay=0 nsec
Rise=20 psec

TLIND
TLD1
Z=49 Ohm
Delay=1.5 nsec

TLIND
TLD2
Z=51 Ohm
Delay=1.5 nsec

R
RL
R=50 Ohm

TRANSIENT

Tran
Tran1
StopTime=20.0 nsec
MaxTimeStep=2.0 psec

A TDR Simulation Model including a 49Ω and a 51Ω Transmission Line Connected in Series with a 50Ω Terminator—Each Transmission Line has a Time Delay of 1.5ns.

Figure 15-8

The simulation results, shown in Figure 15-9, illustrate one of the complexities of interpreting the results of the TDR measurement. The red trace shows the ideal transmission line impedance while the blue trace shows the TDR measurement. The difference is due to the reflections of each transmission line interacting with each other. This simulation also shows that each of the 1.5ns transmission lines results in a 3ns response, confirming that the signal travels the transmission path twice in the TDR measurement. The signal travels the length from the TDR instrument to the end of the transmission length and the reflection travels from the end of the transmission path back to the TDR instrument.

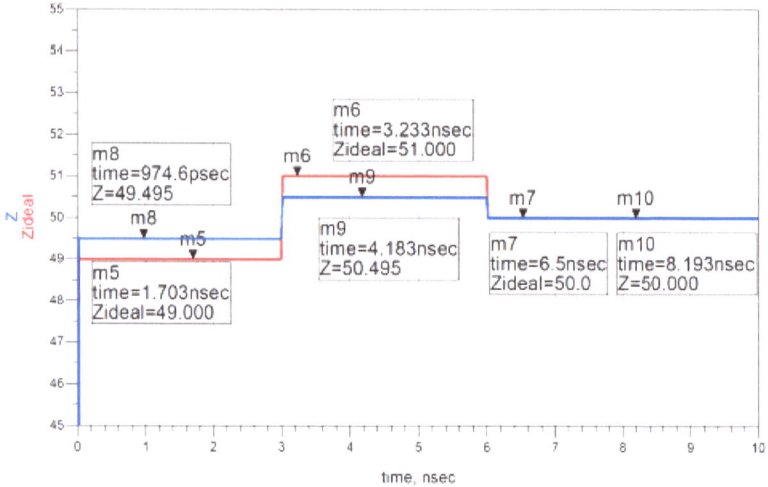

The Red Trace Shows the Ideal Transmission Line Impedance values of 49Ω and 51Ω—The Interaction between Reflections Results in Actual TDR Measurements which are Slightly Different

Figure 15-9

The timing of the TDR measurement can be used to isolate the particular location of a particular artifact if a good estimate of the signal velocity is known. Using the example in Figure 15-8, the 3ns round trip distance is used to determine the 1.5ns signal path. Then with T_d, measured in ns, Equation 15.2 is applied in order to determine the distance from the deskew reference point. This is shown in Equation 15.6.

$$Distance\ from\ Deskew = Signal_{velocity} \cdot T_d \qquad 15.6$$
$$= \frac{1}{\sqrt{\varepsilon_r}} \cdot 11.8\frac{in}{ns} \cdot T_d$$

This same method is used to find the location of an open or a short in a signal path. A similar simulation model, using only a single 50 Ω 1.5ns transmission line, is simulated with the load termination resistor, RL changed from 50Ω to 0Ω (short circuit) and to ∞ (open circuit). The simulation results, shown in Figure 15-10, identify the location of the short or open circuit, which occurs 3ns from the deskew reference point. Again, if a good estimate of the signal velocity is known, then this can be used to determine the location of the short or open circuit. The reference line is the simulation result with RL at 50Ω. The transmission lines in this simulation are set to 52.5Ω for the open circuit 47.5Ω for the short circuit and 50Ω for the 50Ω termination. This is just to separate the traces slightly for clarity.

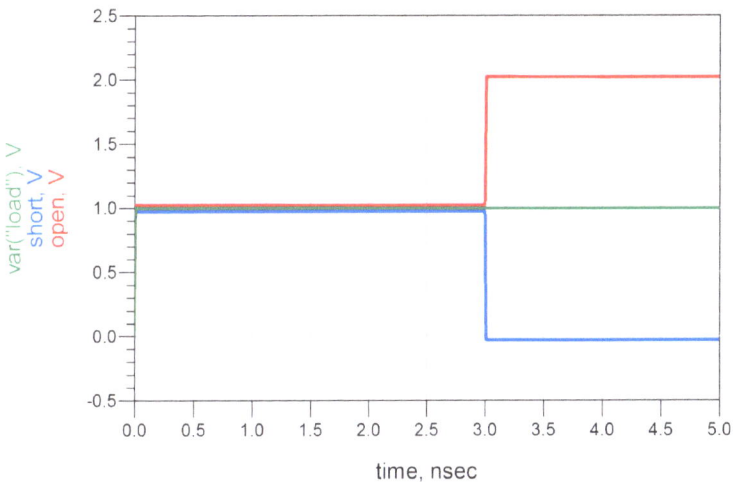

The TDR Simulation Results for a 50 Ω 1.5ns Ideal Transmission Line with a 50Ω Termination (Green Trace), a 0Ω Short-Circuit Termination (Blue Trace) and an $\infty\Omega$ Open Circuit (Red Trace)

Figure 15-10

A second simulation model is created containing two short 50Ω transmission lines, as well as a series inductor between them and a small capacitor located at the load termination, RL.

The simulation model is shown in Figure 15-11.

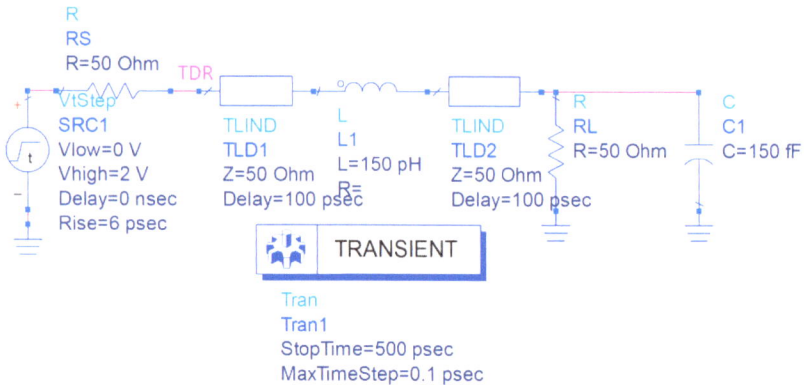

Two Ideal 50Ω 100ps Transmission Lines are Separated by a Series Inductance of 150pH and Terminated into 50Ω in Parallel with 150fF Capacitance

Figure 15-11

As in the prior model we can determine the location of the new elements by the time from the deskew reference to the resulting signal artifact.

In this case, we can see that a series inductor results in a positive artifact while a capacitor results in a negative artifact.

We can interpret the 200ps position of the positive artifact as the round trip distance from the deskew reference to the location of the series inductor, indicating the time from the deskew reference to the inductor is 100ps.

Similarly the 400ps round trip distance to the negative artifact is used to determine that the capacitor is located at 200ps. The simulation results for Γ from Equation 15.4 are

456

shown in Figure 15-12.

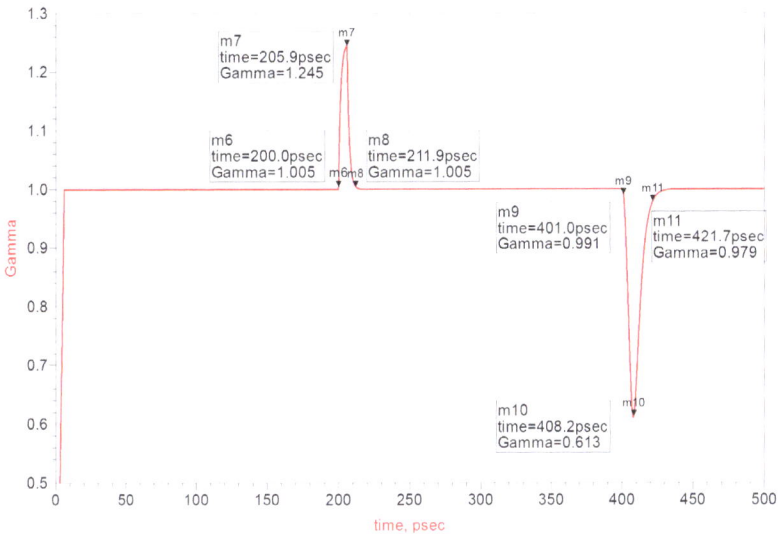

The Simulation Result shows the Positive Artifact at 200ps—Corresponding with a One-Way Transmission Delay of 100ps

Figure 15-12

The artifacts of Figure 15-12 are located after the first transmission line.

The negative artifact occurs at 400ps corresponding with a one way transmission delay of 200ps. This is after the second transmission line.

More complex paths can be interpreted in the same way. The measurement, shown in Figure 15-13, includes several interconnects.

Each artifact can be associated with an inductor for positive excursion or a capacitor for negative excursions. In Figure 15-13, Each element position, referenced to the deskew reference, can be used to determine the physical location of each element.

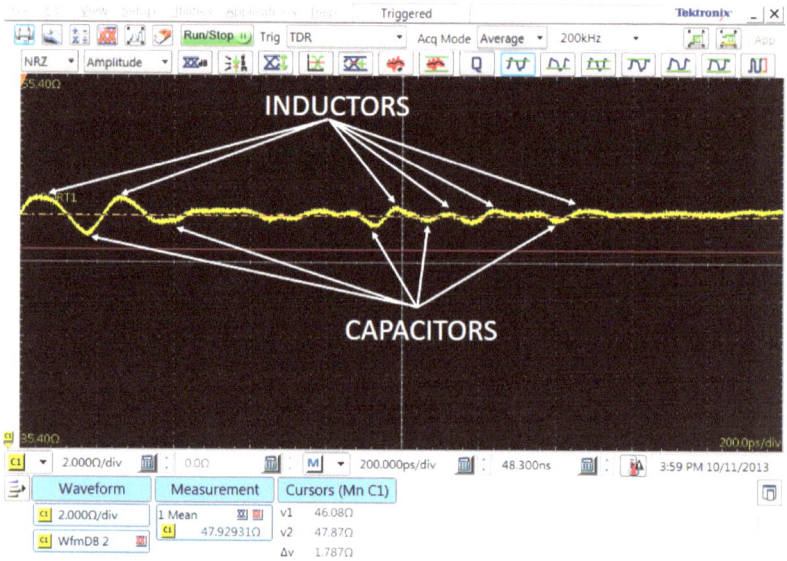

Measurement Result for a Path with Several Interconnects and the Interpretation of the Multiple Inductor and Capacitor Elements

Figure 15-13

Estimating Inductance and Capacitance

The value of the series inductance or shunt capacitance can be estimated from the integration of the reflection artifact.

The inductance, L, is determined from the reference impedance (50Ω) and the integral of the reflection artifact as shown in 15.7.

$$L = 2 \cdot R_{ref} \cdot \int_0^\infty reflection \qquad 15.7$$

Using the measurement in Figure 15-12, three markers are placed on the positive excursion. Two are used to determine the width of the artifact, 11.9ps, while one is used to measure the peak of the excursion, 1.245. Since this peak is above the reference point of 1, the excursion is 0.245. Assuming this waveform is triangular, the inductance is estimated to be 146pH as computed in Equation 15.8.

$$L = 2 \cdot 50\Omega \cdot \frac{11.9ps \cdot 0.245}{2} = 146pH \qquad \text{15.8}$$

This estimate compares very well with the actual value of 150pH used in the simulation model. Similarly, the capacitance can be estimated from the reference impedance (50Ω) and the integral of the reflection artifact as shown in Equation 15.9.

$$C = \frac{2}{R_{ref}} \cdot \int_0^\infty reflection \qquad \text{15.9}$$

Again, using the result in Figure 15-12, three markers are placed on the negative artifact. Two are used to determine the width of the artifact, 20.7ps, and one is used to measure the peak of the excursion, 0.613. Since this peak is below the reference point of 1, the excursion is 0.387. Assuming this waveform is triangular, the capacitance is estimated to be 160fF as computed in Equation 15.10.

$$C = \frac{2}{50} \cdot \frac{20.7ps \cdot 0.387}{2} = 160fF \qquad \text{15.10}$$

This again is very close to the 150fF included in the simulation model.

In order to maintain the edge speed of the TDR pulse, it is important to manage the signal launch, or the interface between the TDR and the DUT. It is ideal to include high quality connectors within the DUT and to connect the TDR instrument using high quality cables.

The rise time is still slowed a bit by the interconnecting cables, so the cables need to be high quality and as short as possible. Additional details of the coaxial cable rise time are included in Chapter 5.

The relationship between the rise time and a single order time constant is derived in Chapter 13, and is copied here in Equation 15.11 for convenience.

15.11

$$T_{rise_{10/90}} = \tau \cdot ln\left(\frac{90\%}{10\%}\right) = 2.19722 \cdot \tau$$

In the case of a cable attachment to the DUT, the time constant is primarily the inductance of the interconnecting leads and the DUT.

Since the TDR instrument presents a 50Ω source impedance, the total series resistance is the sum of the source and DUT resistance terms. The time constant, τ is then:

$$\tau = \frac{L}{R_S + R_{DUT}}$$

15.12

Assuming an interconnect inductance of 1nH, and the DUT to be 50Ω, the time constant using 15.12 is 10ps. Applying Equation 15.11, the rise time due to the interconnect inductance is 21.97ps.

If the DUT is reduced to 1Ω the rise time is increased to 43ps. Reducing the DUT further to 0.1Ω results in 43.85ps. These results are confirmed by simulation. The simulation schematic is shown in Figure 15-14. All three results are simulated simultaneously and the results for the 50Ω and 1Ω DUT resistances are shown in Figure 5-15.

The 0.1Ω DUT simulation result is shown in Figure 15-16. The simulated rise time for 50Ω is (23.48ps-1.557ps) or 21.9ps. The results for 1Ω and 0.1Ω are 43.05ps and 43.87ps, respectively, confirming the calculations.

In Figure 15-14, note that the 50Ω source resistance is included in the inductor model to allow a more compact model.

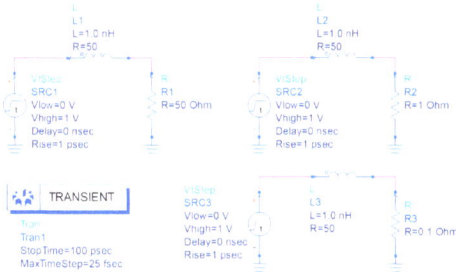

Simulation of Rise Time for 1nH Interconnect including DUT Resistance of 50Ω, 1Ω and 0.1Ω

Figure 15-14

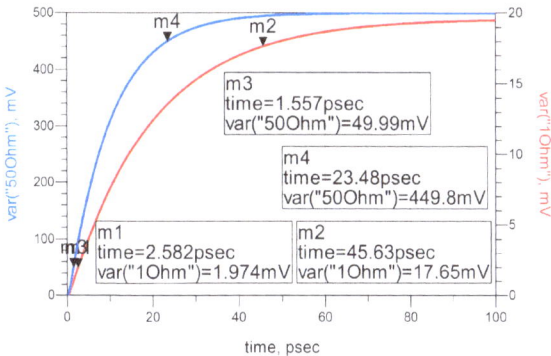

Simulation Rise Time Results for 50Ω and 1Ω DUT Resistance, Resulting in 10%-90% Rise Time of 21.9ps and 43.05ps, Respectively

Figure 5-15

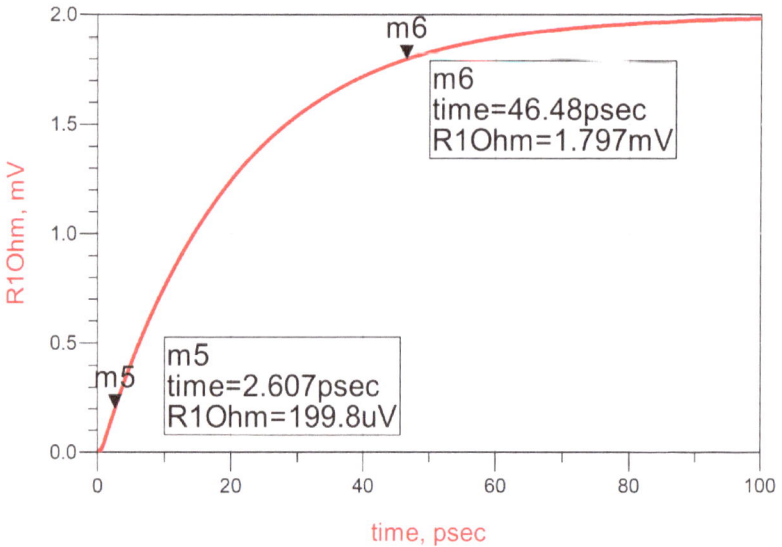

Simulation Rise Time Results for a 0.1Ω DUT Resistance, Resulting in a 10%-90% Rise Time of 43.87ps

Figure 15-16

Since most of the power related measurements have low values of DUT resistance, a good approximation of the rise time due to the interconnecting inductance is 44ps/nH.

This approximation can also be used to estimate the maximum allowable interconnect inductance based on the tolerable degradation of rise time.

The interconnect or launch inductance can also be significant for low quality RF connectors. For this reason high quality connectors should be used, as well as high quality cables.

The setup in Figure 15-17 illustrates the impact of poor quality connectors. In this image, the TDR is deskewed with no cables attached to set the reference plane at the TDR instrument front panel.

A high quality semi-rigid coax is connected to the front panel while an SMA thru connector is used to attach a second high quality cable.

Finally, the second cable is attached to a 50Ω test trace on an FR4 printed circuit board with a low quality SMA edge launch connector.

The measurement results are shown in Figure 15-18.

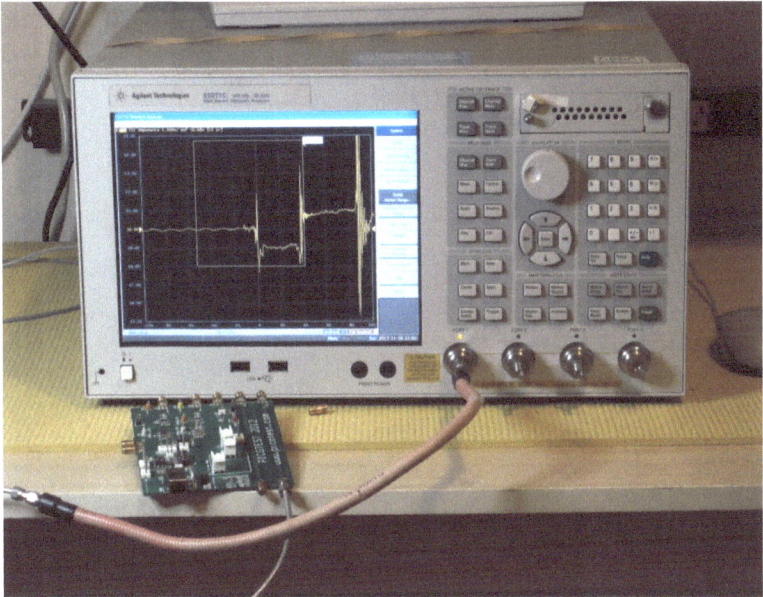

Setup of Two High Quality Cables Connected with an SMA Thru Adapter and Connected to a PCB using a Low-Quality Edge Connector

Figure 15-17

In Figure 15-18, the high quality connectors maintain a very tolerable impedance variation while the low quality edge connector does not.

.

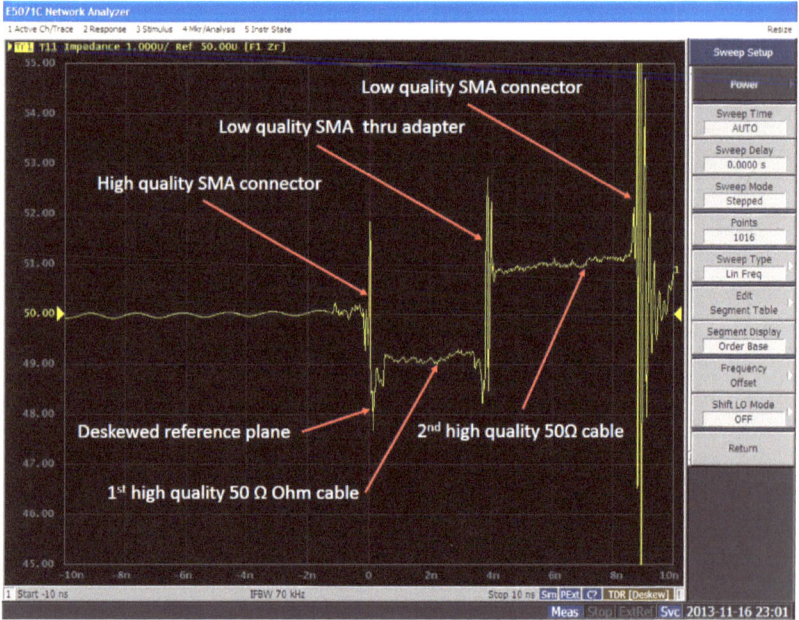

The Measurement Results from Figure 15-17

Figure 15-18

The impedance change seen for the first connection using the high quality SMA connector is within approximately $\pm 2\Omega$.

The first cable has a lower characteristic impedance of 49Ω. The SMA thru connector also keeps the impedance fluctuation to about $\pm 3\Omega$ at the joint between the two high quality cables. The second high quality cable has a characteristic impedance of 51 Ohms. The final connection to the test trace using the low quality connector results in an impedance fluctuation that is off the screen. It is, therefore, much greater than $\pm 5\Omega$. The actual impedance fluctuation is $\pm 15\Omega$ and can be seen in the upper trace of Figure 15-17.

In order minimize the impedance fluctuations resulting from these connector interfaces, it is important to properly torque the connectors. The image in Figure 15-19 shows measurements with a "finger tight" cable and a torqued cable. The properly torqued connector results in an impedance fluctuation of slightly more than 1.6Ω. The same connector, finger tightened, results in an impedance fluctuation of slightly more than 3Ω. This is nearly twice as much fluctuation as compared to the properly torqued connector.

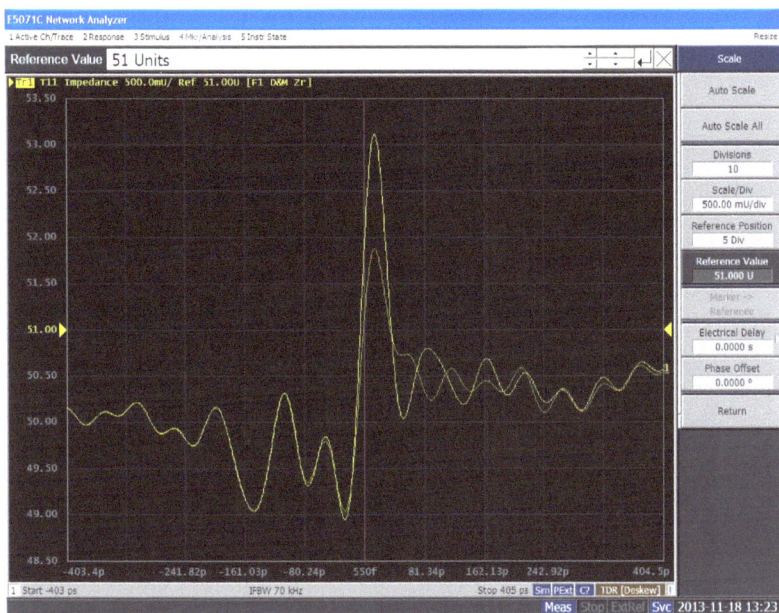

Results of Loose Connector (Bright Trace) and Properly Torqued Connector (Fainter Trace)

Figure 15-19

Not all manufacturers recommend the same torque. The range for SMA connectors is very wide, from approximately 3in-Lbs to 10in-Lbs. Stainless connectors can generally be torqued more than brass. It is recommended that you check with the manufacturer of your TDR instrument and invest in a torque wrench.

Tips and Tricks

1. Use only high quality, low loss cables and connectors.
2. Sometimes the best cable is no cable at all. Small devices can often be attached directly to the instrument.
3. Properly torque connectors. Check your instrument for torque specifications.
4. Perform a calibration and measure a known value to assure the setup and calibration accuracy.
5. Use a rise time that is appropriate for the application.
6. Placing markers on the PCB provides clear distance markers in the measurement.

S-parameter Measurements

Most TDR instruments also offer S-parameter measurements.

In many cases, the S-parameters are determined using a Fourier transformation. The resulting dynamic range is approximately 70dB for these instruments. The Agilent E5071C includes a swept VNA and offers a much higher dynamic range and 20GHz of bandwidth, which is sufficient for all power applications.

The S-parameter measurements are made using the same techniques described in Chapter 7. The cautions about properly torqued high quality connectors and cables and the launch inductance that were previously discussed for TDR measurements also apply to S-parameter measurements at higher frequencies.

One example of a TDR based S-parameter measurement is shown in Figure 15-20. The TDR pulse is shown in the lower right window, while the orange trace shows the insertion loss (S21), the blue trace shows the input return loss (S11) and the pink trace shows the output return loss (S22). The measurement is of a high quality 18GHz cable. Though it is difficult to read the scales, the measurement shows both input and output return loss of approximately 12dB and insertion loss of approximately 2.5dB at 18GHz.

In Figure 15-20, the TDR is the yellow trace in the lower right, while the orange trace in the upper left is the insertion loss (S21). The blue trace is the output return loss (S22) and the pink trace is the input return loss (S11).

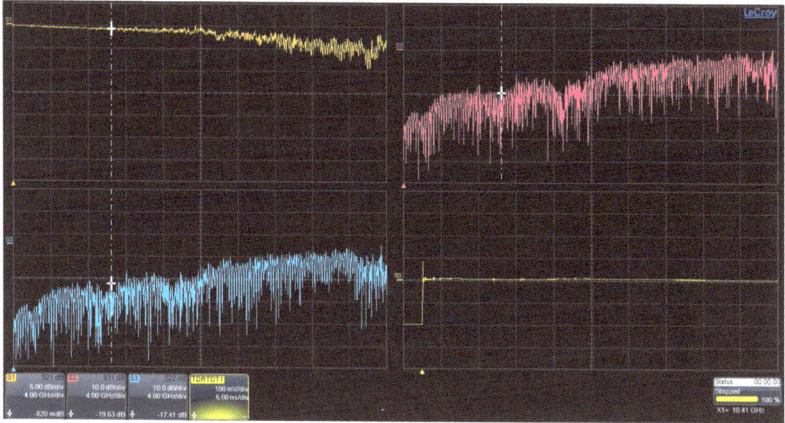

TDR-Based S-Parameter Measurement of an 18GHz High-Quality Coaxial Cable

Figure 15-20

The swept VNA in the Agilent E5071C offers much lower noise and much better dynamic range allowing measurements of much lower impedance values.

The measurement in Figure 15-21 shows a $1m\Omega$ resistor with an S21 result of approximately -86.7dB.

Converting this to a magnitude results in an S21 magnitude of 46.2μ. The measurement is performed from 300kHz to 3GHz and the resulting S21 of -86.7dB results in a resistance of $1.16m\Omega$.

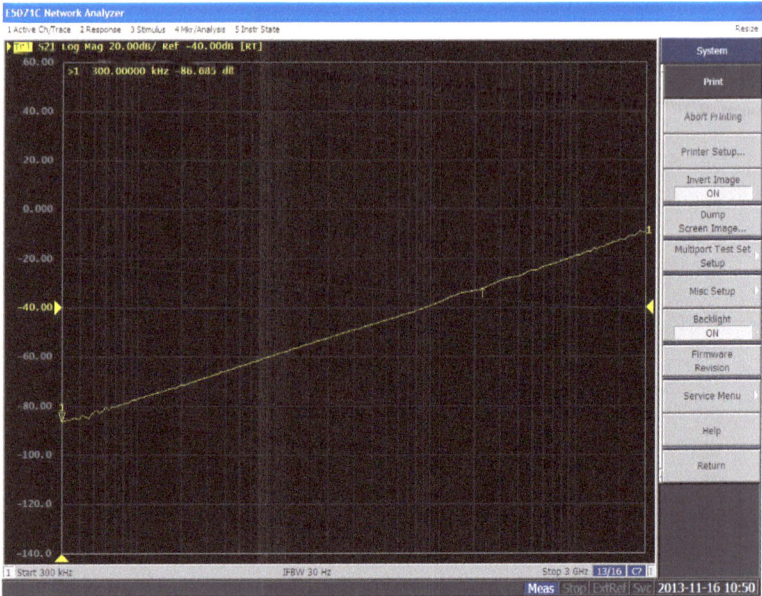

Measurement of a 1mΩ Resistor using the 2-Port Shunt-Thru Method

Figure 15-21

The relationship between the 2-port shunt-thru S-parameter measurement and impedance is derived in Chapter 7 with the result is included here for convenience.

$$Z_{DUT}(S21) = \frac{-S21 \cdot RS \cdot RL}{S21 \cdot RS + S21 \cdot RL - 2 \cdot RL} \qquad 15.13$$

Assuming that both ports, RS and RL are 50Ω this reduces to:

$$Z_{DUT}(S21) = \frac{-S21 \cdot 25}{1 - S21} \qquad 15.14$$

Substituting 46.2u for S21 in Equation 15.14 results in 1.156mΩ. The same measurement is performed over the range of 1kHz to 30MHz using the Agilent E5061B VNA. The measurement result in Figure 15-22 shows 1.16mΩ, which agrees very well. This measurement indicated 1.16mΩ, which is in very good agreement with the E5071C measurement.

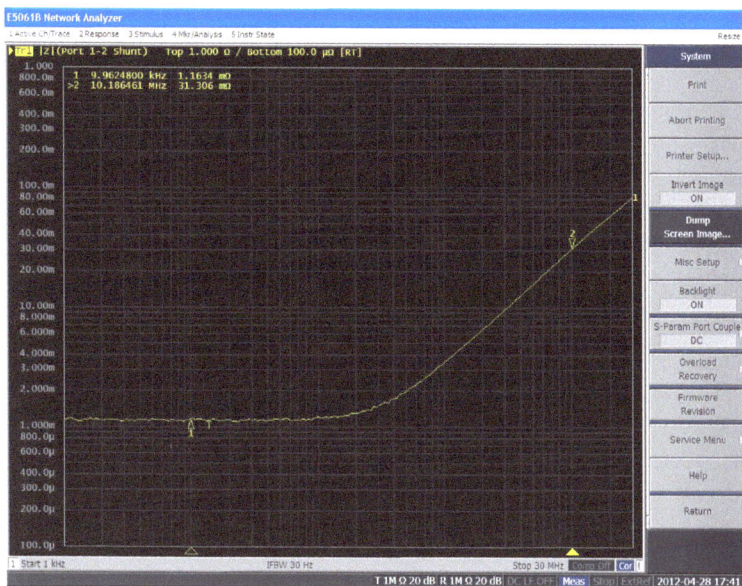

The Same Resistor Measured using the Agilent E5061B Analyzer over the Frequency Range of 1kHz to 30MHz

Figure 15-22

The reduced dynamic range of the TDR based S-parameters allows measurement of resistances as low as 5 or 10mΩ.

Chapter References

1. *Agilent Signal Integrity Analysis Series Part 1: Single-Port TDR, TDR/TDT, and 2-Port TDR*, Agilent Technologies http://cp.literature.agilent.com/litweb/pdf/5989-5763EN.pdf

2. *TDR Impedance Measurements: A Foundation for Signal Integrity*, Tektronix http://www.tek.com/document/fact-sheet/tdr-impedance-measurements-foundation-signal-integrity

3. *Milliohm PDN Measurements with the SPARQ Network Analyzer*, Teledyne Lecroy http://teledynelecroy.com/doc/milliohm-pdn-measurements-with-the-sparq-network-analyzer

4. Michael Steinberger, *TDR: Reading the Tea Leaves,* June 14,2012 http://www.eeweb.com/blog/michael_steinberger/tdr-reading-the-tea-leaves

5. Agilent Technology, *Measure Parasitic Capacitance and Inductance Using TDR* http://literature.agilent.com/litweb/pdf/5988-6505EN.pdf

6. Teledyne Lecroy, *Using 2nd Tier Calibration for Cable Fixture De-embedding,* Apr. 7, 2011 http://teledynelecroy.com/doc/using-2nd-tier-calibration-for-cable-fixture-deembedding

Afterword by Robert Bolanos—Power Integrity in a Changing Landscape

IN THE EVER-EVOLVING world of electronic systems, the importance of power integrity cannot be overstated. Over the past decade, I witnessed remarkable advancements and transformative changes in the field. Amid these changes, Steve Sandler's book, *Power Integrity: Measuring, Optimizing, and Troubleshooting Power-Related Parameters in Electronic Systems*, has not only stood the test of time but has become an indispensable guide for engineers seeking to navigate the challenges of modern electronic design.

The complexity of systems has grown exponentially—

driven by the pursuit of higher performance, increased functionality, lower impedances, miniaturization and higher power densities. As technology advances –including the development of GaN FETs—the power delivery system faces new challenges in providing clean and stable power to electronic systems. Power integrity emerged as critical, demanding meticulous attention to ensure the reliable operation of electronic devices. It is within this context that Steve's expertise, accumulated over four decades, shines through in the pages of *Power Integrity*.

One of the book's key strengths lies in its ability to foster a systems approach to electronic design. Steve emphasizes the interconnectedness of power distribution networks and their impact on system performance. By understanding the intricate relationship between power integrity and overall system reliability, engineers can design more robust and efficient electronic systems. The book provides a comprehensive analysis of power distribution networks, covering their design, analysis, and measurement. It equips engineers with the knowledge and tools necessary to address the evolving challenges in power integrity.

In addition, Steve provides a detailed overview of the different types of equipment that can be used to measure and validate power distribution network models. From the Bode 100, Oscilloscope, FRA and VNAs to TDR analyzers, Steve offers practical guidance on choosing the most appropriate tools for a given application. Moreover, Steve's book goes beyond a simple review of measurement equipment, showing the practical aspects of measurement set-up, calibration, SPICE modeling and validation of the SPICE model (or ADS models). This ensures that engineers not only have access to the right tools, but also know how to use them effectively to yield accurate results.

Steve's passion for teaching and commitment to sharing knowledge are evident throughout the book. His ability to demystify complex concepts and make them accessible to engineers of all levels is truly commendable. The clarity of his explanations, supported by multiple practical examples and decades of real-world insights, empowers engineers to bridge the gap between theory and practice. Whether you are an experienced engineer seeking to expand your knowledge or a student just starting out in the field, *Power Integrity* provides a wealth of practical guidance and insights.

This book is a comprehensive guide empowering engineers to embrace power integrity in a changing landscape. Steve's extensive experience—and dedication to sharing this knowledge—has made a significant impact on the field. As we continue to push the boundaries of electronic design, Steve's work will remain a trusted companion, enabling engineers to achieve excellence in power integrity and drive innovation forward.

In summary, my collaboration with Steve on intricate projects led me to regard him as one of the most brilliant engineers I have ever had the opportunity to work with. I extend my sincere appreciation for his exceptional contributions to the realm of power integrity. His work in this field made a lasting impact, establishing a strong knowledge base and enabling comprehension among many engineers.

With the constantly evolving landscape of electronic design, Steve's accomplishments will serve as a guiding light, empowering engineers to navigate the intricate nature of power distribution networks and guarantee the dependable functionality of electronic systems.

Thank you, Steve

—Robert Bolanos

You can find Robert on LinkedIn:

https://www.linkedin.com/in/robert-bolanos-b12b281b/

INDEX

INDEX

INDEX

INDEX

INDEX

INDEX

Domains, measurement: tips and tricks 76

Dribble effect 122, 396-397

DUT. See Device under test 13, 23, 35

Dynamic range: ADC 43

Dynamic range: defined 43

Dynamic range: ENOBs 45, 47-48

Dynamic range: measurement 44, 46-47, 49

Dynamic range: for oscilloscopes 43

Dynamic range: preamplifiers 60

E

E See Electric field 128, 411-412, 414-415, 417, 425, 431-432, 435, 437,

Edges: BW and rise time relationship 391

Edges: cascading rise time 384

Edges: coaxial cables 379

Edges: high-voltage probe measuring 406, 408

Edges: impact of filtering and bandwidth limiting 388

Edges: interleaved sampling 392-393

Edges: interpolation and 394-395

Edges: measuring 378

Edges: minimum error solver 389

Edges: overview about 378

Edges: passive probes measuring 404

Edges: printed circuit board issues 403

Edges: probe connection criticality 398

Edges: probes 403

Edges: sampling rate 392-393

Edges: TDR 378

Edges: tips and tricks 408

Effective number of bits (ENOBs): dynamic range 43-44, 52

Effective number of bits (ENOBs): sensitivity 40

Electric (E) field probes 414, 425, 431, 435

Electric field orientation 415

Electromagnetic interference (EMI) 71, 74-75, 85, 94, 123, 128-129, 341, 410,

Electromagnetic interference (EMI): compliance testing 129, 410

Electronic load versus current injector 325

Electronic load versus current described 325

EMI. See Electromagnetic interference 71, 74-75, 85, 94, 123, 128-129, 341, 410, 414-415, 420

Emissions: basics 411

Emissions: coefficient in SPICE model of diode example

Emissions: E and H and 411

Emissions: far field boundary 412-413

Emissions: impedance 411, 412

Emissions: transition region 413

ENOBs. See Effective number of bits 40

Equivalent series resistance (ESR) 161, 256

INDEX

INDEX

High-fidelity, interpreting TDR measurements 452

High-fidelity overview 21

High-fidelity and pulse rise time 451

High-fidelity and reference plane 445, 448, 451, 462

High-fidelity, S-parameter measurements 466

High-fidelity and time domain 442

High-frequency loss effects 396

High-voltage loop measurement 241

High-voltage probe 114, 241, 406, 408

I

Impedance. See also Output impedance 19-20, 23, 29, 65, 133, 138, 140, 145-146, 148-151, 169, 172-175, 213-215, 224, 226, 240-241, 245, 253-256, 259, 262-268, 276, 304, 306, 348, 370

Impedance adapters: advantages 215

Impedance adapters: calibrating 217

Impedance adapters: cons 216

Impedance adapters: device setup 216

Impedance adapters: ferrite bead example 220-221

Impedance adapters: low-impedance regulator output impedance example 213

Impedance adapters: Omicron Lab B-SMC and B-WIC 134

Impedance adapters and Omicron Labs Bode 100 216, 224

Impedance adapters: 1.2-nH inductor example 219-220

Impedance adapters: pros 215

Impedance adapters: Semtech SC4437 voltage reference example 214

Impedance adapters: switching regulator output impedance with 10-Ohm load resistor example 214

Impedance adapters: tantalum capacitor example 221-222

Impedance adapters: tips and tricks 216

Impedance adapters: circuit 216

Impedance adapters: control loop stability 130,148-150, 152-153, 271-272

Impedance adapters: current injection measurements, clamp on current probe method 209

Impedance adapters: cons 216

Impedance adapters: device setup 216

Impedance adapters: pros 215

Impedance adapters: solid-state device method 210

Impedance adapters: tips and tricks 216

Impedance adapters: transformer coupled method 210-211

Impedance adapters: domain

Impedance adapters and emissions

Impedance adapters: fixtures 215, 217

INDEX

INDEX

INDEX

INDEX

INDEX

INDEX

INDEX

INDEX

INDEX

INDEX

Preamplifiers, Picotest J2180A, Qi wireless charger evaluation example 426, 427

Preamplifiers, Picotest J2180A and thru calibration 199, 286, 289

Precise 14,21-22,134,357,452

Probes. See also Active probes; 25-26, 104, 113, 116-177, 399-400, 403-405, 433

Probes, differential 104, 117, 122, 211

Probes, low-impedance 104, 113, 118

Probes, passive 104, 107, 114-115, 124, 285, 352-353, 365, 403-404

Probes, specific probe cascading rise time 384

Probes, criticality of voltage connections 398

PSRR. See Power supply rejection ratio 16-18, 23, 65, 132-133, 138, 143-144, 149, 152, 154, 226, 253, 262, 264-265, 271-272, 275, 278, 283-285, 290-294, 297-300, 319, 348

Pulse frequency modulation (PFM) 143

Pulse-width-modulator (PWM) 22, 25

Pulse-width-modulator: locating signal sources example 422

Q

Qi wireless charger evaluation example and microprocessor 426-427

Qi wireless charger evaluation example and MOSFET 426-427

Qi wireless charger evaluation example and noise source locations 426-427

Qi wireless charger evaluation example and preamplifier 426-427

Qi wireless charger evaluation example setup 426-427

Qi wireless charger evaluation example with spectrum analyzer 426-427

R

Random interleaved sampling (RIS) 392-393

RBW. See Resolution bandwidth 49-50, 85, 167, 169, 287, 289, 439

REF-03 voltage reference example 264-265

Reflection coefficient 67, 162-163, 443-444

Resistance and cables and DCR 65, 134, 164, 341

Resistance and cables and ESR 161, 201

Resistance and cables and line injector output 279

Resolution bandwidth (RBW) defined 49

Resolution bandwidth and FFT 50

Resolution bandwidth and measurement 49, 167

Resolution bandwidth and SignalVu 52-54

Return loss (RL) 67, 121, 466-467

INDEX

INDEX

INDEX

INDEX

INDEX

INDEX

INDEX

INDEX

INDEX

INDEX

www.ingramcontent.com/pod-product-compliance
Lightning Source LLC
Chambersburg PA
CBHW040243230326
41458CB00104B/6473